Cosmic Plasmas and Electromagnetic Phenomena

Cosmic Plasmas and Electromagnetic Phenomena

Special Issue Editors

Athina Meli
Yosuke Mizuno
José L. Gómez

MDPI • Basel • Beijing • Wuhan • Barcelona • Belgrade

MDPI

Special Issue Editors

Athina Meli
University of Gent,
Belgium

Yosuke Mizuno
Goethe University Frankfurt,
Germany

José L. Gómez
Instituto de Astrofísica de Andalucía—CSIC,
Spain

Editorial Office
MDPI
St. Alban-Anlage 66
4052 Basel, Switzerland

This is a reprint of articles from the Special Issue published online in the open access journal *Galaxies* (ISSN 2075-4434) from 2018 to 2019 (available at: https://www.mdpi.com/journal/galaxies/special_issues/Plasmas)

For citation purposes, cite each article independently as indicated on the article page online and as indicated below:

LastName, A.A.; LastName, B.B.; LastName, C.C. Article Title. *Journal Name* **Year**, *Article Number*, Page Range.

ISBN 978-3-03921-465-5 (Pbk)
ISBN 978-3-03921-466-2 (PDF)

Contents

About the Special Issue Editors vii

Preface to "Cosmic Plasmas and Electromagnetic Phenomena" ix

Denise Gabuzda
Evidence for Helical Magnetic Fields Associated with AGN Jets andthe Action of a Cosmic
Battery
Reprinted from: *galaxies* **2019**, *7*, 5, doi:10.3390/galaxies7010005 1

Ioannis Contopoulos
Generation and Transport of Magnetic Flux in Accretion–Ejection Flows
Reprinted from: *Galaxies* **2019**, *7*, 12, doi:10.3390/galaxies7010012 15

José-María Martí
Numerical Simulations of Jets from Active Galactic Nuclei
Reprinted from: *Galaxies* **2019**, *7*, 24, doi:10.3390/galaxies7010024 33

Demosthenes Kazanas
MHD Accretion Disk Winds: The Key to AGN Phenomenology?
Reprinted from: *Galaxies* **2019**, *7*, 13, doi:10.3390/galaxies7010013 64

Sergey Bogovalov
Physics of "Cold" Disk Accretion onto Black Holes Driven by Magnetized Winds
Reprinted from: *Galaxies* **2019**, *7*, 18, doi:10.3390/galaxies7010018 78

Roman Gold
Relativistic Aspects of Accreting Supermassive Black Hole Binaries in Their Natural Habitat: A
Review
Reprinted from: *Galaxies* **2019**, *7*, 63, doi:10.3390/galaxies7020063 109

**Ken-Ichi Nishikawa, Yosuke Mizuno, José L. Gómez, Ioana Duţan, Athina Meli,
Jacek Niemiec, Oleh Kobzar, Martin Pohl, Helene Sol , Nicholas MacDonald and
Dieter H. Hartmann**
Relativistic Jet Simulations of the Weibel Instability in the Slab Model to Cylindrical Jets with
Helical Magnetic Fields
Reprinted from: *Galaxies* **2019**, *7*, 29, doi:10.3390/galaxies7010029 127

Markus Böttcher
Progress in Multi-Wavelength and Multi-Messenger Observations of Blazarsand Theoretical
Challenges
Reprinted from: *Galaxies* **2019**, *7*, 20, doi:10.3390/galaxies7010020 147

Asaf Pe'er
Plasmas in Gamma-Ray Bursts: Particle Acceleration, Magnetic Fields, Radiative Processes
and Environments
Reprinted from: *Galaxies* **2019**, *7*, 33, doi:10.3390/galaxies7010033 184

Peter L. Biermann, Philipp P. Kronberg, Michael L. Allen, Athina Meli, Eun-Suk Seo
The Origin of the Most Energetic Galactic Cosmic Rays: Supernova Explosions into Massive
Star Plasma Winds
Reprinted from: *Galaxies* **2019**, *7*, 48, doi:10.3390/galaxies7020048 214

Klaus Michael Spohr, Domenico Doria and Bradley Stewart Meyer
Theoretical Discourse on Producing High Temporal Yields of Nuclear Excitations inCosmogenic
^{26}Al with a PW Laser System: The Pathway to an Astrophysical Earthbound Laboratory
Reprinted from: *Galaxies* **2018**, 7, 4, doi:10.3390/galaxies7010004 . . . . **237**

About the Special Issue Editors

Athina Meli, Ph.D., senior research scientist in the Department of Physics and Astronomy of University of Gent. For over a decade she has been studying the acceleration mechanisms of high energy cosmic-rays and the properties of relativistic extragalactic sources, using Monte-Carlo and RPIC simulation studies of relativistic magnetized extragalactic plasma jets. She has been a member of major international cosmic-ray and cosmic-neutrino observatory experiments such as AUGER, IceCube, Antares, Km3net. She serves as editor and reviewer in several international scientific peer-reviewed journals and she is a co-founder and member of the board of the Cosmoparticle (CosPa) network in Belgium.

Yosuke Mizuno, Ph.D., is a research scientist in the Institute for Theoretical Physics, Goethe University Frankfurt, Germany. He has worked on theoretical plasma astrophysics, relativistic astrophysics, black hole astrophysics, and high-energy astrophysics by using numerical simulations. His research is about black hole, active galactic nuclei, gamma-ray bursts, astrophysical jets, accretion flow, pulsar wind nebulae. His research field covers the fundamental plasma physics including shocks, instabilities, turbulence, magnetic reconnection and particle acceleration. Recently he has been studying the theory of gravity via the black hole shadow.

José L. Gómez, senior research staff at the Instituto de Astrofísica de Andalucía (CSIC), in Granada (Spain).B.S. in Physics/Astronomy from Barcelona University (Spain) and PhD in Astronomy from Granada University (Spain). Postdoctoral research in Boston University (USA) and Manchester University (UK). Main interests include accretion onto supermassive black holes and their associated relativistic jets, extremley high angular resolution observations through millimeter and space VLBI observations, and relativistic magnetohydrodynamic and non-thermal emission simulations.

Preface to "Cosmic Plasmas and Electromagnetic Phenomena"

During the past few decades, plasma science has witnessed a great growth in laboratory studies, in simulations, and in space. Plasma is the most common phase of ordinary matter in the universe. It is a state in which ionized matter (even as low as 1%) becomes highly electrically conductive. As such, long-range electric and magnetic fields dominate its behavior. Cosmic plasmas are mostly associated with stars, supernovae, pulsars and neutron stars, quasars and active galaxies at the vicinities of black holes (i.e., their jets and accretion disks). Cosmic plasma phenomena can be studied with different methods, such as laboratory experiments, astrophysical observations, and theoretical/ computational approaches (i.e., MHD, particle-in-cell simulations, etc.). They exhibit a multitude of complex magnetohydrodynamic behaviors, acceleration, radiation, turbulence, and various instability phenomena. This Special Issue addresses the growing need of the plasma science principles in astrophysics and presents our current understanding of the physics of astrophysical plasmas, their electromagnetic behaviors and properties (e.g., shocks, waves, turbulence, instabilities, collimation, acceleration and radiation), both microscopically and macroscopically. This Special Issue provides a series of state-of-the-art reviews from international experts in the field of cosmic plasmas and electromagnetic phenomena using theoretical approaches, astrophysical observations, laboratory experiments, and state-of-the-art simulation studies

Athina Meli, Yosuke Mizuno, José L. Gómez
Special Issue Editors

galaxies

MDPI

Review

Evidence for Helical Magnetic Fields Associated with AGN Jets and the Action of a Cosmic Battery

Denise Gabuzda

Physics Department, University College Cork, Cork T12 K8AF, Ireland; d.gabuzda@ucc.ie; Tel.: +353-21-490-2003

Received: 9 November 2018; Accepted: 14 December 2018; Published: 27 December 2018

Abstract: Theoretical models for the electromagnetic launching of astrophysical jets have long indicated that this process should generate helical magnetic fields, which should then propagate outward with the jet plasma. Polarization observations of jets are key for testing this idea, since they provide direct information about the magnetic field structures in the synchrotron-emitting radio jets. Together with Faraday rotation measurements, it is possible in some cases to reconstruct the three-dimensional magnetic-field structure. There is now plentiful evidence for the presence of helical magnetic fields associated with the jets of active galactic nuclei, most directly the detection of transverse Faraday-rotation gradients indicating a systematic change in the line-of-sight magnetic field component across the jets. A variety of models involving helical jet magnetic fields have also been used to explain a great diversity of phenomena, including not only the linear polarization and Faraday rotation structures, but also circular polarization, anomalous wavelength dependences of the linear polarization, variability of jet ridge lines, variability of the Faraday rotation sign and polarization angle rotations. A joint consideration of Faraday rotation measurements on parsec and kiloparsec scales indicates a magnetic-field and current structure similar to that of a co-axial cable, suggesting the action of some kind of battery mechanism, such as the Poynting–Robertson cosmic battery.

Keywords: active galactic nuclei; relativistic jets; magnetic fields; radio interferometry

1. Introduction

In normal galaxies such as our Milky Way, the vast majority of the luminosity is associated with thermal emission from stars and glowing gas. Although virtually all galaxies are believed to harbour a supermassive black hole in their central regions, with masses of order 10^6 solar masses, the emission associated with accretion onto this central black hole comprises a small fraction of the overall luminosity of the galaxy. However, the luminosities of a minority of galaxies are dominated by the contribution of non-thermal activity in their nuclei. The strongest of such Active Galactic Nuclei (AGN) are about 10^5-times more luminous than a normal galaxy such as the Milky Way. The activity giving rise to this tremendous energy is believed to be accretion onto a much more massive supermassive black hole ($\sim 10^9$ solar masses) at the galactic centre.

This accretion process sometimes, but not always, creates conditions favourable for the generation and launching of jets of relativistic plasma from the vicinity of the central black hole in the AGN. Although no direct imaging is available, it is natural to suppose that these jets are ejected along the rotational axis of the central black hole. The relativistic electrons present in these jets and the lobes they inflate at their termination points emit radio synchrotron radiation as they move through regions with magnetic (**B**) fields. About 10–15% or so of all AGN, said to be "radio-loud", display substantial amounts of such radio emission (ratio of the 5-GHz flux to the optical B-band flux ≥ 10 [1]). The presence of synchrotron radiation requires both highly relativistic particles and magnetic fields, suggesting that the jets are regions where particle acceleration and magnetic-field amplification can occur.

The jet launching mechanism is widely believed to be electromagnetic in nature. In the Blandford–Payne process [2], a magnetic (**B**) field threading the accretion disk and frozen into the disk plasma is wound up by the rotation of the disk. This gives rise to a magnetic pressure gradient that accelerates material from the disk outward along the field lines. The Blandford–Znajek mechanism [3] provides a means of extracting the rotational energy of the black hole. The rotation of the black hole in the ambient **B** field from the accretion disk gives rise to induced electric fields that accelerate local charged particles, initiating a series of processes that gives rise to jets comprised of electrons and positrons carried outward by an outgoing Poynting flux. A key factor in both mechanisms is rotation of the central black hole and its accretion disk in the presence of a magnetic field, giving rise to a helical **B** field carried outward along with the jets. The generation of such fields has been investigated theoretically in a number of studies over the years. Two of the best-known earlier works in this area are those of Nakamura et al. [4] and Lovelace et al. [5]. Some of the most recent simulations of jet launching and propagation have been done, for example, by Tchekhovskoy and Bromberg [6] and Barniol Duran et al. [7]; many more examples can be found in the literature. The azimuthal/toroidal component of this helical field may well also play a role in collimating the jets. The magnetic field threading the accretion disk is often taken to be more or less randomly oriented, but it has a preferred orientation in some scenarios, such as the "cosmic battery" model described by Contopoulos et al. [8] and Christodoulou et al. [9]. Furthermore, the presence of a toroidal field component unambiguously implies the presence of a current in the jet, whose direction is given by the usual right-hand rule from university physics (Figure 1, adapted from [10]); thus, studies of helical fields associated with AGN jets are intrinsically tied to studies of the currents flowing in AGN jets.

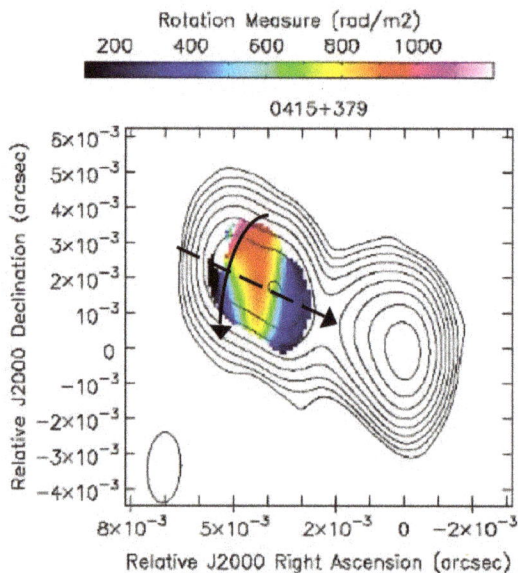

Figure 1. Relationship between an observed transverse Faraday RM gradient (colour scale), the direction of the associated azimuthal **B** field component (solid curved arrow) and the inferred direction of the current, inward in this case (dashed arrow). RM image superposed on intensity contours adapted from [10].

One of the signatures of synchrotron radiation is significant linear polarization. Synchrotron radiation can, in principle, be completely unpolarized if the synchrotron **B** field is fully disordered;

if the magnetic field is partially or fully ordered, however, this will lead to a corresponding degree of linear polarization, up to a maximum of about 75% for a fully-ordered magnetic field and a random pitch angle distribution for the population of radiating electrons [11]. The direction of the observed plane of linear polarization will be orthogonal to the synchrotron magnetic field in the case of optically-thin emission. The plane of linear polarization becomes aligned with the synchrotron **B** field in sufficiently optically-thick regions; however, this transition actually occurs at an optical depth considerably exceeding unity [12]. This means that the polarized emission we observe in radio images of AGN jets is likely all predominantly from optically-thin regions, including polarization from the "core", often interpreted as the location along the jet outflow where the optical depth is near unity [13].

Linear-polarization observations are the only source of direct information about both the orientation of the synchrotron **B** field and the extent to which this field is ordered. For this reason, polarization observations play a key role in studies of both AGN jets and the media through which they propagate. They are also an important means of searching for evidence of the helical **B** fields predicted to form during the launching of the jets.

Very Long Baseline Interferometry (VLBI) provides a means of imaging AGN with very high resolution, in order to obtain information about regions that are as close as possible to the "central engine" and jet-launching region. The jets of AGN are present on the smallest scales it has been possible to directly image with VLBI. At the other end, they can extend many kiloparsecs from the galactic nucleus, often displaying extremely good collimation. Observations of high resolution at multiple wavelengths are required if we wish to obtain information about the spectrum of the radio emission and Faraday rotation of the polarization angles arising at various locations between the emission region and the observer. The detection of a systematic Faraday-rotation gradient across an AGN jet can reveal the presence of a toroidal **B**-field component in the immediate vicinity of the jet, possibly associated with the helical **B** field predicted to form when the jets are launched (see Section 2.2).

In addition to the helical **B** fields that should be carried outward with the jets, there are numerous local phenomena that can also considerably influence the observed jet **B** fields, such as shocks, interactions with the surrounding medium, bending of the jets, turbulence and magnetic reconnection. Thus, one of the observational challenges is to determine which features are mainly associated with the action of local effects and which are mainly a consequence of the intrinsic helical/toroidal magnetic fields of the jets.

2. Observational Evidence for Helical B Fields in AGN Jets

2.1. Linear Polarization Structures

Variety of Observed Structures

The observed polarization in the jets of AGN on parsec scales usually tends to lie either parallel to or perpendicular to the local jet direction, with the inferred direction of the associated jet **B** field being orthogonal to the direction of the polarization, assuming the polarized emission arrives from predominantly optically-thin regions [14]. We should note here that the 90° rotation in the orientation of the observed linear polarization from orthogonal to aligned with the synchrotron **B** field in the transition from optically thin to optically thick occurs at an optical depth τ substantially greater than unity ($\tau \simeq 6$ [12]), so that even polarization observed in the VLBI core region (optical depth $\simeq 1$) is expected to arise in predominantly optically-thin regions. Thus, the assumption that the observed polarization patterns are associated with optically-thin emission is well justified.

Commonly-observed polarization patterns correspond to: (a) extensive regions of **B** field aligned with the local jet direction; (b) localized regions of **B** field oriented perpendicular to the local jet direction, associated with bright, compact features; (c) extensive regions of **B** field oriented perpendicular to the local jet direction; (d) the spine-sheath **B**-field structure across the jet; (e) **B** field aligned with the local jet direction and offset toward one side of the jet; and (f) **B** field aligned with the local jet direction around the outer edge of a bend in the jet. For brevity, we will refer to **B** field (or polarization) aligned with the jet as "longitudinal" and **B** field (or polarization) perpendicular to the jet as "orthogonal." A schematic of these various polarization configurations is shown by Gabuzda [15]. Lyutikov et al. [16], for example, have pointed out that this could be a natural consequence of local cylindrical symmetry displayed by the jets; in this case, the **B** field can always be described as a sum of longitudinal and orthogonal components projected onto the sky.

In fact, helical jet **B** fields could give rise to virtually all of these characteristic polarization patterns, as outlined below.

- (a), (c) Extended regions of longitudinal or orthogonal **B** field could be associated with helical jet **B** fields with comparatively low and high pitch angles (i.e., comparatively "loosely-wound" or "tightly-wound" helical fields), respectively. The pitch angle of the helical field should be due in part to the ratio of the velocities of the rotation and outflow.

- (d) A spine-sheath transverse polarization structure refers to a situation in which the predominant **B** field is orthogonal near the jet axis and longitudinal at one or both edges of the jet. This configuration could be associated with an overall helical field, with the azimuthal component of the field dominant near the jet axis, where its projection is orthogonal to the jet, and the longitudinal component of the helical field dominant near the jet edges, where its projection is along the jet (e.g., [16,17]). In this case, we would also expect to observe an increase in the degree of polarization from the central axis toward the jet edges.

- (e) Depending on the pitch angle of the helical field and the viewing angle of the jet, situations are possible in which the longitudinal **B** field predominates on one side of the jet, while the orthogonal **B** is dominant on the other. If the orthogonal **B** field is too weak to be detected in a particular image, this may also appear as a region of the longitudinal **B** field offset toward one side of the jet on its own (see, e.g., [18]).

- (f) The well-ordered longitudinal **B** field around the outer part of a bend in the jet could come about when the longitudinal component of a helical **B** field is enhanced by the curvature of the jet, stretching the **B** field at the outer edge of the bend.

The only type of pattern that is not naturally explained by the presence of helical jet **B** fields is regions of the orthogonal **B** field associated with bright, compact regions, which are probably more naturally associated with shock compression (see, e.g., [19–22]).

Most of these polarization patterns can also be caused by combinations of other, locally-acting, phenomena, such as interactions with the ambient medium. However, the principle of Occam's razor is relevant here, which can be stated as "if more than one explanation is possible, the simplest should be preferred". Another way of expressing this principle is that, the more assumptions that must be made to explain observations, the more unlikely the explanation. Since helical jet **B** fields are predicted theoretically to be present, Occam's razor suggests an approach in which we initially ask which of the observed polarization characteristics can straightforwardly be explained by the jet's helical field; if all the observed characteristics can be explained by this one theoretically-based assumption, this provides the simplest explanation of the observations. We can then look to other factors such as superposed shocks and shear to explain localized, anomalous or unusual features not easily explained by a helical jet **B** field, guided by additional information, such as variability or spectral data when possible. This approach is most likely to provide a relatively complete picture of the jet itself and the actions of various agents on it as it propagates outward from the AGN.

Another interesting question is whether the jet fluid itself follows helical stream lines, as in the picture of [23], or possibly even rotates as it propagates from the jet base, and what effect this may have on the observed polarization patterns in the jets. Marscher et al. [24] have suggested that one observational manifestation of motion of the jet plasma along helical stream lines could be rapid and smooth rotations of the observed polarization angles, probably upstream of the VLBI core. More direct evidence for the existence of streamline-like structures comes from the detection of oscillatory intensity structures within the overall jet flow in a few well-resolved AGN jets (e.g., [25,26]). However, at present, we have no firm information about how the presence of such helical streamlines in the VLBI jets could potentially affect the observed polarization, although this question is worthy of further study. In terms of possible rotation of the jet plasma, we have no means of detecting this directly, and no indirect evidence for rotation of AGN jets has been reported thus far, making it premature to speculate about possible effects of such rotation on the observed polarization patterns at this time.

It is also of interest to consider how the above polarization patterns could be affected by the angle at which a jet carrying a helical **B** field is viewed. This has been analysed, for example, by Murphy et al. [18], who presented figures summarizing the types of polarization structure observed for various combinations of helical pitch angle and viewing angle (Figures 1 and 2 of [18]). For example, taking into account the typical resolution of centimetre-wavelength VLBI observations, a jet carrying a helical field with a pitch angle in the rest frame of the jet of 80° will display the transverse **B** field all across the jet when the viewing angle in the rest frame of the jet is 80–90°, the longitudinal **B** field on one side and the transverse **B** field on the other if this viewing angle is roughly 60–80° and the longitudinal **B** field at the edges and the transverse **B** field near the jet axis (a "spine + sheath" structure) if this viewing angle is less than about 60°. A technique for deriving the intrinsic speed of the jet and the viewing angle in the observer's frame from (i) the pitch angle of the helical **B** field and the viewing angle in the rest frame of the jet derived from fitting transverse intensity and polarization profiles and (ii) the observed superluminal speeds in the jet is described in [18].

2.2. Faraday Rotation Gradients

An electromagnetic wave can be described as the sum of any two mutually-perpendicular components of the wave's electric field. In radio astronomy, these are often taken to be the Right Circularly-Polarized (RCP) and Left Circularly-Polarized (LCP) components of the **E** field. When such a wave passes through a magnetized medium present along the path from the source to the observer, asymmetry in the interactions between free electrons in the medium and the RCP and LCP components of the polarized electromagnetic wave leads to these two components having different speeds of propagation in the medium. This means that the RCP and LCP components become more and more out of phase as the wave passes through the medium, manifest as a rotation of the plane of polarization of the wave, known as Faraday rotation. The amount of rotation depends on the strength of the ambient magnetic field **B**, the number density of charges in the plasma n, the charge e and mass m of these charges and the wavelength of the radiation λ:

$$\chi = \chi_o + \text{RM}\lambda^2 \qquad\qquad \text{RM} = \frac{e^3}{8\pi^2 \epsilon_o m^2 c^3} \int n\mathbf{B} \cdot d\mathbf{l} \qquad (1)$$

where χ is the observed and χ_o the intrinsic emitted polarization angle, RM the Faraday Rotation Measure, ϵ_o the permittivity of free space and c the speed of light in a vacuum, and the integral is carried out over the line of sight from the source to the observer.

Although Faraday rotation can, in principle, be associated with any free charged particles in the magnetized plasma in which it arises, due to the $1/m^2$ dependence in the expression for the RM, it is usually assumed that the observed RM is due to electrons, rather than protons. Furthermore, because relativistic electrons have appreciably higher effective masses than non-relativistic ones, the electrons giving rise to the Faraday rotation are usually assumed to be non-relativistic (thermal). The RM depends on both the electron density and the Line Of Sight (LOS) **B** field; however, the sign of the RM is determined purely by whether the LOS **B** field is pointed toward (positive) or away from (negative) the observer. This means that the sign of the RM provides a unique diagnostic of the structure of the associated **B** field along the line of sight. In the simplest case when the Faraday rotation occurs outside the emitting volume in the source, it gives rise to a linear dependence of χ on λ^2, as is indicated above, providing a straightforward means of identifying the action of Faraday rotation observationally.

If a jet and its immediate vicinity carry a helical **B** field, this will give rise to a monotonic gradient in the observed Faraday rotation across the jet, due to the monotonic change in the LOS component of the helical **B** field across the jet. Some observers and theoreticians were aware of this effect as long ago as the 1980s (e.g., [27,28]), but it was largely ignored until the detection of a Faraday-rotation gradient across the parsec-scale jet of 3C273, reported in 2002 by Asada et al. [29].

Transverse RM gradients with statistical significances exceeding the 3σ level have now been detected across the jets of about 50 AGNs [30] and also across the kiloparsec-scale jet structures of a number of AGNs and radio galaxies [9,31]). The overall patterns displayed by these transverse RM gradients are discussed in the following subsection.

Note that, in principle, an RM gradient across a jet could come about due to a corresponding gradient in the electron density in a Faraday screen in the vicinity of the jet. However, variations in the electron density in the region of Faraday rotation cannot cause changes in the sign of the RM, whereas a transverse RM gradient due to a helical jet **B** field can encompass RM values of both signs. It is therefore noteworthy that an appreciable number of the transverse RM gradients detected across AGN jets change sign across the jet. Note also that, depending on the pitch angle of the helical field and the jet viewing angle, an RM gradient due to a helical jet **B** field may include RM values of only one sign, so that the lack of a sign change across the jet is not an argument against interpreting an RM gradient as reflecting the presence of a helical magnetic field being carried by the jet.

2.2.1. Overall Patterns of the Transverse RM Gradients on pc and kpc Scales: Evidence for the Action of a Cosmic Battery

The direction of an RM gradient observed across the jet of an AGN implies a direction for the azimuthal (toroidal) magnetic-field component that is giving rise to this gradient, which, in turn, implies the direction for a current associated with this toroidal field (Figure 1). The direction of the azimuthal component of the jet's helical field basically comes about due to the combination of the direction of the rotation of the central black hole and its accretion disk and the direction of the longitudinal field component that is partially transformed into a toroidal component by the rotation.

In this simplest picture, we would expect to see a particular direction for the observed transverse RM gradient all along the jet. This is true in some cases; in addition, transverse RM gradients in the same direction have been observed at multiple epochs in some AGN, indicating that these structures are fairly stable. However, at the same time, reversals in the directions of observed transverse RM gradients have been observed in a number of cases, either with distance from the base of the jet or with time (see [30] and references therein). This can be understood as a consequence of a structure with nested helical fields surrounding the jet, one inside the other, with the directions of the azimuthal field components being opposite in the two [32]. In this case, the line of sight passes through both regions of the helical **B** field, and the direction of the net observed RM gradient is determined by which region (inner or outer) makes the greater contribution to the overall Faraday rotation. If conditions in and around the jet lead to a change in whether the inner or outer region of helical field dominates the observed Faraday rotation, this will lead to a corresponding change in the direction of the associated RM gradient across the jet.

In fact, just such a nested helical-field configuration is expected in the "cosmic battery" model of [8,9]. Further, this model predicts that the orientations of the azimuthal components of the two regions of helical field are not random: the inner helical field should have a counter-clockwise azimuthal component projected onto the sky, associated with an inward current along the jet (as in the example shown in Figure 1), while the outer helical field should have a clockwise azimuthal component projected onto the sky, associated with a more extended region of outward current (essentially present between the two regions of the helical field). This situation is shown schematically in Figure 2.

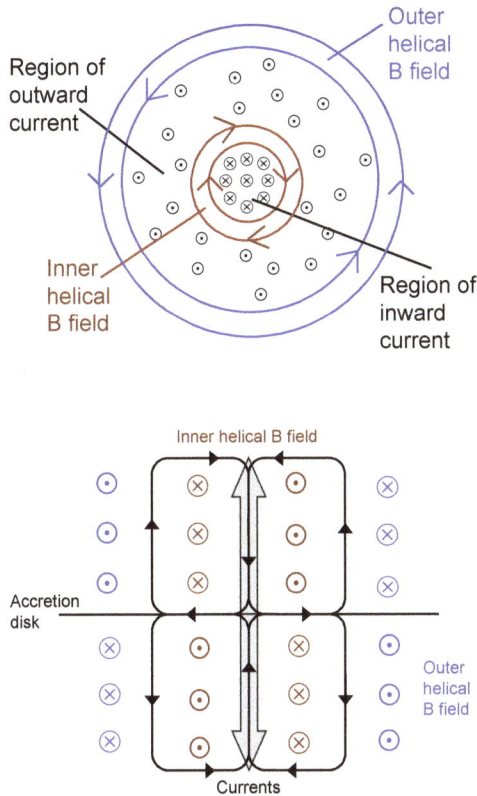

Figure 2. Schematic views of the set of helical **B** fields and associated currents inferred from parsec-scale and kiloparsec-scale Faraday RM gradients looking straight down the jet axis (upper) and viewed at 90° to this, in the plane of the accretion disk (lower). The overall configuration of currents resembles a coaxial cable, with inward current along the jet axis and a more extended region of outward current farther from the axis.

Sufficient statistics to test this picture have recently become available. Gabuzda et al. [30] have demonstrated that the collected data on RM gradients detected across about 50 AGN jets imaged with centimetre-wavelength VLBI indicate a statistically-significant predominance of inward currents (probability of the predominance being spurious $\simeq 0.40\%$). In contrast, the results of [9], supplemented by the more recent results of [31], show the opposite tendency: a strong predominance of outward currents on decaparsec–kiloparsec scales (probability of the predominance being spurious $\simeq 0.05\%$). These observations are consistent with a picture in which the jets have a nested helical-field configuration, with huge current loops oriented roughly orthogonal to the accretion disk, with the current flowing inward along the jet axis, away from the axis in the accretion disk, outward in an extended region around the jet and back toward the jet axis in a region near the termination of the jet. Determining the physical origin and nature of this set of **B** fields and currents will provide key information for our understanding of the formation and propagation of AGN jets.

This is broadly similar to the set of currents and **B** fields in a co-axial cable, but extending over kiloparsec scales. It is important that the data suggest not only a nested helical-field configuration, but specifically one corresponding to inward current along the jet axis and outward current in a more extended region around the jet. It is this clear preference for a particular direction for the currents flowing in and around the jets that strongly suggests the action of some kind of battery, such as the mechanism of [8,9]. Of course, other competing mechanisms may come to light; however, at present, this is the only mechanism specifically predicted to give rise to this configuration of fields and currents.

As was pointed out in [30], it is also interesting that a sizeable minority of the transverse RM gradients that have been detected on parsec scales correspond instead to outward currents. It may be that, in these sources, the outer region of the helical field has already become dominant on scales smaller than those on which the gradient is detected. It may also be that the battery-like mechanism that is leading to a dominance of inward currents along the jet axis on parsec scales is less efficient under some conditions; in this case, a relatively small number of the observed transverse RM gradients could have random directions. One of the most intriguing possibilities is that the cosmic battery mechanism described above may operate in the opposite sense ("in reverse") when the central black hole rotates very rapidly, at more than about 70% of its maximum possible rotational velocity, as suggested by the numerical computations of [33]. The inner edge of the accretion disk will approach closer to the black hole event horizon as the rotational velocity increases. If the rotation is fast enough, the inner edge of the disk is located close enough to the horizon for the rotation of space-time to exert the dominant effect, rather than the disk's rotation. This leads to a reversal in the direction of the radiation force on the electrons in the accretion disk, so that electrons acquire higher, rather than lower, toroidal velocities in the direction of rotation compared to protons in the disk. Searches for differences in the properties of AGNs whose transverse RM gradients correspond to predominant inward and outward currents in their jets on parsec scales would be valuable in this context.

2.2.2. Helical vs. Toroidal Fields

The detection of a transverse RM gradient indicates the presence of an azimuthal **B**-field component, but, on its own, this does not demonstrate that this is one component of an overall helical **B** field. It is therefore important to identify observational diagnostics that can distinguish whether this azimuthal field represents a toroidal field or one component of a helical **B** field. The key to this is whether the intensity and polarization profiles across the jet are symmetric or asymmetric. A toroidal field has no net longitudinal component (because this component is either absent or disordered) and will give rise to symmetric intensity and polarisation profiles across the jet, whereas a helical field has an ordered longitudinal component, which will give rise to asymmetric profiles across the jet. The key point here is that the synchrotron intensity and degree of polarization are both determined by the **B** field component in the plane of the sky.

When helical and toroidal fields are viewed side-on, it becomes apparent that a toroidal field will generally appear symmetric about the central jet axis, whereas a helical field will display asymmetry if viewed at any angle other than exactly 90° to the jet axis (Figure 3). A toroidal field should give rise to symmetric intensity and polarization profiles, independent of the angle at which the jet is viewed. However, the presence of a longitudinal component in the helical field introduces asymmetry. The dominant **B** field along one edge of the jet (top edge of the lower helical field in Figure 3) lies close to the line of sight, and we observe relatively lower degrees of polarization; in contrast, the dominant **B** field along the other edge of the jet (bottom edge of the lower helical field in Figure 3) lies close to the plane of the sky, and we observe relatively higher degrees of polarization. Only for particular viewing angles and helical pitch angles will a helical **B** field give rise to symmetrical intensity and polarization structures (see, e.g., [18,30]).

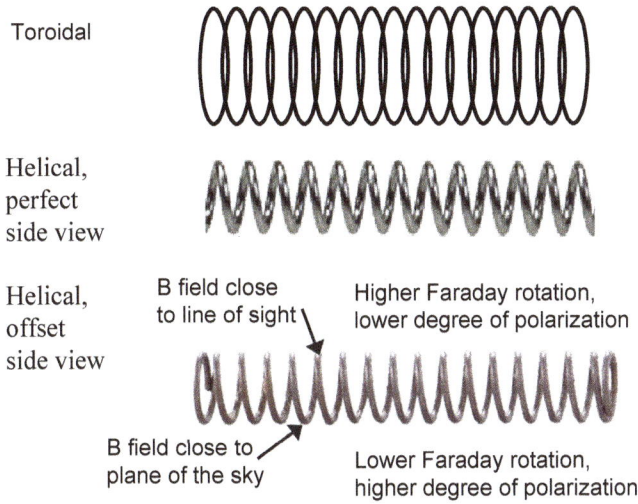

Figure 3. Schematic showing differences in the profiles presented by a toroidal field (upper image), a helical field viewed exactly side-on (middle image) and a helical field viewed slightly offset from exactly side-on (lower image). The jet axes are horizontal in each case and run down the centres of the toroidal and helical configurations shown. The first two cases give rise to symmetrical polarization profiles, whereas the last case gives rise to asymmetrical polarization profiles (in observations with sufficient resolution to detect this effect). In the offset side-view shown for the lower helix, the **B** field is oriented close to the line of sight along the upper edge of the helix and closer to the plane of the sky at the lower edge of the helix. Because Faraday rotation is proportional to the LOS **B** field, we expect the magnitude of the Faraday rotation to be higher at the upper than at the lower edge of this jet. Because the degree of polarization of synchrotron radiation is determined by the **B** field in the plane of the sky (perpendicular to the line of sight toward the observer), we expect the degree of polarization to be higher at the lower than at the upper edge of this jet.

In addition, the fact that Faraday rotation is proportional to the LOS component of the ambient **B** field provides additional potential tests of consistency of observations with the presence of helical **B** fields in jets displaying transverse Faraday RM gradients. The picture depicted in Figure 3 predicts that the side of such a jet with the higher degree of polarization (dominant **B**-field component in the plane of the sky) should display a lower amount of Faraday rotation, whereas the side of the jet with the lower degree of polarization (dominant **B**-field component along the line of sight) should display a higher amount of Faraday rotation. Indeed, some cases where this pattern is observed have been noted [18], but much more work remains to be done.

We should note that it may seem at first inappropriate to speak of viewing a helical field side-on, as in Figure 3), since we believe the jets of core-dominated AGN are oriented close to our line of sight. However, taking into account aberration, a small viewing angle of close to $1/\Gamma$ in the observer's frame (where Γ is the Lorentz factor for the bulk relativistic motion of the jet) will be close to a viewing angle of $90°$ in the jet rest frame, corresponding to the situations shown in Figure 3).

2.3. Variability of the Faraday Rotation Sign

During their monthly monitoring of the AGN Mrk421 at 15, 24 and 43 GHz with the Very Long Baseline Array in 2011, Lico et al. [34] discovered time variability in both the magnitude and the sign of the Faraday RM in the core region. Since the RM sign is determined by the direction of the LOS **B** field, it is not trivial to explain such rapid changes in the RM sign. They suggest that the RM sign reversals observed during their monitoring can be understood if the jet has a nested helical-field configuration such as the one described in Section 2.2.1 and depicted in Figure 2, with opposite directions for the azimuthal field components in the inner and outer regions of the helical **B** field; in this case, the variability of the RM sign could come about due to changes in whether the inner or outer region of the helical field dominates the observed Faraday rotation.

In fact, although the RM signs observed for some AGN can be stable over many years, the RM sign is variable for others. Furthermore, changes in the RM sign are also sometimes observed along the core–jet structure, even after taking into account the Faraday rotation occurring in our own galaxy. Such sign changes reflect changes in the direction of the LOS **B** field relative to our line of sight and so provide information about the three-dimensional **B** fields of these jets. The suggestion of Lico et al. [34] that RM sign changes could be associated with variability in the conditions in a nested-helical-field configuration is worthy of further study and may provide a relatively simple way to explain RM sign changes without invoking dramatic bending of the jet relative to the line of sight.

2.4. Inverse Depolarization

Looking at the wavelength dependence of the degree of polarization of a large sample of AGN in the range 8.1–15.4 GHz, Hovatta et al. [35] found that most of these sources displayed the decrease in their degree of polarization with decreasing frequency that is expected due to the action of Faraday and beamwidth depolarization. However, a small number of sources displayed the opposite tendency: increasing polarization with decreasing frequency, termed "inverse depolarization" by Homan [36]. Homan [36] presented a physical model for this phenomenon in which the structure of the jet magnetic field is such that internal Faraday rotation acts to align the polarization from the far and near sides of the jet; this leads to higher fractional polarization at longer wavelengths, where the superposed polarization from these two regions becomes better aligned. This effect can be produced naturally in both helical **B** fields and randomly-tangled magnetic fields; however, in the latter case, the fields must be tangled on length scales in the jet that are too long to be consistent with the observed levels of fractional polarization, making helical jet **B** fields a more attractive option. Homan [36] also notes that three of the four features that display inverse depolarization are in jets that have transverse RM gradients—a clear signature of the presence of a toroidal field that may represent one component of a helical field—increasing the likelihood that the explanation for this anomalous polarization behaviour lies in helical **B** fields carried by these jets. Incidentally, this is an example of the role of "Occam's razor" in helping us identify the simplest and most likely interpretation of the collected data.

2.5. Variability of Jet Ridge Lines

In their study of the kinematics of ridge lines of the VLBI jet of BL Lac, Cohen et al. [37] identified transverse patterns in the ridge lines, which moved superluminally downstream. They suggested that these patterns were analogous to waves on a whip. This behaviour led them to a model for the observed variability of the jet ridge lines in which the jet carries a tightly-wound helical **B** field, and the transverse patterns represent Alfvén waves propagating downstream along the longitudinal component of this helical field.

2.6. Large Polarization Angle Rotations Associated with Outbursts

Marscher et al. [24] carried out detailed analyses of sequences of VLBI images together with optical polarization measurements of BL Lac in order to probe the acceleration and collimation region upstream of the observed VLBI core. They suggested that rapid, smooth rotations of the polarization could reflect the motion of a polarized region in the jet along a helical stream-line, such as those expected in the model of [23]. Their analysis showed that a bright feature in the VLBI jet caused a double flare observed from optical up to TeV energies, accompanied by rotation of the observed polarization angle and a delayed outburst at radio wavelengths. Marscher et al. [24] concluded that this event was initiated in a region of helical **B** field, which they identified with the acceleration and collimation zone predicted theoretically, on scales upstream from the observed millimetre-VLBI core.

2.7. Double Polarization Angle Rotations

Cohen et al. [38] observed four double polarization angle rotations in OJ 287, in which the polarization angle rotated counter-clockwise for $\simeq 180°$ and then rotated clockwise back to roughly its initial value, over an overall time of 1–3 years. They were able to explain this phenomenon with a model in which two successive polarized outbursts occur, with the polarization angles rotating in each, but in opposite directions, superposed on a constant polarization component corresponding to the underlying quiescent jet. The question then becomes how to explain the occurrence of such pairs of counter-rotating polarized outbursts: Cohen et al. [38] suggested that these can be generated by the supermagnetosonic jet model of Nakamura et al. [39] and Nakamura and Meier [40], in which the jet carries a strong helical magnetic field. The pairs of outbursts are associated with forward and reverse pairs of fast and slow magnetohydrodynamical waves; the plasma inside the two fast/slow pairs rotates in a helical pattern around the axis of the jet, but in opposite directions.

This study supports the earlier suggestion by Cohen [41] that the jet of OJ 287 carries a helical **B** field, based on the evolution of the jet ridge lines, which are twisted and can be interpreted as sections of a rotating helix, as in the model of [39,40].

2.8. Circular Polarization

The circular polarization of synchrotron radiation is intrinsically much less than 1%. At the same time, a substantial minority (about 15%) of radio-loud AGN display low, but significant levels of circular polarization, typically a few tenths of a percent [42]. Thus, some other mechanism is required to explain this, and it is generally believed that the best candidate is the Faraday conversion of linear to circular polarization in a magnetized plasma [43]. Faraday conversion can operate in any situation where a linearly-polarized electromagnetic wave propagates through a region of magnetized plasma with a non-zero component of the ambient magnetic field parallel to the electric field of the wave. Most importantly, Faraday conversion is more efficient at generating circular polarization than the synchrotron mechanism under similar conditions. The presence of a helical jet **B** field provides a natural configuration facilitating Faraday conversion: linearly-polarized radiation emitted at the far side of the jet relative to the observer can be partially converted to circularly-polarized emission as it passes through the part of the helical field that is at the near side of the jet ([44,45]). Thus, transverse RM gradients and circular polarization exceeding the level expected for the synchrotron mechanism are both signs of a helical jet **B** field. When these properties are both detected, they can be analysed together with the jet's linear polarization structure to determine uniquely the direction of the longitudinal component of the helical **B** field, and hence infer the direction of the rotation of the central black hole and its accretion disk. Such an analysis was recently carried out for 12 AGN by Gabuzda [46]. Her results indicate numbers of outward and inward longitudinal B-field components that are equal within the statistical uncertainties and also statistically equal numbers of central black holes rotating clockwise and counter-clockwise projected on the sky. At the same time, the results suggest that the directions of the longitudinal field and of the central rotation are coupled: clockwise central rotation projected onto the sky is preferentially associated with an inward longitudinal **B** field, while counter-clockwise rotation is associated with an outward longitudinal **B** field. Together, this gives rise to a preferred orientation for the toroidal component of the resulting helical **B** field, which corresponds to inward current along the jet axis on parsec scales, just as is predicted by the cosmic battery mechanism of [8,9].

3. Summary

AGN jets are expected theoretically to carry helical **B** fields, due to the joint action of the rotation of the central black hole and accretion disk and the jet outflow. There is now substantial observational evidence that many or all jets do indeed carry helical **B** fields. The most direct evidence comes from the detection of statistically-significant transverse Faraday RM gradients across the jets of some 50 AGN on parsec scales and about a dozen AGN on larger scales of tens to thousands of parsecs. This clearly indicates that these helical fields persist to substantial distances from the jet base.

Furthermore, many of the characteristic polarization/magnetic-field structures observed in AGN jets on parsec scales can be understood as manifestations of a helical jet **B** field. Other evidence comes from studies of a wide variety of phenomena, including circular polarization, inverse depolarization, variability in jet ridge lines and the RM sign and polarization-angle rotations. There is considerable hope for building up links between these observations, since AGN jets carrying helical **B** fields may well display more than one of these phenomena, as well as polarization structures characteristic of helical jet **B** fields and transverse RM gradients. Joint analyses of the linear polarization structure, circular polarization sign and transverse RM gradients are a potentially powerful tools for revealing the full three-dimensional structures of helical **B** fields carried by AGN jets.

At the same time, some of the characteristic polarization structures observed in AGN jets are also consistent with the action of local agents, such as shocks, shear and jet bending. It can be difficult to identify unambiguously the origin of observed polarization structures in particular jets in practice without incorporating additional information about the distribution of the degree of polarisation, the morphology of the jet, the distribution of Faraday rotation in the vicinity of the jet and the possible presence of transverse RM gradients across the jet. In many cases, it is likely that the observed polarization structure is produced by both the intrinsic helical **B** field of the jet and the action of various local factors. One local factor that has been neglected so far in interpretations of observational data, but that may well play an important role, especially in variability, is magnetic reconnection. Work on identifying observational signatures of magnetic reconnection in AGN jets is needed.

The direction of a transverse RM gradient across an AGN jet implies a direction for the azimuthal **B**-field component producing it, which, in turn, implies the direction for the net current flowing inside the region occupied by this toroidal field. The collected data on transverse RM gradients observed on parsec to kiloparsec scales demonstrate a statistically-significant predominance of inward currents along the jet axis on parsec scales (probability of the predominance being spurious $\simeq 0.40\%$) and of outward currents on decaparsec–kiloparsec scales (probability of the predominance being spurious $\simeq 0.05\%$), presumably in a more extended region surrounding the jet. This is fully consistent with the nested-helical-field structure predicted by the "cosmic battery" model described in [8,9]. In this picture, the systems of currents and fields in AGN jets and their accretion disks are essentially similar to those for giant co-axial cables. Simulations of jet launching and propagation taking into account this type of nested-helical-field structure would be very valuable.

Thus, we are essentially in the process of initiating a new paradigm, in which the observed polarization (magnetic-field) structures in AGN jets on essentially all scales are at least partially due to the helical **B** fields carried by these jets. Superposed on the intrinsic polarization patterns associated with these helical fields are the results of local effects, such as shocks, shear, turbulence, jet bending, and magnetic reconnection. Although it may be virtually impossible to determine unambiguously precisely what factors are contributing to the observed polarization structure of a particular AGN jet, joint analyses of multiple types of observations provide hope for giving us some idea of how much of the observed patterns is due to the intrinsic helical jet **B** field and how much is associated with local effects. This will present interesting challenges for future studies in this area.

Funding: This research received no external funding.

Conflicts of Interest: The author declares no conflict of interest.

References

1.   Kellermann, K.I.; Sramek, R.; Schmidt, M.; Shaffer, D.B.; Green, R. VLA observations of objects in the Palomar Bright Quasar Survey. *Astron. J.* **1989**, *98*, 1195–1207. [CrossRef]

2. Blandford, R.D.; Payne, D.G. Hydromagnetic Flows from Accretion Discs and the Production of Radio Jets. *Mon. Not. R. Astron. Soc.* **1982**, *199*, 883–903. [CrossRef]

3. Blandford, R.D.; Znajek, R.L. Electromagnetic Extraction of Energy from Kerr Black Holes. *Mon. Not. R. Astron. Soc.* **1977**, *179*, 433–456. [CrossRef]

4. Nakamura, M.; Uchida, Y.; Hirose, S. Production of Wiggled Structure of AGN Radio Jets in the Wweeping Magnetic Twist Mechanism. *New Astron.* **2001**, *6*, 61–78. [CrossRef]

5. Lovelace, R.V.E.; Li, H.; Koldoba, A.V.; Ustyugova, G.V.; Romanova, M.M. Poynting Jets from Accretion Disks. *Astrophys. J.* **2002**, *572*, 445–455. [CrossRef]

6. Tchekhovskoy, A.; Bromberg, O. Three-dimensional Relativistic MHD Simulations of Active Galactic Nuclei Jets: Magnetic Kink Instability and Fanaroff-Riley Dichotomy. *Mon. Not. R. Astron. Soc.* **2016**, *461*, L46–L50. [CrossRef]

7. Barniol Duran, R.; Tchehovskoy, A.; Giannios, D. Simulations of AGN Jets: Magnetic Kink Instability versus Conical Shocks. *Mon. Not. R. Astron. Soc.* **2017**, *469*, 4957–4978. [CrossRef]

8. Contopoulos, I.; Christodoulou, D.; Kazanas, D.; Gabuzda, D.C. The Invariant Twist of Magnetic Fields in the Relativistic Jets of Active Galactic Nuclei. *Astrophys. J.* **2009**, *702*, L148–L152. [CrossRef]

9. Christodoulou, D.; Gabuzda, D.; Knuettel, S.; Contopoulos, I.; Kazanas, D.; Coughlan, C. Dominance of Outflowing Electric Currents on Decaparsec to Kiloparsec Scales in Extragalactic Jets. *Astron. Astrophys.* **2016**, *591*, A61–A71.

10. Gabuzda, D.C.; Roche, N.; Kirwan, A.; Knuettel, S.; Nagle, M.; Houston, C. Parsec Scale Faraday-rotation Structure Across the Jets of Nine Active Galactic Nuclei. *Astron. Astrophys.* **2017**, *472*, 1792–1801.

11. Pacholczyk, A.G. *Radio Astrophysics*; W. H. Freeman: San Franciso, CA, USA, 1970.

12. Wardle, J.F.C. The Variable Rotation Measure Distribution in 3C 273 on Parsec Scales. *Galaxies* **2018**, *6*, 5. [CrossRef]

13. Blandford, R.D.; König, A. Relativistic Jets as Compact Radio Sources. *Astrophys. J.* **1979**, *232*, 34–48. [CrossRef]

14. Lister, M.L.; Homan, D.C. MOJAVE: Monitoring of Jets in Active Galactic Nuclei with VLBA Experiments. I. First-Epoch 15 GHz Linear Polarization Images. *Astron. J.* **2005**, *130*, 1389–1417. [CrossRef]

15. Gabuzda, D.C. Parsec-Scale Jets in Active Galactic Nuclei. In *The Formation and Disruption of Black Hole Jets*; Astrophysics and Space Science Library; Springer International: Cham, Switzerland, 2015; Volume 414, pp. 117–148.

16. Lyutikov, M.; Pariev, V.I.; Gabuzda, D.C. Polarization and Structure of Relativistic Parsec-scale AGN Jets. *Mon. Not. R. Astron. Soc.* **2005**, *360*, 869–891. [CrossRef]

17. Pushkarev, A.B.; Gabuzda, D.C.; Vetukhnovskaya, Y.N.; Yakimov, V.E. Spine-sheath Polarization Structures in Four Active Galactic Nuclei Jets. *Mon. Not. R. Astron. Soc.* **2005**, *356*, 859–871. [CrossRef]

18. Murphy, E.; Cawthorne, T.V.; Gabuzda, D.C. Analysing the Transverse Structure of the Relativistic Jets of Active Galactic Nuclei. *Mon. Not. R. Astron. Soc.* **2013**, *430*, 1504–1515. [CrossRef]

19. Hughes, P.A.; Aller, H.D.; Aller, M.A. iPolarized Radio Outbursts in Bl Lacertae—Part Two—The Flux and Polarization of a Piston-Driven Shock. *Astrophys. J.* **1985**, *298*, 301–315. [CrossRef]

20. Hughes, P.A.; Aller, H.D.; Aller, M.A. Synchrotron Emission from Shocked Relativistic Jets. II. A Model for the Centimeter Wave Band Quiescent and Burst Emission from BL Lacertae. *Astrophys. J.* **1989**, *341*, 68–79. [CrossRef]

21. Laing, R. A Model for the Magnetic-field Structure in Extended Radio Sources. *Mon. Not. R. Astron. Soc.* **1980**, *193*, 439–449. [CrossRef]

22. Marscher, A.P.; Gear, W.K. Models for High-frequency Radio Outbursts in Extragalactic Sources, with Application to the early 1983 Millimeter-to-infrared Flare of 3C 273. *Astrophys. J.* **1985**, *298*, 114–127. [CrossRef]

23. Vlahakis, N. Disk-Jet Connection. In *Blazar Variability Workshop II: Entering the GLAST Era*; Astronomical Society of the Pacific: San Francisco, CA, USA, 2006; Volume 350, pp. 169–177.

24. Marscher, A.P.; Jorstad, S.G.; D'Arcangelo, F.D.; Smith, P.S.; Williams, G.G.; Larionov, V.M.; Oh, H.; Olmstead, A.R.; Aller, M.F.; Aller, H.D.; et al. The Inner Jet of an Active Galactic Nucleus as Revealed by a Radio-to-γ-ray Outburst. *Nature* **2008**, *452*, 966–969. [CrossRef] [PubMed]

25. Lobanov, A.; Hardee, P.; Eilek, J. Double Helix in the Kiloparsec-Scale Jet in M 87. In *Future Directions in High Resolution Astronomy: The 10th Anniversary of the VLBA*; Astronomical Society of the Pacific: San Francisco, CA, USA, 2005; Volume 340, pp. 104–106.

26. Lobanov, A.P.; Zensus, J.A. A Cosmic Double Helix in the Archetypical Quasar 3C273. *Science* **2001**, *294*, 128–131. [CrossRef] [PubMed]

27. Blandford, R.D. *Astrophysical Jets*; Cambridge University Press: Cambridge, UK, 1993; p. 26.

28. Perley, R.A.; Bridle, A.H.; Willis, A.G. High-resolution VLA Observations of the Radio Jet in NGC 6251. *Astrophys. J. Suppl.* **1984**, *54*, 291–334. [CrossRef]

29. Asada, K.; Inoue, M.; Uchida, Y.; Kameno, S.; Fujisawa, K.; Iguchi, S.; Mutoh, M. A Helical Magnetic Field in the Jet of 3C 273. *Publ. Astron. Soc. Jpn.* **2002**, *54*, L39–L43. [CrossRef]

30. Gabuzda, D.C.; Nagle, M.; Roche, N. The Jets of AGN as Giant Coaxial Cables. *Astron. Astrophys.* **2018**, *612*, A67–A79.

31. Knuettel, S.; Gabuzda, D.C.; O'Sullivan, S.P. Evidence for Toroidal B-Field Components in AGN Jets on Kiloparsec Scales. *Galaxies* **2017**, *5*, 61. [CrossRef]

32. Mahmud, M.; Coughlan, C.P.; Murphy, E.; Gabuzda, D.C.; Hallahan, D.R. Connecting Magnetic Towers with Faraday Rotation Gradients in Active Galactic Nuclei Jets. *Mon. Not. R. Astron. Soc.* **2013**, *431*, 695–709. [CrossRef]

33. Koutsantoniou, L.; Contopoulos, I. Accretion Disk Radiation Dynamics and the Cosmic Battery. *Astrophys. J.* **2014**, *794*, 27–38. [CrossRef]

34. Lico, R.; Gomez, J.L.; Asada, K.; Fuentes, A. On the Time Variable Rotation Measure in the Core Region of Markarian 421. *Galaxies* **2017**, *5*, 57. [CrossRef]

35. Hovatta, T.; Lister, M.L.; Aller, M.F.; Aller, H.D.; Homan, D.C.; Kovalev, Y.Y.; Pushkarev, A.B.; Savolainen, T. MOJAVE: Monitoring of Jets in Active Galactic Nuclei with VLBA Experiments. VIII. Faraday Rotation in Parsec-scale AGN Jets. *Astron. J.* **2012**, *144*, 105–138. [CrossRef]

36. Homan, D.C. Inverse Depolarization: A Potential Probe of Internal Faraday Rotation and Helical Magnetic Fields in Extragalactic Radio Jets. *Astrophys. J.* **2012**, *757*, L24–L28. [CrossRef]

37. Cohen, M.H.; Meier, D.L.; Arshakian, T.G.; Clausen-Brown, E.; Homan, D.C.; Hovatta, T.; Kovalev, Y.Y.; Lister, M.L.; Pushkarev, A.B.; Richards, J.L.; et al. Studies of the Jet in Bl Lacertae. II. Superluminal Alfvén Waves. *Astrophys. J.* **2015**, *803*, 3. [CrossRef]

38. Cohen, M.H.; Aller, H.D.; Aller, M.F.; Hovatta, T.; Kharb, P.; Kovalev, Y.Y.; Lister, M.L.; Meier, D.L.; Pushkarev, A.B.; Savolainen, T. Reversals in the Direction of Polarization Rotation in OJ 287. *Astrophys. J.* **2018**, *862*, 1. [CrossRef]

39. Nakamura, M.; Garofalo, D.; Meier, D.L. A Magnetohydrodynamic Model of the M87 Jet. I. Superluminal Knot Ejections from HST-1 as Trails of Quad Relativistic MHD Shocks. *Astrophys. J.* **2010**, *721*, 1783–1789. [CrossRef]

40. Nakamura, M.; Meier, D.L. A Magnetohydrodynamic Model of the M87 Jet. II. Self-consistent Quad-shock Jet Model for Optical Relativistic Motions and Particle Acceleration. *Astrophys. J.* **2014**, *785*, 152–157. [CrossRef]

41. Cohen, M.H. OJ 287 as a Rotating Helix. *Galaxies* **2017**, *5*, 12. [CrossRef]

42. Homan, D.C.; Lister, M.L. iMOJAVE: Monitoring of Jets in Active Galactic Nuclei with VLBA Experiments. II. First-Epoch 15 GHz Circular Polarization Results. *Astron. J.* **2006**, *131*, 1262–1279. [CrossRef]

43. Jones, T.W.; O'Dell, S.L. Transfer of Polarized Radiation in Self-absorbed Synchrotron Sources. I. Results for a Homogeneous Source. *Astrophys. J.* **1977**, *214*, 522–539. [CrossRef]

44. Ensslin, T.A. Does Circular Polarisation Reveal the Rotation of Quasar Engines? *Astron. Astrophys.* **2003**, *401*, 499–504.

45. Wardle, J.F.C.; Homan, D.A. The Nature of Jets: Evidence from Circular Polarization Observations. In *Particles and Fields in Radio Galaxies*; Astronomical Society of the Pacific: San Francisco, CA, USA, 2001; pp. 152–163.

46. Gabuzda, D.C. Determining the Jet Poloidal B Field and Black-Hole Rotation Directions in AGNs. *Galaxies* **2018**, *6*, 9. [CrossRef]

galaxies

MDPI

Review

Generation and Transport of Magnetic Flux in Accretion–Ejection Flows

Ioannis Contopoulos [1,2,†]

[1] Research Center for Astronomy and Applied Mathematics, Academy of Athens, 11527 Athens, Greece; icontop@academyofathens.gr; Tel.: +30-210-6597167
[2] National Research Nuclear University (Moscow Engineering Physics Intitute), 31 Kashirskoe Highway, 115409 Moscow, Russia
[†] Current address: Research Center for Astronomy and Applied Mathematics, Academy of Athens, 4 Soranou Efessiou Str., 11527 Athens, Greece.

Received: 9 November 2018; Accepted: 3 January 2019; Published: 9 January 2019

Abstract: Astrophysical accretion flows are associated with energetic emission of radiation and outflows (winds and jets). Extensive observations of these two processes in X-ray binary outbursts are available. A convincing understanding of their dynamics remains, however, elusive. The main agent that controls the dynamics is believed to be a large scale magnetic field that threads the system. We propose that during the quiescent state, the field is held in place by a delicate balance between inward advection and outward diffusion through the accreting matter. We also propose that the source of the field is a growing toroidal electric current generated by the aberrated radiation pressure on the innermost plasma electrons in orbit around the central black hole. This is the astrophysical mechanism of the Cosmic Battery. When the return magnetic field outside the toroidal electric current diffuses through the surrounding disk, the disk magnetic field and its associated accretion rate gradually increase, thus leading the system to an outburst. After the central accretion flow approaches equipartition with radiation, it is disrupted, and the Cosmic Battery ceases to operate. The outward field diffusion is then reversed, magnetic flux reconnects with the flux accumulated around the central black hole and disappears. The magnetic field and the associated accretion rate slowly decrease, and the system is gradually driven back to quiescence. We conclude that the action (or inaction) of the Cosmic Battery may be the missing key that will allow us to understand the long-term evolution of astrophysical accretion–ejection flows.

Keywords: black holes; accretion disks; X-ray binaries; active galactic nuclei; magnetic fields

---

1. Accretion–Ejection Flows around Astrophysical Black Holes

Active galactic nuclei (AGN), X-ray binaries (XRB) and other energetic astrophysical sources are believed to be powered by the infall (accretion) of gaseous matter (plasma) into a central black hole. The infall proceeds as a rotating disk along which matter gradually releases enormous amounts of gravitational energy in the form of energetic outflows (winds and jets) and radiation across the full electromagnetic spectrum.

In their seminal paper [1] (hereafter SS73), Shakura and Sunyaev proposed that the structure and radiation spectrum of the accretion disk depend mainly on the matter accretion rate $\dot{M}_{\mathrm{disk}}$ in the disk. They went on to calculate the radiation spectrum as a superposition of black-body spectra emitted as matter locally converts gravitational energy into radiation at all radii in the disk. Their model serves as the standard model of accretion flows to the present day. It is interesting that SS73 realized that high-energy radiation can evaporate the gas and counteract the matter inflow in the disk. They concluded that, at high accretion rates, most of the gas will outflow along the way from the outer to the central regions of the disk, and only a small fraction will accrete into the central black hole. The observational fact that accretion disks indeed generate outflowing winds and jets from their surfaces nevertheless remains rather surprising.

Morover, it has been observed that, in many cases, the kinetic power of the outflow exceeds the total luminosity of the disk by several orders of magnitude [2–4]. This may be explained by the presence of an external agent that removes angular momentum and thus also gravitational energy from the accretion flow. In the SS73 picture, angular momentum is removed through viscous stresses in the disk, and, therefore, gravitational energy is released locally at all radii as high-energy radiation. Recent observations suggest that angular momentum is removed via a large scale magnetic field that threads the accretion and ejection flows. In doing so, gravitational energy is transformed into the kinetic energy of the outflow, thus there is no need for it to be radiated locally in the disk. This action of the large scale magnetic field suppresses the total radiation from the disk and modifies its high-energy spectrum [5–7].

The study of accretion flows is very complicated by itself. Now that we have concluded that accretion flows co-exist and as it seems to depend strongly on ejection flows, one needs to study accretion and ejection as one coherent process. Over the past couple of decades, several works contributed to the development of our understanding of accretion–ejection flows (e.g., [8–16]). This may not be the common view of the community, but we consider the work of Ferreira and collaborators as the most comprehensive. Due to the high degree of complexity of the problem, their treatment of the accretion flow has been similar to that of SS73. Accretion proceeds in the form of a more-or-less standard Shakura–Sunyaev-type disk, and ejection takes place as a gradual small perturbation above its surface. Let us denote with $\dot{M}_{\mathrm{disk}}(r)$ the accretion rate at radius r in the disk, and $\dot{M}_{\mathrm{wind}}(r)$ the total outflow rate enclosed within radius r in the wind. We assume that the accretion–ejection structure extends from an internal boundary near the black hole horizon at radius r_{in} out to an outer boundary at radius r_{out}. We also assume that the densities $\rho_{\mathrm{disk}}(r)$ in the disk and $\rho_{\mathrm{wind}}(r)$ at the base of the outflow vary as power laws with distance, namely,

$$\rho_{\mathrm{wind}}(r) \propto \rho_{\mathrm{disk}}(r) \propto r^{-p} \tag{1}$$

(in a radially self-similar configuration all densities have the same power-law dependence), and that both the ejection velocity v_z at the base of the wind and the accretion velocity v_r in the disk follow Keplerian profiles

$$v_z(r) \propto v_r(r) \propto r^{-1/2} \tag{2}$$

Finally, we assume that the disk scale-height $h \equiv rc_s/v_K$ is proportional to r (c_s is the speed of sound at the midplane of the disk, $v_K \equiv (GM/r)^{1/2}$ is the Keplerian velocity, and M is the mass of the central black hole). Putting all the above together, we obtain that

$$\dot{M}_{\mathrm{disk}} = 2\pi r h v_r \rho_{\mathrm{disk}} \propto r^{1.5-p} \tag{3}$$

$$\dot{M}_{\mathrm{wind}} = \int_{r_{\mathrm{in}}}^{r_{\mathrm{out}}} 2\pi r v_z \rho_{\mathrm{wind}} \, dr \tag{4}$$

and since mass conservation requires that $d\dot{M}_{\mathrm{wind}}/dr = d\dot{M}_{\mathrm{disk}}/dr$, we obtain

$$\xi \equiv 1.5 - p = \frac{\rho_{\mathrm{wind}}}{\rho_{\mathrm{disk}}} \frac{r}{h} \frac{v_z}{v_r} \tag{5}$$

where ξ is the local ejection efficiency parameter of [8]. In general, we expect that v_z at the base of the wind is comparable to v_r in the disk. When $\dot{M}_{\mathrm{disk}} \approx \mathrm{const.}$, $\xi \approx 0$, $\rho_{\mathrm{wind}} \ll \rho_{\mathrm{disk}}$, and, therefore, the flow is dominated by a strong accretion disk, a small fraction of which outflows to infinity in the form of a wind. On the contrary, when ξ differs significantly from zero, ρ_{wind} may be a significant fraction of ρ_{disk}, and, therefore, $\dot{M}_{\mathrm{wind}}(r_{\mathrm{out}}) \approx \dot{M}_{\mathrm{disk}}(r_{\mathrm{in}})(r_{\mathrm{out}}/r_{\mathrm{in}})^{\xi} \gg \dot{M}_{\mathrm{disk}}(r_{\mathrm{in}})$. Ferreira and collaborators concluded that the ejection efficiency parameter must be very small [8,9]. More specifically, they concluded that, in cold flows, the disk winds are only a small perturbation of the accretion process. For $\zeta \approx 0$, the wind above the disk may be described by the Blandford and Payne (hereafter BP82) radially self-similar wind solution with $B \propto r^{-5/4}$ and $\rho \propto r^{-3/2}$ [17]. This has been the canonical model of disk winds for almost three decades now.

It is very interesting that BP82-type solutions do not extend to infinity, but are diverted toward the axis beyond some distance [18,19]. Several years later, the scaling of the density in the wind with radial distance was generalized to $\rho_{wind} \propto r^{-p}$ [19,20]. Contopoulos showed that, among radially self-similar solutions, only the ones with $p \leq 1$ extend to infinity in the form of cylindrically collimated jets. What is most interesting though is that the analysis of observations of X-ray absorption lines in extended winds from AGN and XRB by Fukumura and collaborators [21–25] have shown that such extended disk winds may be modeled best by power-law exponents p closer to 1 than 1.5 (e.g., $p \approx 1.2$ in the wind of GRO J1655-40 [25]). This observational result prompted [26] to propose that disk winds may actually be generated by a "warm" (instead of "cold") disk. On the other hand, ref. [27] even considered the possibility that the magnetic field in the disk is not held in place by a balance between inward advection of the field by the accreting matter and outward diffusion of the field through the disk. Our present understanding is that a value of p different from 1.5 does not necessarily imply flux imbalance. We expect that a density scaling with $p \neq 1.5$ may be compatible with $\dot{M}_{disk} \approx$ const. if $h \propto r^{p-1/2}$ (and not $\propto r$). In the present work, we assume that $p = 1.5$, but our conclusions should remain valid even for a different density scaling in the wind (although this has not been formally verified).

The issue of the transport of magnetic flux through the disk is not new. The assumed presence of the large-scale magnetic field threading the disk required the efficient inward transport of magnetic flux. This issue has been strongly debated over the years (e.g., [28–30]). What is even more perplexing is that the most recent state-of-the-art numerical simulations of magnetized accretion–ejection flows performed by the Harvard group (e.g., [31,32]) reach a so-called Magnetically Arrested Disk (MAD) state where, seemingly, accretion proceeds continuously, whereas the total magnetic flux accumulated over the central black hole saturates to a limiting maximum value. However, if one studies these numerical simulations more closely, one realizes that every parcel of accreting matter brings the magnetic flux associated with it very close to the black hole horizon. Then, right before that parcel of matter plunges into the horizon, it gets rid through reconnection of the magnetic flux that it carried all along. Matter continues to accrete, but magnetic flux accumulates in the vicinity of the black hole, continuously reducing the average density of the accreting matter. It is straightforward to see that, over time in these simulations, $\dot{M}_{disk}(r_{in})$ and the total magnetic flux Ψ_{BH} accumulated onto the central black hole indeed both remain unchanged on average. However, the density of the surrounding disk continuously decreases with time as one can see by the color becoming more and more yellow in the top panel of Figure 1 in [31]. The duration of these simulations may be very long in numerical terms (several tens of thousands of dynamical times GM/c^3), but in actual physical terms, the time scale is tiny. To our understanding, the issue of magnetic flux transport in accretion disks remains open, and the numerical simulations of the Harvard group did not produce a convincing answer.

The origin of the magnetic field itself remains an open question. Ref. [32] argued that "observations do show patches of coherent magnetic flux surrounding astrophysical systems that can feed black holes". They reached the conclusion that even if only about 10 percent of it manages to accrete via the accretion disk, it would be enough to generate all the interesting accretion–ejection phenomena that we are investigating (see references in their paper). Here, we discuss another possibility, namely that constant-polarity flux is generated by the Poynting–Robertson drag effect on the plasma electrons in a so-called Cosmic Battery around the central black hole [33,34]. As we show below, the astrophysical implications of this effect, first proposed by Contopoulos and Kazanas more than twenty years ago, may be the missing key needed to understand the long time-evolution of accretion–ejection structures.

2. The Origin of the Magnetic Field: A Cosmic Battery

Let us discuss first how radiation acts on a plasma. In non-relativistic dynamics, radiation is introduced as an extra term in the equation of motion of the plasma, namely

$$\rho \frac{d\mathbf{v}}{dt} = \cdots + \frac{\rho}{m_p} \mathbf{f}_{\text{rad}} , \tag{6}$$

where ρ and $\mathbf{v}$ are the plasma matter density and velocity, respectively, and m_p is the mass of the proton (for ease of presentation, we assume a simple electron–proton plasma). It is well known that radiation acts on the plasma electrons, and much less on the plasma protons, and in fact $\mathbf{f}_{\text{rad}}$ is the radiation force per electron. Thus, how is it possible that radiation contributes to the dynamics of the plasma as a whole, e.g., how is it possible that radiation holds stars from collapsing under their own weight? The answer is that, in the presence of radiation, an inductive electric field $\mathbf{E}$ develops in the interior of the plasma. Why is this electric field important becomes clear when we consider the equations of motion not only of the protons, but also of the electrons. In the presence of radiation, the equation of motion of the protons contains an extra term

$$m_p \frac{d\mathbf{v}_p}{dt} = \cdots + e\mathbf{E} , \tag{7}$$

and the equation of motion of the electrons also contains the radiation force term

$$m_e \frac{d\mathbf{v}_e}{dt} = \cdots + \mathbf{f}_{\text{rad}} - e\mathbf{E} . \tag{8}$$

Here, $\mathbf{v}_p, \mathbf{v}_e$ are the velocities of the protons and the electrons, respectively, m_e is the mass of the electron, and e is the magnitude of the electron charge. The velocities of the electrons and the protons do not differ much and $m_e \ll m_p$, therefore, Equation (8) is equal to zero to a very good approximation. Thus,

$$\mathbf{E} = \frac{\mathbf{f}_{\text{rad}}}{e} . \tag{9}$$

This is how the radiation force appears in the equation of motion for the protons and the plasma as a whole. This effect is often ignored by the younger generation of researchers. Another way to explain this result is that, without the electric field, the radiation force would have disturbed the motion of the electrons so dramatically that an enormous electric current and an associated magnetic field would have appeared in the interior of the highly conducting astrophysical plasma. This is, however, not the case. It is not possible to "simply turn on" an electric current inside a highly conducting plasma. Maxwell's equations and in particular Faraday's induction law teach us that the plasma will react, and an induction electric field will appear in the direction opposite to the direction of the growing electric current that will counteract its growth. It is thus not correct to study the growth of the electric current and the associated magnetic field by considering the velocity difference between the electrons and the protons, and their number densities in the plasma. This approach ignores the inductive reaction of the plasma, and leads to wrong conclusions.

The radiation force per electron is calculated in the rest frame of the electron. Even if the radiation field is isotropic, the moving electron absorbs photons coming from the direction opposite to its direction of motion, and then re-emits them isotropically, losing momentum in the process. This is how the electrons feel the radiation force. In relativistic tensor notation, the spatial components of the radiation force are equal to

$$f^i_{\text{rad}} = \frac{\sigma_T F^i_{\text{rad}}}{c} \tag{10}$$

where σ_T is the electron Thomson cross-section and F_{rad}^i is the radiation flux (flow of radiation energy per unit surface) as seen in the frame of the electron. The radiation flux components are given by the projection of the stress-energy tensor of the radiation $T_{rad}^{\mu\nu}$ in the frame of the electron as

$$F_{rad}^i = h_\nu^i T_{rad}^{\mu\nu} u_\mu , \qquad (11)$$

where, u^μ is the electron 4-velocity which almost coincides with that of the plasma, and $h_\nu^\mu \equiv -\delta_\nu^\mu - u^\mu u_\nu$ is a tensor that projects opposite to the target electron 4-velocity [35].

The calculation of the radiation field in the vicinity of an astrophysical source of X-rays is complex and involves radiation transfer with absorption, emission, and detailed ray tracing. Another simpler approach is to consider the radiation field as a fluid, but this is not as accurate (see [36] for details). Ref. [37] performed ray-tracing calculations over several optically thick equatorial distributions of matter (thin and thick disks, thick torus) extending beyond the Innermost Stable Circular Orbit (ISCO) around the central spinning black hole. Koutsantoniou (unpublished work) recently extended this calculation over an optically thin radiation-emitting torus.[1] In Figure 1, we show in Mollweide projection[2] the view of the torus as seen from the frame of the plasma rotating with Keplerian velocities at various distances in the equatorial plane. The central black hole and its associated "light ring" can be discerned. The direction of rotation is to the right of the black hole. The region to the left of the black hole is the region opposite to the direction of the rotating plasma. It is interesting to notice here that inside about $3GM/c^2$, the radiation field becomes stronger along the direction of motion. This effect is due to the rotation of spacetime which forces most photons near the black hole horizon to reach the rotating plasma from behind. The normalized azimuthal radiation force per electron at the ISCO around a 5 $M_\odot$ Kerr black hole with various spin parameters a are listed in the third column of Table 1.

We can obtain a crude estimate of the radiation force per electron in the idealized case of a central point radiation source emitting with luminosity L, and an electron in circular motion around it at distance r. In that case,

$$\mathbf{f}_{rad} = \frac{L\sigma_T}{4\pi r^2 ce}\hat{r} - \frac{L\sigma_T}{4\pi r^2 ce}\left(\frac{v^\phi}{c}\right)\hat{\phi} . \qquad (12)$$

The latter is the same expression as the well known Poynting–Robertson azimuthal drag force on dust grains in orbit around the sun, only in that case the Thomson cross-section is replaced by the geometric cross-section of the grains [38,39]. Obviously, the radiation field around an accreting rotating black hole is far more complex, as suggested by the calculated sky maps shown in Figure 1.

We will now discuss the growth of the magnetic field generated by the intense radiation and rotational velocity fields in the innermost accretion flow around an astrophysical black hole. The correct way to do this is via the induction equation, which in the case of an astrophysical accretion disk threaded by a large scale poloidal magnetic field **B** takes the form

$$\frac{\partial \mathbf{B}}{\partial t} = -\nabla \times (-\mathbf{v} \times \mathbf{B} + \mathbf{E}c + \eta\nabla \times \mathbf{B}) , \qquad (13)$$

where η is the magnetic diffusivity in the interior of the disk. If $\nabla \times \mathbf{E} = 0$, as is the case in stellar interiors, radiation does not generate electric currents neither magnetic fields. In stars, radiation pushes the electrons outwards leading to a surplus of electrons in the outer layers of the star, and a depletion in its center. As a result, stars become electrically polarized along their radius and develop

---

[1] The distribution of matter in that torus is ad hoc, not a result of a numerical simulation. The torus has a circular cross section between r_{ISCO} and $2r_{ISCO}$, and a density distribution that drops exponentially such that the optical depth from the center to the surface is taken to be equal to 2. The motion is everywhere Keplerian.

[2] The Mollweide or homalographic projection is a common map projection generally used for global maps of the world or the sky. The equator is represented as a straight horizontal line perpendicular to the central meridian in the direction of the black hole.

an electric potential difference between the center and the surface (in the case of the sun, this is on the order of one volt), but no magnetic fields are generated. As we now see, rotation introduces electric fields with non-zero rotation (curl).[3]

Combining the simple expression obtained in Equations (12) and (9), and the integrated form of Equation (13), we obtain

$$\frac{\partial \Psi}{\partial t} \approx 2\pi r \left((\mathbf{v} \times \mathbf{B})^\phi - \frac{c f_{\mathrm{rad}}^\phi}{e} + \eta (\nabla \times \mathbf{B})^\phi \right) , \tag{14}$$

where Ψ is the magnetic flux contained inside radius r (see also [41] for a generalized expression). The second term in the right-hand side of the above equation generates a poloidal magnetic field in the direction of the angular velocity vector ω in the disk (f_{rad}^ϕ is negative in most parts of the disk except possibly in its innermost part just above the horizon, as suggested by the results obtained in Figure 1). The magnetic flux thus generated closes in the outer parts of the disk not reached by radiation from the center where f_{rad}^ϕ drops to zero. This poloidal magnetic flux will be advected inwards by the ideal accretion flow represented by the first term in Equation (14). We assume that, inside the ISCO, ideal MHD conditions apply, and flux accumulated inside the inner edge of the disk will keep growing. The growth will cease and the accumulated magnetic flux will saturate if the flow also carries along the return polarity of the magnetic field [33,41,42]. However, if the return polarity lies in a region with significant magnetic diffusivity so that the third term in Equation (14) dominates over the first [28–30] (see also, however, [43]), the growth of the accumulated magnetic flux will proceed unimpeded. This latter point was emphasized by Contopoulos and Kazanas (see Figure 1b in [33]) and was missed by [41]. As we show in the next section, actual accretion disks may favor both types of field evolution at different stages of their evolution (outward flux diffusion followed by the generation of poloidal magnetic field loops around the disk's inner edge, and inward flux advection followed by field reconnection with the flux accumulated inside the inner edge of the disk).

The field growth cannot proceed beyond equipartition. There are various definitions of equipartition (e.g., balance of gravity by radiation pressure, balance of gravity by magnetic forces, etc.). Whatever the definition, it is clear that the field cannot keep growing steadily beyond its equipartition value B_{eq} at the inner edge of the disk, because the innermost accretion region will be severely disrupted, and our analysis will break down. Nevertheless, we can estimate a rough timescale τ_{CB} for the innermost magnetic field to grow to astrophysically significant magnetic field values of order B_{eq} if we assume that the system radiates continuously at roughly its Eddington value. That timescale may be obtained from a dimensional analysis of Equation (14) as

$$\tau_{\mathrm{CB}} \sim \frac{e B_{\mathrm{eq}} r_{\mathrm{ISCO}}}{c f_{\mathrm{rad}}^\phi} . \tag{15}$$

The values of τ_{CB} obtained for a $5 M_\odot$ black hole with various black hole spin parameters and $B_{\mathrm{eq}} = 10^7$ G are shown in the fourth column of Table 1. These characteristic timescales vary from about one hour (for maximally rotating stellar mass black holes) to several days (for slowly rotating ones). These times scale roughly proportionally to $M^{3/2}$ with black hole mass, and therefore, the corresponding times to reach equipartition range from one to ten billion years for $10^8 \, M_\odot$ supermassive black holes. We emphasize that these rough estimates have been obtained for accretion flows that radiate continuously at close to their Eddington limit. As we show in the next section, in realistic astrophysical sources such as XRB outbursts, this is not the case.

---

[3] Notice, however, that the contribution of rotation to the growth of the solar magnetic field by the Poynting–Robertson effect is insignificant [40].

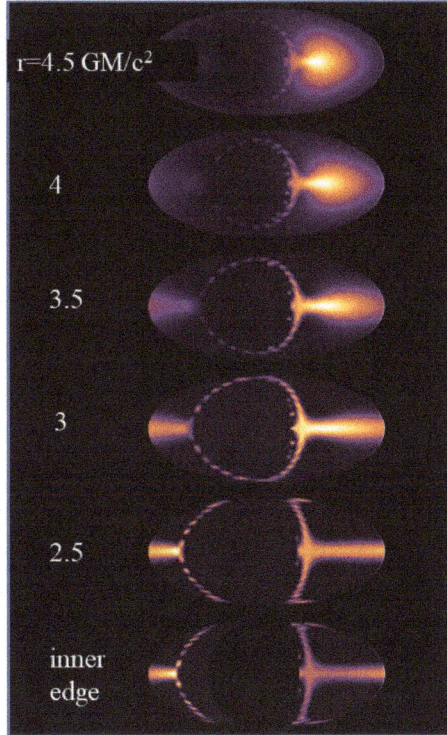

Figure 1. Sky maps in Mollweide projection obtained by ray-tracing of the radiation field emitted by an equatorial torus as seen from an equatorial Keplerian observer at various distances r in units of GM/c^2 (see Footnote 1 for details). The central black hole and its associated "light ring" can be discerned. Black hole spin parameter $a = 0.9\,M$. The direction of rotation is to the right of the black hole. The color scale corresponds to the intensity of radiation.

Table 1. Azimuthal Radiation Force per electron at the ISCO and Cosmic Battery Timescales.

a/M	$r_{ISCO}/(GM/c^2)$	$f^{\phi}_{rad}/(GMm_p/r^2_{ISCO})$	τ_{CB} in Hours
0	6	−0.11	39
0.1	5.7	−0.12	31
0.2	5.3	−0.13	25
0.3	5.0	−0.15	19
0.4	4.6	−0.18	15
0.5	4.2	−0.21	11
0.6	3.8	−0.27	8
0.7	3.4	−0.38	5
0.8	2.9	−0.59	3
0.9	2.3	+1.19	2
0.92	2.2	+1.46	2
0.94	2.0	+1.87	1
0.96	1.8	+2.58	1
0.98	1.6	+4.17	1

Observational confirmation of the Cosmic Battery in astrophysical systems (or any other kind of battery mechanism) may be found in observations of magnetic field asymmetries [44–47]. As is well known, the equations of motion in MHD involve only quadratic terms in the magnetic field through the Lorentz and electric forces $J \times B \sim (\nabla \times B) \times B \sim B^2$ and $\rho_e E \sim (\nabla \cdot E)E \sim E^2 \sim B^2$, respectively.

In other words, the dynamics of the flow do not depend on the direction of the magnetic field, unless some kind of battery mechanism is in action. Ref. [47] compared the direction of galactic rotation and the direction of the line-of-sight magnetic field in the central regions of nine spiral galaxies seen edge on. The observations were found to be in agreement with the magnetic asymmetry $B \parallel \omega$ predicted by the Cosmic Battery. Another asymmetry has to do with the helical structure of the magnetic field in the jet, and, in particular, with the direction of its axial electric current. The Cosmic Battery predicts that the axial electric current contained in the electron–proton disk wind/jet always flows away from the black hole (see Figure 2 of [46] for details). A definite direction of axial electric current is related to a steady Faraday rotation measure gradient across the wind/jet. We were able to observe steady gradients across parts of the kpc-scale wind/jet in only 18 cases, and, in all of them, the direction of the axial electric current was found to be outwards, in agreement with the prediction of the Cosmic Battery. For a more detailed review, the reader may consult [34].

3. A New Paradigm

We hope that we managed to convince the reader that the Cosmic Battery is one plausible origin for the large-scale magnetic field that threads the accretion–ejection flows around astrophysical black holes. Other options may be found in the recent work on mean-field dynamos in disks [48–50]. In fact, the Cosmic Battery could be considered as the source for the mean field dynamo in these simulations.

We now argue that the Cosmic Battery may be precisely the missing key that controls the general evolution of these systems. The ideal laboratory to test our ideas are X-ray binary outbursts where a multitude of temporal and spectral observations still awaits the development of a coherent self-consistent picture. Let us first consider the angular momentum conservation equation in the disk, namely

$$\dot{M}_{\text{disk}} \frac{\partial(r v_K)}{\partial r} + \frac{\partial}{\partial r}(4\pi r^2 t_{r\phi} h) + r^2 B_\phi B_z = 0 . \tag{16}$$

Here, $t_{r\phi}$ is the tangential viscous stress and B_ϕ, B_z are the toroidal and vertical components of the magnetic field at the base of the wind (notice that the product $B_\phi B_z$ is negative in a magnetic field configuration that removes angular momentum from the disk). Recent numerical simulations (see Figure 2 of [51]) suggest that the origin of the disk viscosity may be the Magneto-Rotational Instability (MRI) [52], and that

$$t_{r\phi} \sim \frac{B^2}{4\pi} , \tag{17}$$

where B is the average value of the magnetic field in the disk midplane. This numerical result allows us to simplify our analysis and to assume that the main factor that controls the removal of angular momentum in order for accretion to proceed is the large scale magnetic field that threads the accretion disk. In fact, Equation (16) may be rewritten as

$$\dot{M}_{\text{disk}} \sim \frac{2r^2 B_\phi B_z}{v_K} + \frac{2}{v_K} \frac{\partial}{\partial r}(r^2 B^2 h) = \frac{\epsilon r^2 B^2}{v_K} , \tag{18}$$

where ϵ is a factor of order unity. Ref. [53] argued that B^2 (calculated in the interior of the disk) may be much larger than $|B_\phi B_z|$ (calculated on the surface of the disk). In the present paper, we make the more natural assumption that the two are of the same order. In what follows, we also drop factors of order unity in our calculations.

In their seminal paper SS73, Shakura and Sunyaev were the first to emphasize the role of the magnetic field in the transport of angular momentum in astrophysical disks. They considered a turbulent magnetic field in the disk and incorporated its contribution to the disk viscosity in their now famous α-parameter. Many years later, several researchers realized that a similar effect may be due to the action of a large scale magnetic field that threads the accretion disk, only now magnetic torques remove angular momentum to large distances ("infinity") above and below the disk, and not radially through the disk. There exist many solutions in the literature in which the magnetic breaking mechanism is shown to work (e.g., [9,54]), but it seems that the astrophysical community still tries to "do everything" through the α-parameter without specifically mentioning the magnetic field.

In a slight departure from the work of Ferreira and collaborators, we consider the possibility that the distribution of magnetic field through the disk may evolve due to a slight imbalance between inward advection and outward diffusion through the disk, with one winning over large or small parts of the disk over the other. The timescale of this evolution is much longer than the dynamical timescale in the disk, and, therefore, the conclusions of previous works where the magnetic field was considered to be perfectly balanced in its position across the disk remain valid to a very good approximation.

Let us now discuss the evolution of a typical black-hole X-ray binary outburst, in particular GX 339-4 during its 2002–2003 outburst, through its associated Hardness-Intensity Diagram (HID). For this system,

$$M \approx 14 \, M_\odot \, , \, r_{in} \approx 6GM/c^2 = 1.3 \times 10^7 \text{cm} \, , \text{ and } r_{out} = 1.5 \times 10^9 \text{ cm} = 100 \, r_{in} \, . \tag{19}$$

Five positions in the HID are very important during the system's evolution (see Figures 2–4).

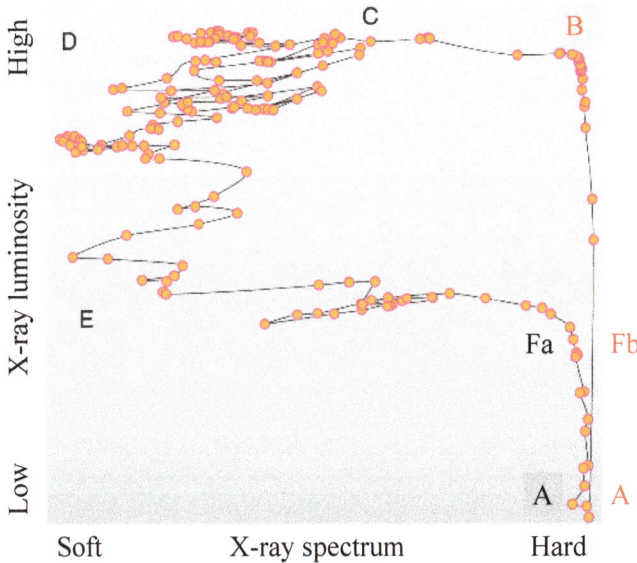

Figure 2. Typical XRB Hardness-Intensity Diagram (HID) (shown the one for XRB GX 339-4 during its 2002–2003 outburst). A: Quiescent state; B: High Hard state with fully developed compact jet; C: Bright Intermediate state where the compact jet is destroyed; D: Soft High state with no jet; E: Faint Intermediate state; Fa: Intermediate state traversed downwards (compact jet reappears but decreases in size); Fb: Intermediate state traversed upwards (compact jet grows in size). Red fonts: states where the Cosmic Battery operates (adapted from [55]).

A: The Quiescent state, which lasts for several months to a few years. In that state, the accretion disk is threaded by a large scale weak magnetic field that generates weak radio emission, probably also a weak magnetically driven wind, but no compact jet. In such a configuration, the mass loss in the wind is very weak, $\dot{M}_{disk} \approx$ const., $\zeta \approx 0$, thus Equation (18) yields

$$B(r) \approx B_{in} \left(\frac{r}{r_{in}} \right)^{-5/4} \, , \tag{20}$$

where $B_{in} \equiv B(r_{in})$. This is the canonical radial magnetic field scaling first proposed in BP82. In that case, the total magnetic flux threading the accretion disk that extends from $r = r_{in}$ to r_{out} is equal to

$$\Psi_{disk} = \int_{r_{in}}^{r_{out}} 2\pi r B(r) dr \approx \frac{8\pi}{3} B_{in} r_{in}^2 \left(\frac{r_{out}}{r_{in}}\right)^{3/4} \approx 260 \, B_{in} r_{in}^2 \tag{21}$$

In the present work, we assume that ζ remains always very close to zero, and that the radial field configuration always remains close to the canonical BP82 one. Notice that the investigation of BP82-type Magnetized Accretion–Ejection Structures (MAES) by Ferreira and collaborators required that the magnetic field stays in place and does not diffuse inwards or outwards through the accreting flow. In reality, the total magnetic flux threading the disk may slowly change as the system evolves in the HID, and, therefore, to keep the BP82 radial scaling, the field will re-arrange itself as the system evolves during the XRB outburst. We conclude that, in practice, the delicate balance between inward advection and outward diffusion is not one hundred percent satisfied. We show below why this is very important during an XRB outburst, and how it may be related to the action of the Cosmic Battery around the central black hole.

B: The High Hard state. At some point in time, without any prior indication, the system "decides" to flare up. It takes a few months for the outburst to rise to its highest luminosity. During that time, a compact jet appears, whose radius at its base seems to increase with time as the system evolves from the quiescent to the high-hard state [56]. Quoting [5], "whenever a disk is capable of driving jets, these will carry away a fraction of the released accretion (gravitational) energy, and the disk luminosity will be quenched (it radiates only a small fraction of the accretion power)". It is well known that we can interpret the observed spectra and spectra variations during an XRB outburst with a low radiative efficiency ADAF disk model inside an evolving transition radius r_{tr} (e.g., [57–60]). Ferreira and collaborators argued that it is possible to interpret the observations equally well with their JED model [5–7]. We prefer the latter explanation since, as we show below, it requires a minimum number of assumptions and free parameters. Notice that the rise from the Quiescent to the High state must be explained.

C: The Bright Intermediate state. The system transitions from the Hard to the Soft High state, and the radius of the compact jet quickly diminishes in size at its base [56]. We associate this with a reduction of the size of the inner JED. At some point, the jet disappears abruptly and episodically in the form of a micro-quasar. According to the theory of JEDs, this corresponds to an abrupt reduction of the magnetization in the jet. Obviously, if the system is threaded by a large scale unidirectional magnetic field, there is no way to make the magnetic flux disappear. On the contrary, if magnetic flux is carried inwards by the shrinking JED, the magnetic field and its associated magnetization are both expected to increase and not decrease. This sudden reduction of the magnetization in the inner accretion disk must be explained.

D: The Soft High state. In that state, the JED has disappeared, and the soft emission originates in a disk of Shakura–Sunyaev type. The system stays in the soft state but gradually decreases in luminosity in the course of several months. This decrease in luminosity is not monotonic, and the system makes several unsuccessful efforts to return to the hard state. This non-monotonic decrease in luminosity must be explained.

E: The Faint Intermediate state. After several months, the system transitions to a state beyond which the system is "ready" to return to the Hard state without hesitating. It is interesting that a particular bursting XRB system may reach differing levels of peak luminosity in different outbursts, but always returns to approximately the same Faint Intermediate state before it decides to return to the quiescent state in the HID. This observation must be explained.

F: The Low Hard state. The return point. The system reaches the Low Hard state and then decides to turn downwards in the HID and return to the quiescent state. Position Fa during the descending phase is very close to position Fb during the ascending crossing of the Low Hard state. The system configuration must be very similar at Points Fa and Fb (similar spectra and luminosities, similar disk types, similar radio emissions, and similar distributions of large-scale magnetic field through the disk), yet in the former the system traverses the HID downwards, whereas in the latter it traverses it upwards. Why that happens remains unexplained.

We believe that the answer to the latter question holds the key to understanding the dynamics of Magnetized Accretion–Ejection Structures. We propose that the answer has to do with the delicate balance (or better imbalance) between inward flux advection and outward flux diffusion. In our physical picture, the action (or inaction) of the central Cosmic Battery plays a key role as follows:

The Cosmic Battery generates the magnetic field in the immediate vicinity of the central black hole. One polarity of the field is held by the accreting matter infalling onto the central black hole, whereas the return polarity diffuses outward through the outer accretion disk. Let us denote the rate of return field generation by the Cosmic Battery as $\dot{\Psi}_{CB}$. Obviously, $\dot{\Psi}_{CB}$ depends on the innermost disk luminosity L, which depends on the accretion rate in the disk (see below). The pressing question is to understand why the system beyond Fa evolves in the ascending part, whereas beyond Fb in the descending. We propose that the answer lies with the direction of magnetic flux redistribution: in the ascending part of the HID, outward diffusion slightly wins over inward advection, and, therefore, the disk is slowly filled with magnetic flux generated by the CB. According to the discussion in the previous section, the CB yields an approximate rate of generation of magnetic flux in the disk

$$\dot{\Psi}_{CB} \approx \frac{L\sigma_T}{4\pi r_{in}^2 ec} \left(\frac{v_K(r_{in})}{c} \right) 2\pi r_{in} c = f_{CB} \frac{\dot{M}_{disk}c^2 \sigma_T}{2 r_{in} e} , \tag{22}$$

where the factor in front is defined as $f_{CB} \equiv (L/\dot{M}c^2)(v_K(r_{in})/c)$. Combining Equations (18), (21) and (22) then yields

$$\frac{\partial B_{in}}{\partial t} \approx \frac{f_{CB}c^2 \sigma_T}{8er_{in}^{1/4} r_{out}^{3/4} v_K(r_{in})} B_{in}^2 . \tag{23}$$

This equation has the solution

$$B_{in}(t) = \frac{B_{quiesc}}{1 - t/\tau_{asc}} , \tag{24}$$

where the time t is measured from the beginning of the outburst in the quiescent state, and the characteristic ascending timescale τ_{asc} is equal to

$$\tau_{asc} \approx \frac{8er_{in}v_K(r_{in})}{f_{CB}c^2 \sigma_T B_{quiesc}} \left(\frac{r_{out}}{r_{in}} \right)^{3/4} = \frac{1}{f_{CB}} \left(\frac{B_{quiesc}}{10^6 \, G} \right)^{-1} \text{yr} . \tag{25}$$

This timescale is much longer than the characteristic timescales of Table 1 because according to our present model, most of the time, the system accretes at rates two orders of magnitude below equipartition, thus most of the time, the Cosmic Battery operates very inefficiently. Notice that the imbalance between inward flux advection and outward flux diffusion is just a very small fraction of each one of these terms, namely

$$\frac{\dot{\Psi}_{CB}}{2\pi r_{in} v_r(r_{in}) B_{in}} = f_{CB} \frac{B_{in}c^2 \sigma_T}{2\pi e v_K(r_{in}) v_r(r_{in})} \lesssim 10^{-8} \left(\frac{B_{in}}{10^7 \, G} \right) . \tag{26}$$

It is impossible to simulate such a small effect in global numerical simulations like the ones performed by [31]. In the ascending phase of the HID (the growing phase of the outburst), the disk is slowly filled with magnetic flux, and the field at its inner edge becomes stronger and stronger with time according to Equation (24). As a result, the disk accretion rate and luminosity also rise as

$$L(t) \propto \dot{M}_{\mathrm{disk}}(t) \sim \frac{r_{\mathrm{in}}^2 B_{\mathrm{quiesc}}^2}{v_{\mathrm{K}}(r_{\mathrm{in}})(1 - t/\tau_{\mathrm{asc}})^2}. \tag{27}$$

At the same time, the magnetic flux Ψ_{BH} accumulated around the central black hole continuously increases. The accretion rate becomes stronger and stronger according to Equation (18). At some point, the accretion will approach equipartition with radiation and the innermost accretion flow will be disrupted. Equipartition corresponds to

$$B_{\mathrm{eq}} \approx \left(\frac{4\pi G M m_{\mathrm{p}} v_{\mathrm{K}}^2(r_{\mathrm{in}})}{f_{\mathrm{CB}} r_{\mathrm{in}}^2 \sigma_T c^2} \right)^{1/2} \approx 7 \times 10^6 \, f_{\mathrm{CB}}^{-1/2} \, \mathrm{G} \, . \tag{28}$$

Notice that, according to Equation (27), as the luminosity rises by two orders of magnitude with respect to the quiescent state, the magnetic field rises only by one order of magnitude. Therefore, we naturally expect that the innermost value of the magnetic field in the disk during quiescence is on the order of

$$B_{\mathrm{quiesc}} \approx \frac{B_{\mathrm{eq}}}{10} \sim 10^6 \, \mathrm{G} \, . \tag{29}$$

Beyond that point, the inner accretion flow is disrupted, the Cosmic Battery ceases to operate, and no new magnetic flux is generated. The delicate balance between inward flux advection and outward flux diffusion is now reversed, and the magnetic flux in the disk is gradually brought to the center where it reconnects with the magnetic flux accumulated around the central black hole. As a result, in the descending return part of the HID, inward advection slightly wins over outward diffusion, and the total flux in the disk decreases. The overall field in the disk becomes weaker and weaker, and, as a result, the disk accretion rate and luminosity also decrease. During the descending phase, the system remains close to equipartition. This keeps disrupting the innermost accretion flow. At the same time, the field annihilation proceeds through a series of field reconnection events, and is therefore, not clearly monotonic. This explains why the decrease of the outburst luminosity is rather irregular and often non-monotonic.

As in the ascending phase, the magnetic field in the descending phase is strong enough to be able to support a JED at the inner part of the disk. However, because the innermost accretion flow is disturbed so dramatically around equipartition, the conditions for restarting the inner compact jet are not favorable. Eventually, the central magnetic field decreases well below equipartition so that the JED forms without further disruption. Why this takes place at that particular value of the disk luminosity that corresponds to Point E in the HID is not yet clear to us. We need to investigate in greater detail what happens to the overall accretion–ejection around equipartition, possibly with numerical simulations.

After the compact jet reappears, the disk magnetic flux continues to "drip" towards the inner edge of the disk, and, therefore, the Cosmic Battery does not operate to halt the field decay and reverse the decrease of the disk accretion rate. This is why at Point Fa the system continues to traverse the HID downwards towards the quiescent state, and does not start a new outburst as is the case at Point Fb where the system traverses the HID upwards.

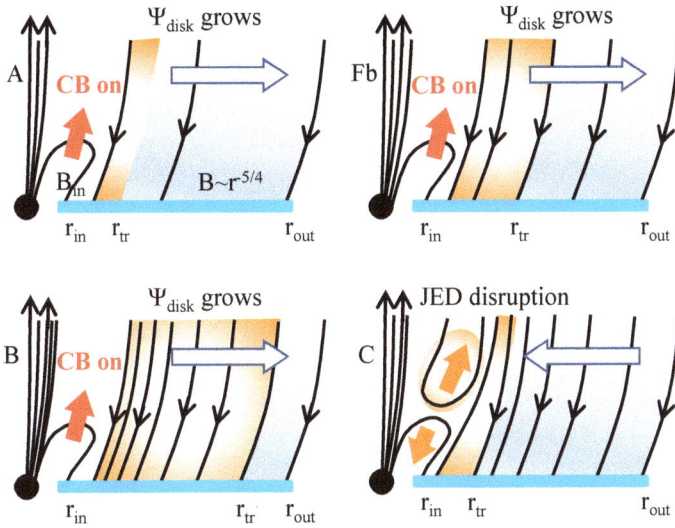

Figure 3. Schematic of our proposed model. Ascending phase of the HID. Blue strip: accretion disk; Black circle: central black hole; Solid lines with arrows: magnetic field. The direction of the angular velocity vector ω is along the z-axis. Magnetic field directions shown according to the predictions of the Cosmic Battery: B parallel to ω around the black hole, B anti-parallel to ω in the disk. Blue arrows: opening up of poloidal loops generated by the Cosmic Battery; White arrows: direction of flux redistribution in the disk; Orange arrows: reconnecting magnetic flux; Light blue: weak disk wind; Light orange regions: compact jet from JED; Light orange oval: ejected blobs of matter as the inner part of the JED is destroyed.

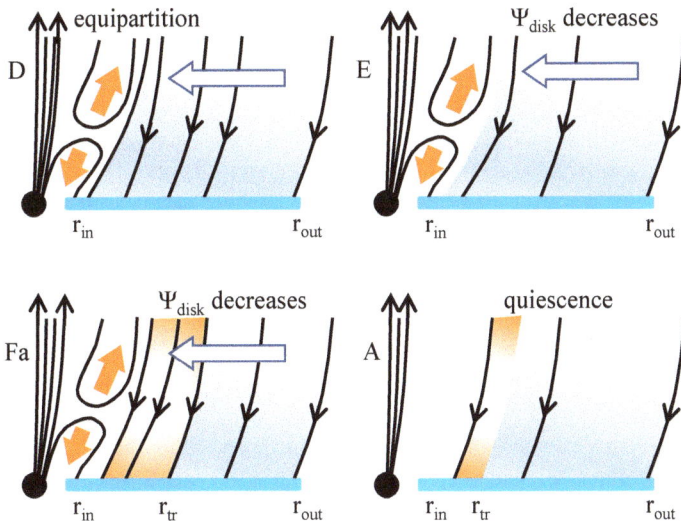

Figure 4. Descending phase of the HID (similar to Figure 3). Notice how similar the intermediate states Fa and Fb are (they differ only in the direction of field redistribution through the disk). Notice also that the direction of flux redistribution during the ascending and descending phases is opposite to that in the model for XRB state transitions of [61].

4. Summary and Conclusions

We believe that the key element missing from most previous efforts to understand the dynamics of accretion–ejection flows in X-ray binary outbursts may be the action of the Cosmic Battery in the innermost accretion region around the central black hole. Our model is plausible, but has not yet been conclusively proven by simulations (see however [36]). The Cosmic Battery offers the possibility to generate poloidal magnetic field loops around the inner edge of the accretion disk. Due to differential rotation, these loops open up above and below the disk. As a result, one polarity of the field (the one where B and ω are parallel) is brought to the center and inundates the black hole horizon, whereas the return field polarity (the one where B and ω are anti-parallel) threads the surrounding disk.

What is crucial for the distribution of magnetic flux in the disk is the delicate balance between inward flux advection and outward flux diffusion. We propose that the field reaches such a balance in the disk and therefore attains a large scale configuration similar to the one proposed by [17]. The balance, however, is not perfect and is influenced by what happens around the center, namely by the action of the Cosmic Battery.

During the ascending part of the XRB outburst, the Cosmic Battery is in operation and magnetic flux is introduced to the accretion disk at its inner edge. As a result, the total magnetic flux that threads the disk continuously increases at a particular rate given by Equation (22). Similarly, the disk accretion rate and its associated luminosity also increase according to Equation (27). Eventually, the accretion rate reaches equipartition with radiation, at which point the innermost accretion flow is dramatically disturbed. The outward field diffusion is reversed, and magnetic flux is advected inwards towards the central black hole. The Cosmic Battery ceases to operate, and the total magnetic flux that threads the accretion disk is gradually lost via reconnection with the magnetic field that is accumulated around the center. At the same time, the energetic jet is destroyed and the accretion disk transitions to a Shakura–Sunyaev type. The total magnetic flux that threads the disk continuously decreases, as does the disk accretion rate and its associated luminosity. Notice that the directions of flux redistribution during the ascending and descending phases of the outburst that we propose in the present work are opposite to those proposed by [61] in their model for XRB state transitions. In both models, however, the accumulated magnetic field increases/decreases during the ascending/descending phase of the outburst. The difference between the two models is precisely the operation of the Cosmic Battery around the central black hole.

It is interesting that, during the descending phase of the outburst, the system may make several attempts to generate a compact jet, i.e. to form a JED in its innermost part. This may happen several times as the luminosity decreases, and a transient compact jet may form several times before the system eventually reaches the so-called Faint Intermediate state (point E in the HID) where it makes a final transition to a JED and subsequently returns to the quiescent state. According to [5], for a JED to form, the magnetization μ in the disk, defined as the ratio of magnetic to gas pressures must be greater than unity. In our present formulation, and with the help of Equation (18),

$$\mu \equiv \frac{B^2/8\pi}{\rho_{\text{disk}} c_s^2} \sim \frac{\dot{M}_{\text{disk}} v_K/(\epsilon r^2)}{(\dot{M}_{\text{disk}}/2\pi r 2 h v_r)(h v_K/r)^2} \approx \frac{v_r}{v_K} \cdot \frac{r}{h} \tag{30}$$

The magnetization parameter μ may exceed unity (within detailed numerical factors that are missing from our crude calculation), either when the accretion velocity approaches the Keplerian rotational velocity and $h \sim r$, either when $h \ll r$. The latter is a possibility in a Shakura–Sunyaev-type disk. We are more interested here in the former possibility where $v_r \sim v_K$. It seems to us that the system can always support an inner JED, but when it reaches equipartition, the JED is destroyed. However, whenever it finds the opportunity, as in the multiple attempts of the system to cross the so-called jet line in the HID, the field is strong enough for a JED to form. Why the disk changes type and generates a persistent (not transient) JED in its innermost parts at the particular luminosity that corresponds to point E in the HID is not yet clear to us (see, however, also [61,62] for interesting ideas).

We acknowledge that much more work must be done to understand all the details of the dynamics of bursting XRB systems. We have not investigated ourselves how the various spectral states of the HID arise in our picture of an inner JED and an outer Shakura–Sunyaev-type disk, and only refer to the work of other groups. The assumption that accretion proceeds only via magnetic torques due to the large scale magnetic field that threads the disk is clearly simplistic. This allows us, however, to make very definite simple predictions about the rise and fall of the outburst that can directly be compared to observations. Finally, our treatment of the Cosmic Battery is too simplistic since it ignores general relativity and the complicated geometry of the central source of radiation, and assumes a point central source of radiation as in the Poynting–Robertson effect in our solar system.

We would like to note that our results are also applicable to AGN with or without jets. It is possible that, as in the case of bursting XRB sources, jets appear only during a small fraction of the system's lifetime. This may be the answer why only a small fraction of AGN develop large scale jets. The scales of those systems are much longer, namely $M \sim 10^9 \, M_\odot$, $r_{in} \sim 10^{15}$ cm, and therefore, $B_{eq} \sim 10^3$ G. If $B_{quiesc} \sim 10^3$ G, then $\tau_{asc} \sim 10^{11}$ yr, which is longer than the Hubble time. Obviously, the model that we used in GX 339-4 (where most of the time the system accretes several orders of magnitude below equipartition and the Cosmic Battery operates very inefficiently) does not directly apply here. We expect that AGN with jets accrete at rates close to equipartition. More work needs to be done to understand the role of the Cosmic Battery in AGN with jets.

In summary, we hope that we have convinced the reader that the action (or inaction) of the Cosmic Battery in the central regions of XRB and AGN may be the missing key in the study of the dynamics of outbursts in these systems.

Funding: This research received no external funding.

Acknowledgments: Most of the ideas presented in this work arose from discussions with Nick Kylafis, Jonathan Ferreira, Demos Kazanas, and Sergei Bogovalov. The material in Section 2 was provided by Leela Koutsantoniou.

Conflicts of Interest: The author declares no conflict of interest.

References

1. Shakura, N.I.; Sunyaev, R.A. Black holes in binary systems. Observational appearance. *Astron. Astrophys.* **1973**, *24*, 337–355.

2. Ghisellini, G.; Tavecchio, F.; Maraschi, L.; Celotti A.; Sbarrato, T. SDSS J114657.79+403708.6: The third most distant blazar at z = 5.0. *Mon. Not. R. Astron. Soc. Lett.* **2014**, *440*, L111–L115. [CrossRef]

3. Striani, E.; Vercellone, S.; Tavani, M.; Vittorini, V.; D'Ammando, F.; Donnarumma, I.; Pacciani, L.; Pucella, G.; Bulgarelli, A.; Trifoglio, M.; et al. The Extraordinary Gamma-ray Flare f the Blazar 3C 454.3. *Astrophys. J.* **2010**, *718*, 455–459. [CrossRef]

4. Ackermann, M.; Ajello, M.; Baldini, L.; Ballet, J.; Barbiellini, G.; Bastieri, D.; Bechtol, K.; Bellazzini, R.; Berenji, B.; Blandford, R.D.; et al. Fermi Gamma-ray Space Telescope Observations of Gamma-ray Outbursts from 3C 454.3 in 2009 December and 2010 April. *Astrophys. J.* **2010**, *721*, 1383–1396. [CrossRef]

5. Ferreira, J.; Petrucci, P.-O.; Henri, G.; Sauge, L.; Pelletier, G. A unified accretion-ejection paradigm for black hole X-ray binaries. I. The dynamical constituents. *Astron. Astrophys.* **2006**, *447*, 813–825. [CrossRef]

6. Marcel, G.; Ferreira, J.; Petrucci, P.O.; Henri, G.; Belmont, R.; Clavel, M.; Malzac, J.; Coriat, M.; Corbel, S.; Rodriguez, J.; et al. A unified accretion-ejection paradigm for black hole X-ray binaries. II. Observational signatures of jet-emitting disks. *Astron. Astrophys.* **2018**, *615*, A57. [CrossRef]

7. Marcel, G.; Ferreira, J.; Petrucci, P.O.; Belmont, R.; Malzac, J.; Clavel, M.; Henri, G.; Coriat, M.; Corbel, S.; Rodriguez, J.; et al. A unified accretion-ejection paradigm for black hole X-ray binaries. III. Spectral signatures of hybrid disk configurations. *Astron. Astrophys.* **2018**, *617*, A46. [CrossRef]

8. Ferreira, J.; Pelletier, G. Magnetized accretion-ejection structures. I. General statements. *Astron. Astrophys.* **1993**, *276*, 625–636.

9. Ferreira, J.; Pelletier, G. Magnetized accretion-ejection structures. III. Stellar and extragalactic jets as weakly dissipative disk outflows. *Astron. Astrophys.* **1995**, *295*, 807–832.

10. Casse, F.; Keppens, R. Magnetized Accretion-Ejection Structures: 2.5-dimensional Magnetohydrodynamic Simulations of Continuous Ideal Jet Launching from Resistive Accretion Disks. *Astrophys. J.* **2002**, *581*, 988–1001. [CrossRef]

11. Zanni, C.; Ferrari, A.; Rosner, R.; Bodo, G.; Massaglia, S. MHD simulations of jet acceleration from Keplerian accretion disks. The effects of disk resistivity. *Astrophys. J.* **2007**, *469*, 811–828.

12. Murphy, G.C.; Ferreira, J.; Zanni, C. Large scale magnetic fields in viscous resistive accretion disks. I. Ejection from weakly magnetized disks. *Astron. Astrophys.* **2010**, *512*, 82–95. [CrossRef]

13. Sheikhnezami, S.; Fendt, C.; Porth, O.; Vaidya, B.; Ghanbari, J. Bipolar Jets Launched from Magnetically Diffusive Accretion Disks. I. Ejection Efficiency versus Field Strength and Diffusivity. *Astrophys. J.* **2012**, *757*, 65–87. [CrossRef]

14. Stepanovs, D.; Fendt, C. An Extensive Numerical Survey of the Correlation Between Outflow Dynamics and Accretion Disk Magnetization. *Astrophys. J.* **2016**, *825*, 14–31. [CrossRef]

15. Qian, Q.; Fendt, C.; Noble, S.; Bugli, M. rHARM: Accretion and Ejection in Resistive GR-MHD. *Astrophys. J.* **2017**, *834*, 29–48. [CrossRef]

16. Qian, Q.; Fendt, C.; Vourellis, C. Jet Launching in Resistive GR-MHD Black Hole-Accretion Disk Systems. *Astrophys. J.* **2018**, *859*, 28–48. [CrossRef]

17. Blandford, R.D.; Payne, D.G. Hydromagnetic flows from accretion discs and the production of radio jets. *Mon. Not. R. Astron. Soc.* **1982**, *199*, 883–903. [CrossRef]

18. Contopoulos, J. Magnetically Driven Jets and Winds: Exact Solutions. Ph.D. Thesis, Cornell University, Ithaca, NY, USA, 1992.

19. Contopoulos, J.; Lovelace, R.V.E. Magnetically Driven Jets and Winds: Exact Solutions. *Astrophys. J.* **1994**, *429*, 139–152. [CrossRef]

20. Konigl, A.; Kartje, J.F. Disk-driven hydromagnetic winds as a key ingredient of active galactic nuclei unification schemes. *Astrophys. J.* **1994**, *434*, 446–467. [CrossRef]

21. Fukumura, K.; Kazanas, D.; Contopoulos, I.; Behar, E. Magnetohydrodynamic Accretion Disk Winds as X-ray Absorbers in Active Galactic Nuclei. *Astrophys. J.* **2010**, *715*, 636–650. [CrossRef]

22. Fukumura, K.; Kazanas, D.; Contopoulos, I.; Behar, E. Modeling high-velocity QSO absorbers with photoionized magnetohydrodynamic disk winds. *Astrophys. J. Lett.* **2010**, *723*, L228–L232. [CrossRef]

23. Fukumura, K.; Tombesi, F.; Kazanas, D.; Shrader, C.; Behar, E.; Contopoulos, I. Stratified Magnetically Driven Accretion-disk Winds and Their Relations to Jets. *Astrophys. J.* **2014**, *780*, 120–131. [CrossRef]

24. Fukumura, K.; Tombesi, F.; Kazanas, D.; Shrader, C.; Behar, E.; Contopoulos, I. Magnetically Driven Accretion Disk Winds and Ultra-fast Outflows in PG 1211+143. *Astrophys. J.* **2015**, *805*, 17–27. [CrossRef]

25. Fukumura, K.; Kazanas, D.; Shrader, C.; Behar, E.; Tombesi, F.; Contopoulos, I. Magnetic origin of black hole winds across the mass scale. *Nat. Astron.* **2017**, *1*, 62. [CrossRef]

26. Chakravorty, S.; Petrucci, P.O.; Ferreira, J.; Henri, G.; Belmont, R.; Clavel, M.; Corbel, S.; Rodriguez, J.; Coriat, M.; Drappeau, S.; et al. Magneto centrifugal winds from accretion discs around black hole binaries. *Astron. Nachr.* **2016**, *337*, 429–434 [CrossRef]

27. Contopoulos, I.; Kazanas, D.; Fukumura, K. Magnetically advected winds. *Mon. Not. R. Astron. Soc. Lett.* **2017**, *472*, L20–L24. [CrossRef]

28. Van Ballegooijen, A.A. Magnetic fields in the accretion disks of cataclysmic variables. In *Accretion Disks and Magnetic Fields in Astrophysics*; Belvedere, G., Ed.; Kluwer: Dordrecht, The Netherlands; Boston, MA, USA, 1989; pp. 99–106.

29. Lubow, S.H.; Papaloizou, J.C.B.; Pringle, J.E. Magnetic field dragging in accretion discs. *Mon. Not. R. Astron. Soc.* **1994**, *267*, 235–240. [CrossRef]

30. Lovelace, R.V.E.; Romanova, M.M.; Newman, W.I. Implosive accretion and outbursts of active galactic nuclei. *Astrophys. J.* **1994**, *437*, 136–143. [CrossRef]

31. Tchekhovskoy, A.; Narayan, R.; McKinney, J.C. Efficient generation of jets from magnetically arrested accretion on a rapidly spinning black hole. *Mon. Not. R. Astron. Soc. Lett.* **2011**, *418*, L79–L83. [CrossRef]

32. McKinney, J.C.; Tchekhovskoy, A.; Blandford, R.D. General relativistic magnetohydrodynamic simulations of magnetically choked accretion flows around black holes. *Mon. Not. R. Astron. Soc.* **2012**, *423*, 3083–3117. [CrossRef]

33. Contopoulos, I.; Kazanas, D. A Cosmic Battery. *Astrophys. J.* **1998**, *508*, 859–863. [CrossRef]

34. Contopoulos, I. A Cosmic Battery around Back Holes. In *The Formation and Disruption of Black Hole Jets*; Contopoulos, I., Gabuzda, D., Kylafis, N., Eds.; Springer: Basel, Switzerland; New York, NY, USA, 2015; pp. 227–244, ISBN 978-3-319-10355-6.

35. Miller, M.C.; Lamb, F.K. Motion of Accreting Matter near Luminous Slowly Rotating Relativistic Stars. *Astrophys. J.* **1996**, *470*, 1033–1051. [CrossRef]

36. Contopoulos, I.; Nathanail, A.; Sadowski, A.; Kazanas, D.; Narayan, R. Numerical simulations of the Cosmic Battery in accretion flows around astrophysical black holes. *Mon. Not. R. Astron. Soc.* **2018**, *473*, 721–727. [CrossRef]

37. Koutsantoniou, L.E.; Contopoulos, I. Accretion Disk Radiation Dynamics and the Cosmic Battery. *Astrophys. J.* **2014**, *794*, 27–39. [CrossRef]

38. Poynting, J.H. Radiation in the solar system: Its effect on temperature and its pressure on small bodies. *Mon. Not. R. Astron. Soc.* **1903**, *64*, A1–A5. [CrossRef]

39. Robertson, H.P. Dynamical effects of radiation in the solar system. *Mon. Not. R. Astron. Soc.* **1937**, *97*, 423–437. [CrossRef]

40. Cattani, D.; Sacchi, C. A theory on the creation of stellar magnetic fields. *Nuovo Cimento B* **1966**, *46*, 258–272. [CrossRef]

41. Bisnovatyi-Kogan, G.S.; Lovelace, R.V.E.; Belinski, V.A. A Cosmic Battery Reconsidered. *Astrophys. J.* **2002**, *580*, 380–388. [CrossRef]

42. Contopoulos, I.; Kazanas, D.; Christodoulou, D.M. The Cosmic Battery Revisited. *Astrophys. J.* **2006**, *652*, 1451–1456. [CrossRef]

43. Lovelace, R.V.E.; Rothstein, D.M.; Bisnovatyi-Kogan, G.S. Advection/Diffusion of Large-Scale B Field in Accretion Disks. *Astrophys. J.* **2009**, *701*, 885–890. [CrossRef]

44. Contopoulos, I.; Christodoulou, D.M.; Kazanas, D.; Gabuzda, D.C. The Invariant Twist of Magnetic Fields in the Relativistic Jets of Active Galactic Nuclei. *Astrophys. J. Lett.* **2009**, *702*, L148–L152. [CrossRef]

45. Kylafis, N.D.; Contopoulos, I.; Kazanas, D.; Christodoulou, D.M. Formation and destruction of jets in X-ray binaries. *Astron. Astrophys.* **2012**, *538*, 5–11. [CrossRef]

46. Christodoulou, D.M.; Gabuzda, D.C.; Knuettel, S.; Contopoulos, I.; Kazanas, D.; Coughlan, C.P. Dominance of outflowing electric currents on decaparsec to kiloparsec scales in extragalactic jets. *Astron. Astrophys.* **2016**, *591*, 61–72. [CrossRef]

47. Lynden-Bell, D. Magnetism Along Spin. *Observatory* **2013**, *133*, 266–269.

48. Stepanovs, D.; Fendt, C.; Sheikhnezami, S. Modeling MHD Accretion-Ejection: Episodic Ejections of Jets Triggered by a Mean-field Disk Dynamo. *Astrophys. J.* **2014**, *793*, 31–52. [CrossRef]

49. Dyda, S.; Lovelace, R.V.E.; Ustyugova, G.V.; Koldoba, A.V.; Wasserman, I. Magnetic field amplification via protostellar disc dynamos. *Mon. Not. R. Astron. Soc.* **2018**, *477*, 127–138. [CrossRef]

50. Fendt, C.; Gassmann, D. Bipolar Jets Launched by a Mean-field Accretion Disk Dynamo. *Astrophys. J.* **2018**, *855*, 130–140. [CrossRef]

51. Salvesen, G.; Simon, J.B.; Armitage, P.J.; Begelman, M.C. Accretion disc dynamo activity in local simulations spanning weak-to-strong net vertical magnetic flux regimes. *Mon. Not. R. Astron. Soc.* **2016**, *457*, 857–874. [CrossRef]

52. Balbus, S.A.; Hawley, J.F. A Powerful Local Shear Instability in Weakly Magnetized Disks. I. Linear Analysis. *Astrophys. J.* **1991**, *376*, 214–233. [CrossRef]

53. Bogovalov, S.V. Ratio of kinetic luminosity of the jet to bolometric luminosity of the disk at the "cold" accretion onto a supermassive black hole. *J. Mod. Phys.* **2019**, in press.

54. Li, Z.-Y. Magnetohydrodynamic disk-wind connection: Self-similar solutions. *Astrophys. J.* **1995**, *444*, 848–860. [CrossRef]

55. Fender, B.; Belloni, T. Stellar-Mass Black Holes and Ultraluminous X-ray Sources. *Science* **2012**, *337*, 540–544. [CrossRef] [PubMed]

56. Kylafis, N.D.; Reig, P. Correlation of time lag and photon index in GX 339-4. *Astron. Astrophys.* **2018**, *614*, L5–L9. [CrossRef]

57. Narayan, R.; McClintock, J.E.; Yi, I. A New Model for Black Hole Soft X-Ray Transients in Quiescence. *Astrophys. J.* **1996**, *457*, 821–833. [CrossRef]

58. Hameury, J.-M.; Lasota, J.-P.; McClintock, J.E.; Narayan, R. Advection-dominated Flows around Black Holes and the X-Ray Delay in the Outburst of GRO J1655-40. *Astrophys. J.* **1997**, *489*, 234–243. [CrossRef]

59. Esin, A.A.; Narayan, R.; Cui, W.; Grove, J.E.; Zhang, S.-N. Spectral Transitions in Cygnus X-1 and Other Black Hole X-Ray Binaries. *Astrophys. J.* **1998**, *505*, 854–868. [CrossRef]

60. Esin, A.A.; McClintock, J.E.; Drake, J.J.; Garcia, M.R.; Haswell, C.A.; Hynes, R.I.; Muno, M.P. Modeling the Low-State Spectrum of the X-Ray Nova XTE J1118+480. *Astrophys. J.* **2001**, *555*, 483–488. [CrossRef]

61. Begelman, M.C.; Armitage, P.J. A Mechanism for Hysteresis in Black Hole Binary State Transitions. *Astrophys. J. Lett.* **2014**, *782*, L18–L22. [CrossRef]

62. Kazanas, D. X-ray Binary Phenomenology and their Accretion Disk Structure. In *The Formation and Disruption of Black Hole Jets*; Contopoulos, I., Gabuzda, D., Kylafis, N., Eds.; Springer: Basel, Switzerland; New York, NY, USA, 2015; pp. 207–225, ISBN 978-3-319-10355-6.

galaxies

MDPI

Review

Numerical Simulations of Jets from Active Galactic Nuclei

José-María Martí

Departament d'Astronomia i Astrofísica and Observatori Astronòmic, Universitat de València, 46100 Burjassot (València), Spain; jose-maria.marti@uv.es

Received: 12 November 2018; Accepted: 14 January 2019; Published: 22 January 2019

Abstract: Numerical simulations have been playing a crucial role in the understanding of jets from active galactic nuclei (AGN) since the advent of the first theoretical models for the inflation of giant double radio galaxies by continuous injection in the late 1970s. In the almost four decades of numerical jet research, the complexity and physical detail of simulations, based mainly on a hydrodynamical/magneto-hydrodynamical description of the jet plasma, have been increasing with the pace of the advance in theoretical models, computational tools and numerical methods. The present review summarizes the status of the numerical simulations of jets from AGNs, from the formation region in the neighborhood of the supermassive central black hole up to the impact point well beyond the galactic scales. Special attention is paid to discuss the achievements of present simulations in interpreting the phenomenology of jets as well as their current limitations and challenges.

Keywords: active galactic nuclei; relativistic jets; magneto-hydrodynamics; plasma physics; numerical methods

1. Introduction

Theoretical models of jets from active galactic nuclei (AGN) motivated by observations have been the subject of thorough testing by numerical simulations for almost forty years, since the simulations of supersonic jets performed in the early 1980s by Norman and collaborators [1] confirmed the viability of the *beam model* proposed by Scheuer [2], and Blandford and Rees [3] one decade before to explain the powering of lobes of extended radio sources and quasars (*classic doubles* or *symmetric doubles*, as these sources were known at those dates).

According to our present understanding [4], jets are produced on scales of a few gravitational radii of the central black hole (BH) powering the AGN ($R_{BH} \approx 10^{-4} (M_{BH}/10^9 M_\odot)$ pc, where M_{BH} is the BH mass) but extend to hundreds of kiloparsecs. This disparity in scales is exacerbated when considering those microphysical processes involved in the formation of the jet or the particle acceleration. The viscosity of the accretion disc from which the jet originates is thought to be generated by the magneto-rotational instability [4] affecting the magnetic field in the disc on scales orders of magnitude smaller than R_{BH}. Besides that, the acceleration of particles in the jet is ultimately governed by processes occurring on scales of the gyroradius of mildly relativistic electrons/protons, again several orders of magnitude smaller than the radius of the central BH, for typical jet non-thermal particle energies and magnetic fields. Interestingly, however, the fact that the gyroradius (that establishes the effective collisional mean free paths by suppressing particle diffusion perpendicular to the magnetic fields) and the Debye length (governing the scales of effective charge neutrality) are much smaller than the jet width, R_j, supports a continuum (i.e., *magneto-hydrodynamical*) approximation of the otherwise collisionless plasma forming the jet [5,6].

Incorporating all the relevant microphysics (shock acceleration, magnetic reconnection, radiative processes, etc.) into a global (general relativistic) magneto-hydrodynamical simulation describing the formation of the jet from an accreting BH and its propagation across the interstellar and intergalactic media along hundreds of kiloparsecs represents a daunting task for present-day computational tools and numerical methods. However, numerical simulations have been accumulating important (although partial) triumphs along their decades of development. This review is aimed to summarize them as well as their current limitations and challenges.

The book edited by Böttcher, Harris and Krawczynski [7] is a basic reference covering many theoretical, observational and numerical aspects of the physics of AGN jets of relevance for the present review. A major break-through in the field of Computational Relativistic Astrophysics (and, in particular, in the simulation of jets from AGN) was accomplished in the 1990s when the so-called *high-resolution shock-capturing* (HRSC) methods were applied to integrate the equations of relativistic hydrodynamics (RHD), and few years later, those of relativistic magneto-hydrodynamics (RMHD) overcoming the traditional difficulties of standard finite-differencing methods in simulating flows with high Lorentz factors. Two recent reviews [8,9] summarize the implementation of these methods in special-relativistic and general-relativistic magnetohydrodynamics (GRMHD) codes and a description of relevant numerical applications in different areas of Relativistic Astrophysics.

The review is organized as follows. In Section 2, we give an overview of the status of the observations and theoretical models of jets that establish the framework for the numerical simulations. Simulations have traditionally divided the study of the jet phenomenon into separate problems. Section 3 goes into the domain of the simulations of jet formation. Section 4 focuses on simulations of jets at parsec scales up to the smallest scales accessible to observations with present-day instruments. Section 5 analyzes the present status of the simulations of jets at the largest spatial and temporal scales including the connection with the formation and evolution of galaxies and clusters of galaxies. The contribution ends with a summary of the achievements, limitations and challenges of contemporary numerical simulations in Section 6.

2. Observations and Theoretical Models

2.1. Models of Jet Formation

Models proposed to explain the origin of the relativistic jets found in several astrophysical scenarios involve accretion in the form of a disc onto a compact central object. In the case of the jets emanating from AGNs, the central object is a rotating supermassive BH fed by interstellar gas and gas from tidally disrupted stars. There is a general agreement that MHD processes are responsible for the formation, collimation and acceleration up to relativistic speeds of the outflows. In the models of magnetically driven outflows [10–13], poloidal magnetic fields anchored at the basis of the accretion disc generate a toroidal field component and consequently a poloidal electromagnetic flux of energy (Poynting flux) that accelerates the magnetospheric plasma and plasma from the disc along the poloidal magnetic field lines, converting the Poynting flux into kinetic energy of bulk motion reaching relativistic speeds at extended ($\approx$pc) scales [14–17]. Energy can also be extracted from rotating BHs via the Blandford–Znajek (BZ) mechanism [18,19]. Several parameters are potentially important for powering the jets: the BH's mass and spin, the accretion rate, the type of accretion disc, the properties of the magnetic field, and the environment of the source [20].

Since in any version of a central engine powering an AGN the accretion disc is magnetized, we expect a wind to be driven from its surface (close to the BH) shrouding the BZ jet, producing transversely stratified outflows in both composition (outer electron–proton wind and inner electron–positron jet) and speed. Furthermore, the disc wind can provide the initial confinement of the jets.

Even at the highest angular resolutions achievable today (tens of micro-arcseconds with space very long baseline interferometry (VLBI) imaging using *RadioAstron* [21]; $1\,\mu\text{as} \approx 1\,(D/20\,\text{Mpc})(10^9\,M_\odot/M_{BH})R_{BH}$, where D is the distance to the source), the scales of jet launching are still inaccessible to observations (although they are the target of the *Event Horizon*

Telescope (EHT) [22], for the M 87 jet −3.9 μas BH). In these conditions, numerical simulations boosted by the development of specific HRSC techniques to solve the GRMHD equations in the last two decades still provide the main guide for further theoretical advances.

2.2. Parsec-Scale Jets and Superluminal Radio Sources

At parsec scales, AGN jets, imaged via their synchrotron emission at radio frequencies with VLBI networks, appear to be highly collimated with a bright spot (the *core*) at one end of the jet and a series of components which separate from the core, sometimes at superluminal speeds [23]. In the standard model of Blandford and König [24], these speeds originally predicted by Rees [25] are the consequence of relativistic bulk motion in jets propagating at small angles to the line of sight with Lorentz factors up to 20 or more. Moving components in these jets, usually appearing after outbursts in emission at radio wavelengths, are interpreted in terms of traveling shock waves [24]. The evolution of the continuum spectrum of the ejected components fits nicely with this interpretation [26].

An ongoing, important debate is concerned with the nature of the radio core. Whereas in the standard Blandford and König's conical jet model the core corresponds to the location near the BH where the jet becomes optically thin, recent multi-wavelength observations of several sources (e.g., 3C 120 [27], BL Lac [28], 3C 111 [29], and 1803+784 [30]) suggest that the radio core can be a physical feature in the jet (as, e.g., a recollimation shock in Daly–Marscher's [31] model) placed probably parsecs (i.e., tens of thousands of gravitational radii of the central BH) away from the central engine.

In the acceleration region the width of the jet broadens rather slowly with distance (with a parabolic-like surface [14,16,17]) and the jet collimates. The collimation and acceleration continues until the flow ceases to be Poynting dominated. At this point, the flow has reached its terminal speed and becomes inertial. If the jet is not confined by the external pressure, then the flow becomes conical. Signatures of the acceleration/collimation region can be investigated in those jets where the BH position is matched with the radio core. Parabolic/conical transitions in the jet shape have been observed in M 87 at scales of 10^5–10^6 R_{BH} [32] and Cyg A at scales $\approx 2.5 \times 10^4$ R_{BH} [33].

Observations of jets at subparsec scales also show signs of transversal stratification which could be revealing the duality of the jet formation mechanism. The transversal structure could also be the observational counterpart of the growth of different kinds of instabilities triggered at the jet base and/or at the jet/wind or jet/ambient medium interfaces. In 3C 84, the ridge-to-limb brightening change in a scale of years [34] is interpreted as a change in the viewing angle of a jet with a highly relativistic spine and a mildly relativistic sheath. A fast central spine with a slow sheath is also inferred in Cyg A [33] and M 87 [35]. In this last case, the limb brightened structure points towards a subrelativistic layer associated either with an instability pattern speed or an outer wind, and a fast, accelerating stream. Moreover, the systematic difference of the apparent speeds in the northern and southern limbs of the jet provides evidence for jet rotation about the jet axis. The angular velocity of the magnetic field line associated with this rotation suggests that the jet in M 87 is launched in the inner part of the disc, at a distance ≈ 5 R_{BH} of the central engine. In 3C 273, the double-rail structure has been consistently interpreted as created by helical Kelvin–Helmholtz instabilities [36].

The composition of jets is still a major undecided issue. The presence of protons in the jet can be inferred from a weak component of circular polarization (CP) in the synchrotron emission (which would be exactly zero in the case of a pure pair plasma jet due to the cancellation of the electron and positron separated contributions). However, the CP can be contaminated from Faraday conversion of linear polarization. Recently, VLBI imaging of both circular and linear polarization have been carried out for a few blazars on sub-parsec scales [37,38]. Gabuzda and collaborators [37] detected CP in eight AGN, the most likely origin being Faraday conversion in helical magnetic fields. Homan and collaborators [38] obtained the full polarization spectra of 3C 279 and modeled them by radiative transfer simulations to constrain the magnetic field and particle properties of the pc-scale jet and concluded that the jet is kinetically dominated by electron–proton plasma with a non-negligible contribution of pair plasma to the jet radiation. However, this conclusion is based on an estimate of the low-energy cutoff of the power-law electron distribution, a parameter prone to large uncertainties.

Another way to determine the jet composition is fitting its multiwavelength spectrum. Potter and Cotter [39] developed a multizone inhomogeneous fluid jet emission model to fit with unprecedented accuracy the entire multiwavelength spectrum of a large sample of quiescent blazar spectra with a pure electron–positron plasma trhough their synchrotron, self-Compton and external Compton emission. The model can be viewed as providing an estimate of the power contained in the magnetic and non-thermal electrons in the jet hence establishing a lower limit of the jet power.

Finally, hadronic emission models [40] currently seem to require implausibly large proton kinetic luminosities to match the data (although the recent observation of a highly energetic neutrino from the blazar TXS 0506+056 is best explained by a hadronic emission process [41]).

The quest for the signatures of helical magnetic fields (as those invoked by the jet formation mechanisms) in the inner regions of extragalactic jets represents another observational challenge. Observational signatures for the existence of such helical magnetic fields can be obtained by looking for Faraday rotation measure (RM) gradients across the jet width, produced by the systematic change in the net line-of-sight magnetic field [42] and have been reported for 3C 273 [43,44] and more recently for 3C 454.3 [45] and BL Lac [21]. The paper by Gabuzda in this volume reviews the observational evidences for helical magnetic fields associated with AGN jets.

2.3. Kiloparsec-Scale Jets

At kiloparsec scales, jets divide into two main classes following the classification established by Fanaroff and Riley [46] in the 1970s according to their large scale morphology. Fanaroff–Riley type I (FR I) jets are decollimated at kiloparsec scales and display extended lobes of diffuse emission (e.g., 3C 31 [47]). On the contrary, jets from Fanaroff–Riley type II (FR II) objects (e.g., Cyg A [48]) remain highly collimated until the hot-spots, the regions where the jet flow impacts with the ambient medium. Whereas current models [49] interpret FR I morphologies as the result of a smooth deceleration from relativistic to non-relativistic, transonic speeds on kpc scales, flux asymmetries between jets and counter-jets in FR II radio galaxies and quasars indicate that relativistic motion extends up to kpc scales in these sources [50].

Beyond the original divide in power between FR I and FR II sources [46,51], and the differences in their hosts, the widely accepted explanation for the FR I jet deceleration is the entrainment of cold and dense gas through a mixing-layer [52–54] or from stars within its volume [55]. Parameterized models assuming intrinsically symmetrical, axisymmetric, relativistic, stationary jets (see [49] and references therein) have been successful in interpreting the brightness distributions of FR I jets characterized by an extended flaring region. However, the ultimate validation/rejection of the different models must rely on long-term (M)HD simulations to account for the non-linear processes leading to the flow deceleration and decollimation.

2.4. AGN Jets in the Cosmological Context

Observational evidence is growing that the baryonic part of the low-redshift Universe has been shaped by the energy and momentum output of BHs, through AGN feedback, with profound implications for our understanding of galaxy, group, and cluster evolution [56]. In the so-called *kinetic* or *radio-jet* mode, the powerful jets emerging from the AGN push the galactic halo (as inferred from the observed anti-correlation between the radio lobes formed by the jets and the X-ray emission from the cluster gas) and are responsible for inhibiting radiative cooling, which would otherwise lead to unrealistically high star formation rates. The proof of this connection between radio jet heating and radiative cooling suppression is provided by the fact that most cool core clusters (i.e., those with the shortest central cooling times) contain radio lobes [57] and by the direct correlation found between the estimated energy content of the lobes and the cooling rate/luminosity of the intra-cluster medium (ICM) [58].

The morphology, particularly in the closest clusters, of the X-ray *holes* has led to the interpretation of these cavities as bubbles of relativistic gas [57] blown by the AGN and rising against the cluster's potential well by buoyancy. Whereas this description does apply to the so-called *ghost cavities*, recent observations of shocks with low Mach number surrounding the lobes in powerful radio sources (e.g., Her A [59], Hydra A [60], MS0735.6+7421 [61], and HCG 62 [62]), tell us that these cavities have not yet reached the buoyancy stage. Hence, although the gross energetics of the AGN/ICM feedback process is roughly understood, the details are not.

On the other hand, the modeling of the feedback processes induced by AGNs is a crucial ingredient to understand the mass distribution of galaxies and their morphologies, the star formation rates and some spatial distributions of the stellar populations (such as the radial gradients of age, metallicity and velocity dispersion).

3. Simulations of Jet Formation

3.1. Jet Formation Mechanisms on the Test Bench

With the advances in the numerical methods in RMHD incorporated into general relativistic codes in the late 1990s, the possibility to explore for the first time the formation mechanism of relativistic jets opened. The first papers considered the problem of jet formation from Schwarzschild and Kerr BHs surrounded by accretion discs. In the case of Schwarzschild black holes [63–65], jets were formed via Blandford–Payne's mechanism [12] with a two-layered concentric structure with an inner fast gas pressure-driven jet and an outer slow magnetically-driven outflow both being collimated by the global poloidal magnetic field that penetrates the disc. For Kerr BHs [66], qualitative differences arise between corotating and counter-rotating disc cases. Whereas corotating discs around Kerr BHs produce almost the same kind of outflows as those from Schwarzschild BHs, counter-rotating discs lead to powerful (although still subrelativistic) magnetically-driven jets accelerated by the magnetic field anchored in the ergospheric disc inside the gas pressure-driven jets. Although there is an agreement about the ultimate source of energy for the outflow (the rotational energy of the BH), whether this extraction was the result of a kind of *Penrose process* [19] that uses the magnetic field to extract rotational energy of the BH at the cost of swallowing plasma with negative *energy at infinity*, or a purely electromagnetic mechanism, was in dispute [67,68]. The debate now seems to have been settled on the basis of recent theoretical and numerical work [69] which establishes the absorption of negative energy and angular momentum as the necessary and sufficient conditions for arbitrary fields or matter (in the case under study, the magnetic field piercing the BH ergosphere) to extract the BH rotational energy (*BZ/Penrose mechanism*).

Despite the ground-breaking character of the simulations discussed in the previous paragraph, none of them was able to generate sustained, relativistic outflows. Simulations lasted typically for less than a couple of rotation cycles at the accretion disc's innermost stable circular orbit (ISCO). Besides that, in these simulations, both the accretion discs and the magnetic fields were prescribed in the initial conditions. As a next step, a couple of studies [70,71] focused on the influence of the initial magnetic field configuration around the rotating BH on the outflow properties and considered monopole magnetospheres as in the original BZ mechanism. Koide [70] obtained outflows with Lorentz factors of ~2 but again the simulation was extremely short (a fraction of a rotation at the ISCO). In a longer simulation [71], the numerical solution evolved towards a stable steady-state solution very close to the force-free solution found by Blandford and Znajek. For the first time, numerical solutions showed the development of a Poynting-flux dominated ultrarelativistic particle wind (Lorentz factor ~15). The wind was mostly radial and had the largest Lorentz factors on the equatorial plane. Finally, direct numerical simulations of the BZ mechanism were performed [72,73] by solving the time-dependent equations of force-free electrodynamics in a Kerr BH magnetosphere.

3.2. Long-Term Simulations of Jet Formation

In parallel, the first simulations of *self-consistent* jet production (i.e., without assuming a large-scale magnetic field right from the beginning) from accretion discs orbiting Kerr BHs in 2D (axisymmetric) [74] and 3D [75–77] were performed. In all the cases, the outflows (formed at the edge of a funnel about 0.5 rad wide around the BH's rotation axis) were sub-relativistic. However, in the axisymmetric case, McKinney [78] succeeded in generating a collimated, long-lived (~1000 orbital periods at the BH horizon), super-fast magnetosonic, relativistic Poynting-flux dominated jet by tuning the floor model used to refill the evacuated funnel. Interestingly, the flow accelerates along paraboloidal field lines in agreement with analytic models [14,16,17] reaching up to a Lorentz factor ~10 beyond 10^3 R_{BH} (although still only a factor ~10^{-2} to 10^{-3} of the estimated asymptotic value). Komissarov and collaborators [79] revisited McKinney's work with a simplified setup where the low-speed outflow wrapping around the relativistic Poynting-flux dominated jet was replaced by a rigid boundary of a prescribed shape in order to reduce the effects of the numerical dissipation, and fixed injection conditions at the jet base. In all cases, the outcome was a steady state characterized by a spatially extended acceleration region and an efficient transfer of Poynting flux into kinetic energy (much more efficient than that of McKinney's work).

The extended-acceleration scenario raised the question of the stability of the jets against the development of kink instabilities at large distances where the toroidal magnetic field dominates over the poloidal one [78]. Starting with a realistic setup, McKinney and Blandford [80] simulated the generation and propagation of a relativistic highly magnetized jet in 3D and explored both the stability of the jet against the development of the highly-disruptive non-axisymmetric helical kink mode (angular frequency of the radial perturbation, $m = 1$), and the stability of the jet formation process itself during accretion of dipolar and quadrupolar fields. In their dipolar model, despite strong non-axisymmetric disc turbulence, the jet reaches Lorentz factors of ~10 with an opening half-angle ~5° at $10^3 R_{BH}$ without significant disruption (see Figure 1).

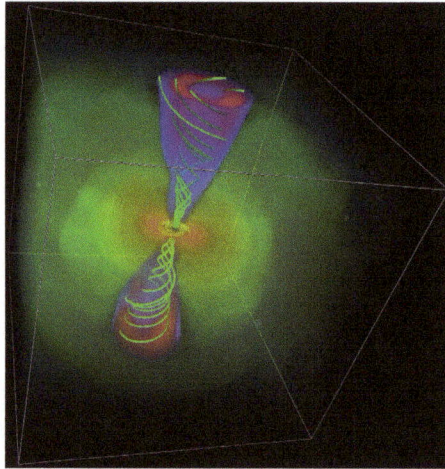

Figure 1. Inner ±100 M_{BH} cubical region at $t = 4000$ M_{BH} of a 3D global GRMHD simulation of the formation of a jet starting with an equilibrium magnetized matter torus, whose angular momentum is aligned with the BH spin. The figure shows the BH, accretion disc (pressure, yellow isosurface), outer disc and wind (log rest-mass density, low green, high orange, and volume rendering), relativistic jet (Lorentz factor $\Gamma \leq 4$, low blue, high red, volume rendering) and magnetic field lines (green) threading BH. (Figure 1 of [80].)

A series of papers [81–83] concentrate on the process of accretion onto rapidly rotating BHs and the role of magnetic fields in regulating both the accretion process and the jet production. The chosen numerical setup led to *magnetically arrested discs* [84,85] where the excess in magnetic flux near the BH impedes the accretion of gas hence maximizing the magnetic flux threading the BH per unit of accreted mass. As a result, very powerful outflows are produced, which in the case of rapidly rotating BHs reach efficiencies (power flowing out of the BH over rest-mass energy flux flowing into the BH) greater than 100%, demonstrating the net energy extraction from spinning BHs via the BZ/Penrose mechanism (see Figure 2). Recent numerical simulations along this line [86,87] have examined the interplay among the properties of the magnetized disc flow, the spin of the central BH, and the jet in producing warped discs and precessing jets as well as in controlling the alignment/misalignment properties of the jet with the disc/BH rotation axes. Let us note that the simulations presented in [87] (which qualify as the simulations of BH accretion discs at the highest resolutions to date) were carried out with a 3D-GRMHD code accelerated by GPUs performing 10 times faster than on last-generation multi-core CPUs.

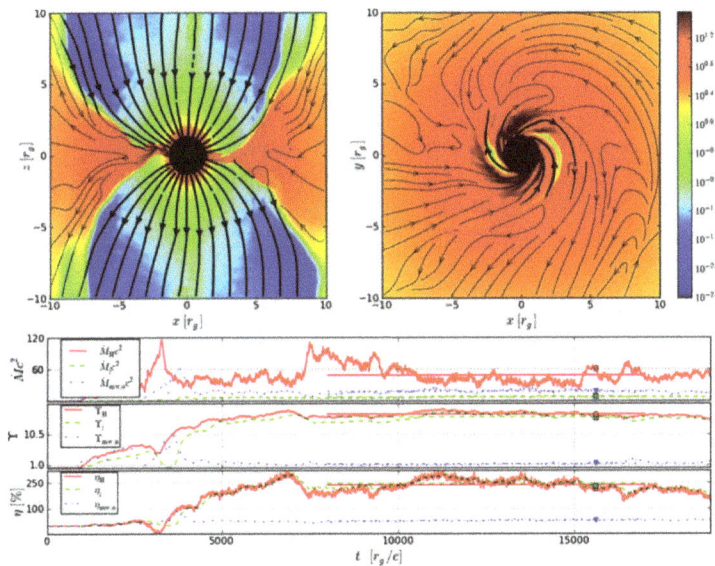

Figure 2. Evolved snapshot of an initially weakly magnetized thick accretion disc around a rotating BH at $t = 15{,}612\, r_g/c$ ($r_g \equiv GM_{\mathrm{BH}}/c^2$, where M_{BH} is the BH mass). **Top panels** show the logarithm of rest-mass density in color (see the legend on the right-hand side) in both the z–x plane at $y = 0$ (**top left-hand panel**) and the y–x plane at $z = 0$ (**top right-hand panel**). The black lines trace field lines, where the thicker black lines show where field is lightly mass-loaded. **The bottom panel** has three subpanels. **The top subpanel** shows $\dot{M}$ through the BH ($\dot{M}_{\mathrm{H}}$), out in the jet ($\dot{M}_{\mathrm{j}}$, at $r = 50\, r_g$), and out in the magnetized wind ($\dot{M}_{\mathrm{mw,o}}$, at $r = 50\, r_g$) with legend. **The middle subpanel** shows the magnetic flux (Y) for similar conditions. **The bottom subpanel** shows the efficiency (η) for similar conditions. Horizontal lines of the same colors show the averages over the averaging period, while square/triangle/circle tickers are placed at the given time and values. A sustained efficiency for η_{H}, the power flowing out of the BH over rest-mass energy flux flowing into the BH, demonstrates the net energy extraction from spinning BHs via the BZ/Penrose mechanism. (Figure 4 of [83].)

3.3. Current-Driven Kink Instability and Magnetic Energy Dissipation

The noticeable large-scale stability of the newly-born fully three-dimensional jets emanating from accreting, rapidly rotating BHs in the first long-term simulations [78,80] prompted a series of studies addressed to re-evaluate the development of 3D instabilities in relativistic magnetized jets under controlled conditions. We concentrate on these studies here and leave for Section 4.2 a more general description of the instabilities arising in relativistic, magnetized flows and their role in the interpretation of the phenomenology in AGN jets at larger scales. Mizuno, Hardee and Nishikawa [88] focused on the stability of magnetized relativistic precessing spine-sheath jets with purely poloidal magnetic fields proving the stabilization effect of weakly relativistic sheaths against Kelvin–Helmholtz instabilities (KHI). In a series of papers, Mizuno and collaborators studied the development of the current-driven kink instability in relativistic force-free jets under idealized conditions (temporal analysis within the periodic box): comoving plasma columns [89]; sheared jets [90]; and differentially rotating jets [91]. The last papers in this series [92,93] have concentrated on the spatial growth of the current-driven instability (CDI) in relativistic jets. On its own, Porth [94] studied the stability of jets from rotating magnetospheres along the initial acceleration and collimation region by means of high-resolution adaptive mesh refinement (AMR) simulations in 3D. His analysis showed that the $m = 1$–5 modes saturate at a height of $\sim$20 inner disc radii.

The models of jet acceleration discussed thus far have considered a gradual acceleration of the jet by magnetic forces. However, dissipation via reconnection of alternating magnetic fields [95,96] has been suggested as an alternative energy conversion mechanism. On the other hand, the large-scale magnetic field may dissipate if the regular magnetic structure is destroyed as a result of a global MHD instability, the CD kink instability being the most plausible candidate. O'Neill and collaborators [97] considered the development of CDI in comoving plasma columns in initial radial force balance through various combinations of magnetic, pressure and rotational forces, and examined the resulting flow morphologies and energetics. Models in which the initial magnetic field is force-free deform, but do not become disrupted. Systems that achieve initial equilibrium by balancing pressure gradients and/or rotation against magnetic forces, however, tend to shred, mix and develop turbulence. Consistently with this result, CDI-driven kinetic energy amplification is slower and saturates at a lower value in force-free models. Singh and collaborators [93] also explored the possible link between CD kink instability and magnetic reconnection in their study of the spatial grow of CDI in relativistic jets. It is interesting to note however that, in these papers, based on (ideal) RMHD simulations, none of the solutions in the non-linear regime can be regarded as *converged*. For convergence, a physical dissipation scale provided by either, e.g., a Navier–Stokes viscosity or Ohmic resistivity (in most astrophysical scenarios, many orders of magnitude smaller than the numerical dissipation scales attainable by magneto-fluid dynamical codes with present-day computational resources) would have to be included in the problem. Additionally, magnetic reconnection can play a major role in regulating the accretion process and hence the production of jets [98–100].

3.4. General Relativistic Radiative Transfer Simulations

Impelled by the forthcoming EHT observations of Sgr A* and the radio core of M 87, spectra and images of the accretion disc/BH/jet system are currently being performed by post-processing GRMHD simulations using relativistic radiative transfer (for radiative processes such as synchrotron emission, self-absorption, and inverse-Compton processes) and ray-tracing including general relativistic effects ([101] and references therein; see also Figure 3) after modeling the location-dependent distribution functions of the emitting particles. However, it is important to note that, despite the huge advances experienced by the numerical simulation of the jet formation process, they still suffer from important theoretical uncertainties. A remarkable example is the mass-loading of the jets, caused by the known failure of up-to-date GRMHD codes inside the highly magnetized funnel [102], with important implications in the jet composition issue (see Section 2.2) and the interpretation of the future observations from the EHT [103]. O'Riordan and collaborators [104] investigated the observational signatures of mass loading in the funnel by performing general-relativistic radiative transfer calculations on a range of 3D-GRMHD simulations of accreting BHs by removing the contribution to the spectrum of the artificially supplied floor material, i.e., restricting the analysis to the case where the funnel material is highly magnetized. Conversely, Broderick and Tchekhovskoy [105] considered the creation and acceleration of pairs at the stagnation surface of Poynting-dominated jets and the subsequent inverse Compton cascade as the mechanism for filling the jets with non-thermal particles near the horizon to explain the compact radio emission in M 87.

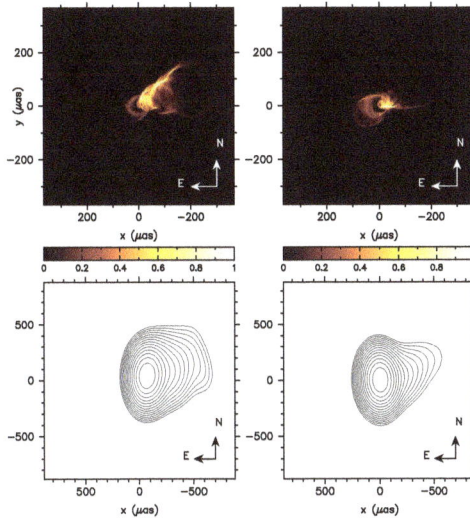

Figure 3. Top panels: Intensity maps ($\lambda = 7$ mm) for a viewing angle of 20° (**left**) and 160° (**right**) from a time-dependent model of a radiatively inefficient accretion flow onto a BH based on a fully 3D GRMHD simulation and scaled to the M 87 disc/BH/jet system. **Bottom panels:** Contour maps of the model convolved with a telescope beam to simulate current observations of M 87. For details about the numerical simulation as well as the emission model and the electron distribution function in the disc and in the jets, see [106]. (Figure adapted from Figures 4 and 6 of [106], reproduced with permission ©ESO.)

4. Simulations of Jets at Parsec and Subparsec Scales

4.1. (Magneto-)Hydrodynamical Simulations

The first simulations of compact jets studied the structure of shocks of steady, relativistic jets propagating through pressure decreasing atmospheres [107,108] and the evolution of relativistically moving perturbations travelling down the underlying steady flow [109–112]. To account for the relativistic effects (relativistic Doppler boosting, light aberration, and time delays) dominating the emission of parsec-scale jets and compare with observations, calculations of the synchrotron radiation transfer were performed by post-processing the hydrodynamical models. These calculations led to the first synthetic radio images of compact jets reproducing the basic phenomenology of these sources (with steady quasi-periodic as well as superluminal radio components), and the first multifrequency (radio) light curves. In particular, Gómez and collaborators [110] simulated the interaction of standing shocks and relativistic perturbations mimicking the ejection of superluminal components from the VLBI core as in the Daly–Marscher's [31] model, and studied the dependence of the structure of radio components with frequency [113]. Next simulations [114–117] aimed to explain the complex behavior observed in many sources as, e.g., the dragging of steady components in 3C 279 [118], the presence of trailing components in 3C 120 [113,115] and 3C 111 [119], and the tangled evolution of components in 3C 111 [117]. In a very recent work, Fromm and collaborators [120] analyzed the influence of the thermal absorption of the obscuring torus in the appearance of radio jets.

41

The preceding simulations were based upon pure (relativistic) hydrodynamical models and the synchrotron emissivity was computed after adding an *ad-hoc* magnetic field with ordered and random components. Hence, as a natural improvement, subsequent simulations incorporated consistent magnetic fields from RMHD models of jets. Broderick and McKinney [121] used jets formed from rapidly rotating, accreting BHs through GRMHD simulations [80] extrapolated to parsec scales, and analyzed the RM generated in the disc wind wrapping the jet, containing ordered toroidally dominated magnetic fields. Porth and collaborators [122] studied the synchrotron emission from RMHD jets with large-scale helical magnetic fields along the acceleration and collimation region. Both simulations [121,122] reproduce the expected gradients in RM across the jet width due to the toroidal component of the helical magnetic field [42]. In addition, Porth et al. [122] provided detailed insights regarding the polarization structure throughout the jet, which depends strongly on the helical magnetic field pitch angle, jet viewing angle, Lorentz factor, and opacity.

Several papers have concentrated on the observational signatures of recollimation shocks in relativistic, magnetized, overpressured jets, commonly associated with the stationary components often seen in parsec-scale VLBI observations of AGN jets [21,123–125]. Mizuno and collaborators [126] considered kinematically-dominated jets with axial, toroidal, and helical force-free magnetic fields to investigate the effects of different magnetic field topologies and strengths on the recollimation structures. Martí, Perucho and Gómez [127] extended the previous study to jet models with (non-force-free) helical magnetic fields and different energy compositions (internal, kinetic, and magnetic) including models with high magnetizations. The total-pressure mismatch between the jet and the ambient medium combines with the radial Lorentz force to generate jets with a complex superposition of periodical recollimation shocks and gentle expansions and compressions along the jet, and steep transversal gradients. In [128], the authors modeled the optically thin total and linearly polarized synchrotron emission from a selection of models of [127] (see Figure 4), with the ultimate goal to connect the properties of the magneto-hydrodynamical jets with the structures observed in extragalactic jets at parsec scales. In particular, these simulations show a top-down emission asymmetry along the jet produced by the helical magnetic field and a noticeable spine brightening for highly magnetized jets. The bright stationary components associated with the recollimation shocks (whose separation and strength depend on the magneto-hydrodynamical parameters) present relative intensities modulated by the Doppler boosting ratio between the pre-shock and post-shock states. Models at small viewing angles display a roughly bimodal distribution in the polarization angle, due to the helical structure of the magnetic field.

Complementing the above studies on overpressured jets propagating through homogeneous ambient media, the large-scale structure of initially free-expanding jets through pressure-decreasing atmospheres, $p_a \propto z^{-\kappa}$, has been tackled using various approximate analytical approaches and numerical simulations for both purely hydrodynamical and magnetized relativistic jets (see, e.g., [129,130] and references therein). Values of $\kappa \simeq 1 - 2$ produce reconfinement shocks reaching the jet axis, whereas values of $\kappa \geq 4$ keep the jets expanding freely. Most importantly, κ governs the global stability of the jets: beyond $\kappa = 2$, the jet suffers such a rapid lateral expansion that the causal communication across the jet is completely lost and hence global instabilities of any type become totally suppressed [131]. Fromm and collaborators [132] have computed synthetic radio maps of hot, mildly magnetized jets propagating down a pressure-decreasing atmosphere with $\kappa = 1$. Most interestingly, these authors incorporated the characteristics of the observing array (VLBA in this case) and of the observing experiment, and the imaging algorithm (*Difmap*'s [133] *CLEAN* deconvolution [134]) in the computation of the radio maps.

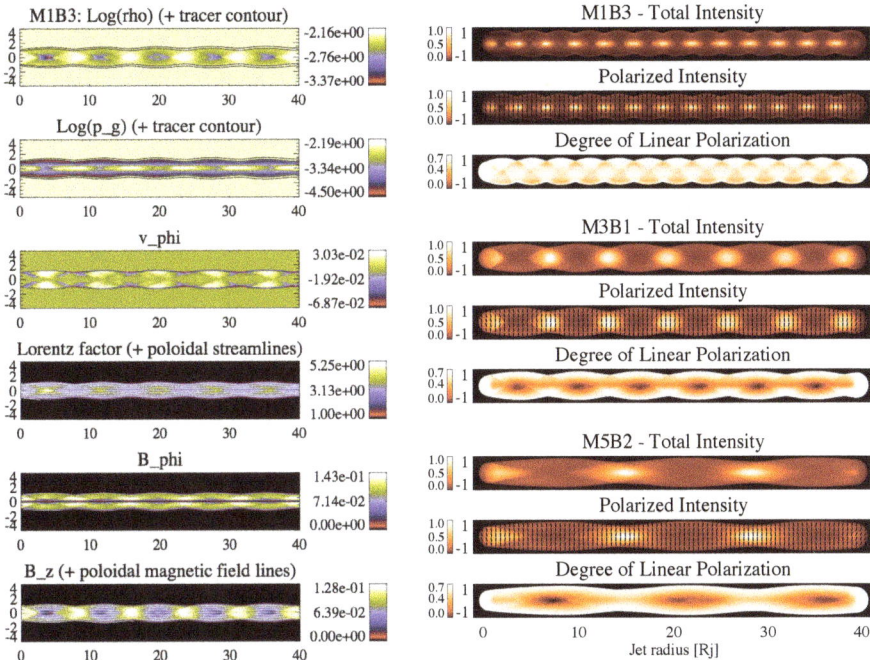

Figure 4. (**Left**) Steady structure of a magnetically dominated relativistic jet model. From top to bottom: Distributions of rest-mass density, gas pressure, toroidal flow velocity, flow Lorentz factor, and toroidal and axial magnetic field components. Poloidal flow and magnetic field lines are overimposed onto the Lorentz factor and axial magnetic field panels, respectively. (**Right**) Synthetic total and polarized synchrotron intensities and degree of linear polarization from RMHD jet models at 20° to the line of sight. **Top panel** (M1B3 model): magnetically dominated jet corresponding to the RMHD models on the left; **middle panel** (M3B1): hot jet; **bottom panel** (M5B2): kinetically dominated jet. (Figure adapted from Figures 2 and 14 of [128] ©AAS. Reproduced with permission.)

4.2. Instabilities

Magneto-hydrodynamic flows can exhibit KH as well as CD instabilities. Whereas KHI grow in the contact discontinuity or shear layer of flows with relative speeds (and can develop in both magnetized and unmagnetized flows), the growth of CDI modes is driven by the distribution of the electric density current flowing along the jet in magnetized models. Strictly speaking, it is not possible to unambiguously separate their effects in a magnetized jet, although, in general, CDI prevail in magnetically dominated jets, whereas kinetically dominated jets are subject to KHI. In the case of CDI, the stability conditions can be predicted theoretically in the force-free approximation. In kinetically dominated jets, the KHI conditions follow from the linearized (R)MHD equations. Hardee [135] provided a condensed review of both KHI and CDI including references to papers on linear stability analyses (see also [8]). The analysis of the stability against these types of perturbations allows one to constrain the space of flow parameters leading to a successful jet launch and propagation. On the other hand, instability modes can grow to finite amplitudes without destroying the jet and leave observational imprints (in the form of radio knots, transverse structure, bends, etc.) whose properties can be used to pinpoint the jet flow parameters. This analysis has been successfully applied to probe the physical conditions in the jets of several sources (e.g., M 87 [136–138], S5 0836+710 [139–142], 3C 273 [36,143], 3C 120 [144–146], and Bl Lac [147]).

Beyond the linear regime, the analysis requires numerical (hydrodynamic or magneto-hydrodynamic) simulations. Here, the main purpose is to assess the stability, collimation, and mass entrainment properties of jets at large (temporal and spatial) scales. We restrict ourselves to the relativistic case. In a series of papers, Perucho and collaborators studied the effects of relativistic dynamics and thermodynamics on the development of KH instabilities in relativistic slab jets—both in the vortex-sheet approximation [148,149] and for sheared flows [150]—covering the linear, saturation, and non-linear phase of the evolution by means of hydrodynamic simulations. In [150,151], the authors considered the effects of very high-order reflection modes on sheared relativistic (both kinematic and thermodynamic) slab jets, and in full 3D for initially cylindrical jets, in [152]. Other authors [153–155] have focused on the properties of the fully-developed turbulence flow driven by the KHI and check its consistency (velocity power spectrum, flow intermittency) with the classical Kolmogorov's model in the *inertial range* in which the turbulent eddies interact without external forcing or dissipation.

In the context of the magnetic KHI, Mizuno, Hardee and Nishikawa [88] focused on the stability of magnetized relativistic precessing spine-sheath jets with purely poloidal magnetic fields (see Section 3.3), whereas other papers [156,157] have been devoted to elucidate the basic properties (dynamo amplification of magnetic fields, power spectra) of the KHI-driven RMHD turbulence. In particular, Beckwith and Stone [157] presented preliminary results on the turbulent amplification of the magnetic field in the cocoon formed along the propagation of a kpc-scale jet. Let us note that, instead of starting from a shearing flow, Zrake and MacFadyen [158] studied the development of RMHD turbulence from initial spatially-uniform conditions.

The non-linear development of the CDI have concentrated on the most disruptive mode, the kink mode. The state of the art of these numerical studies as well as of the CDI-driven RMHD turbulence are summarized in Section 3.3 in the context of the magnetically-dominated flows arising at the jet formation region.

The KHI and CDI are by far the best studied instabilities in the context of (relativistic) jets. However, in the last years, other fluid instabilities have deserved the attention of theorists and numerical simulators in the context of reconfining jets. These instabilities are the Rayleigh–Taylor (RTI) and the centrifugal instabilities (CI). The radial force arising from the pressure mismatch between the jet and the surrounding ambient medium in the expanding/compressing jet flow is the driving force of the RTI [159–161]. In the case of the CI considered in [162], the driving force is the centrifugal force experienced by the fluid parcels moving along the poloidally curved trajectories in the course of the reconfinement process. The transverse structure of the jet is remarkably deformed by the non-linear growth of these radial oscillation-induced instabilities, which can evolve into turbulence leading to the complete jet disruption. These instabilities could be behind the division of AGN jets (although see Section 5.2) into two morphological types in the Fanaroff–Riley classification (see Figure 5). It is also worth noting that the CI is also present in the rotating flows forming the jets near the central engine [163–165]. Further work is needed to untangle the competition between the KHI and CI in the context of the FR division [162], and to analyze the effect of including strong magnetic fields which may inhibit the growth of CI and KHI modes and promote CDI [166].

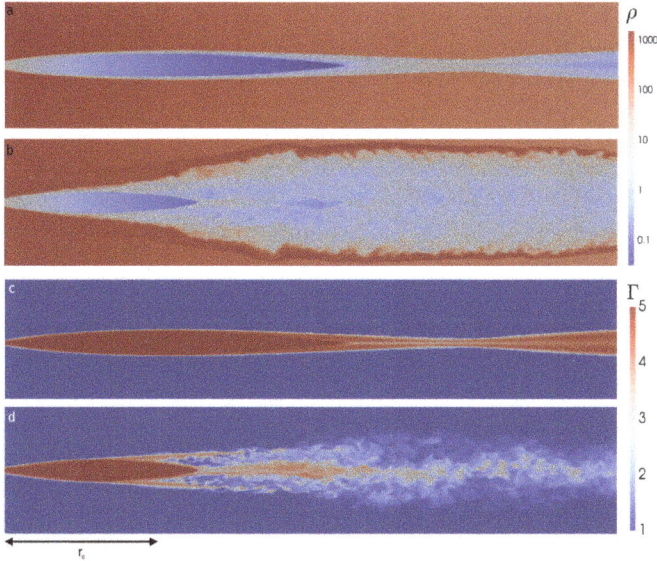

Figure 5. (**a**,**b**) Jet density (ρ) in arbitrary units; and (**c**,**d**) Lorentz factor (Γ) for a model corresponding to a CI-driven FR I jet making its way through a galactic corona. A section of the jet, containing the jet axis is shown for ρ (**a**) and Γ (**c**) in the steady state solution and ρ (**b**) and Γ (**d**) in the final 3D solution. The results correspond to simulation C1 in [162]. The jet to ambient density ratio is 1.6×10^{-2} and the initial jet Lorentz factor is 5. The jet kinetic power is 2×10^{44} ergs/s. (Figure adapted from [162] kindly provided by the authors.)

4.3. Microphysics

The simulations discussed previously rely on a pure macroscopic (magneto-)fluid dynamical modelization of the jet that accounts only for the evolution of the thermal plasma. However, the emission in these sources is produced by non-thermal particles (NTPs) accelerated at shocks [167] or in magnetic reconnection events [168] that require a microscopic description whose implementation in macroscopic fluid dynamical models is a challenge for present-day scientific computing.

Current models of blazar jet emission (e.g., [39] and references therein) focus on the microscopic processes (acceleration of NTPs, emission processes, and adiabatic losses) and a simplified jet structure model (in this case, a 1D time-independent relativistic fluid flow with a variable shape and bulk Lorentz factor) with the goal of fitting simultaneous multifrequency observations. Marscher [169] presented a parameterized model for variability of the flux and polarization of blazars in which turbulent plasma flowing at a relativistic speed down a jet crosses a standing conical shock representing the mm-wavelength core of these sources.

On the opposite side, RMHD simulations produce self-consistent models for the underlying jet and implement simplified recipes to describe the NTPs. Several papers [109–111,122,128,170] set the energy density of the NTPs proportional to the pressure of the thermal plasma. Since the pressure is a good tracer of shocks, this approach mimics the emission if the process of shock acceleration were dominant. Alternatively, other works [121,122,128] consider the energy density of the NTPs as proportional to the magnetic energy density more suitable to account for the particle acceleration in the case of magnetic reconnection. Anantua and collaborators [171] suggested some other similar recipes where the energy density of the NTPs is assumed proportional to a power of the magnetic energy density.

An alternative approach in the numerical modeling of non-thermal emission from astrophysical jets treats the population of NTPs as a separate population advected by the fluid. Transport of

NTP populations in classical jets in radio galaxies were carried out in [172–174]. Mimica and collaborators [175] improved the approach of Gómez et al. [109,110] by incorporating a transport algorithm for the advection of the NTPs including synchrotron losses to study the spatial and temporal variations of the spectral index of the radio components. The same numerical treatment of the NTP transport and the synchrotron emission is used in [176] to interpret the light-curve and the VLBI observations of the 2006 flare of CTA 102 within the shock-shock interaction model devised in [110]. Vaidya and collaborators [177] incorporated a particle module in an RMHD code that includes particle acceleration at shocks and synchrotron and (external) inverse Compton emission.

Particle-in-cell (PIC) methods appear as the most realistic way to simulate the kinetic dynamics of the plasma forming the jet. PIC codes model plasmas as a collection of charged *macroparticles* (representing many physical particles) that are moved by integration of the Lorentz force. Currents associated with the macroparticles are deposited on a grid on which Maxwell's equations are discretized. Electromagnetic fields are then advanced via Maxwell's equations, with particle currents as the source term. Finally, the updated fields are extrapolated to the particle locations and used for the computation of the Lorentz force, so the loop is closed self-consistently. To ensure that kinetic effects are resolved in the simulation, it is necessary that the grid spacing be much smaller than the plasma skin depth, c/ω_p, and that the timestep be much smaller than the corresponding timescale ω_p^{-1}, where ω_p is the plasma frequency ($\omega_p \approx 5.3 \times 10^5/\sqrt{n_e}$ cm; n_e is the electron density). PIC methods are described in this volume by Nishikawa. This approach is capable of treating all effects present in collisionless plasmas, including particle acceleration at shocks [178] and through magnetic reconnection [179] and incorporate them into global simulations of relativistic jets [180,181]. Finally, there are efforts to combine the dynamical evolution of the thermal and non-thermal components via MHD-PIC equations [182–184] (see Figure 6). In this approach, it is not necessary to resolve the plasma skin depth and it is enough to resolve the much larger Larmor (gyration) radius.

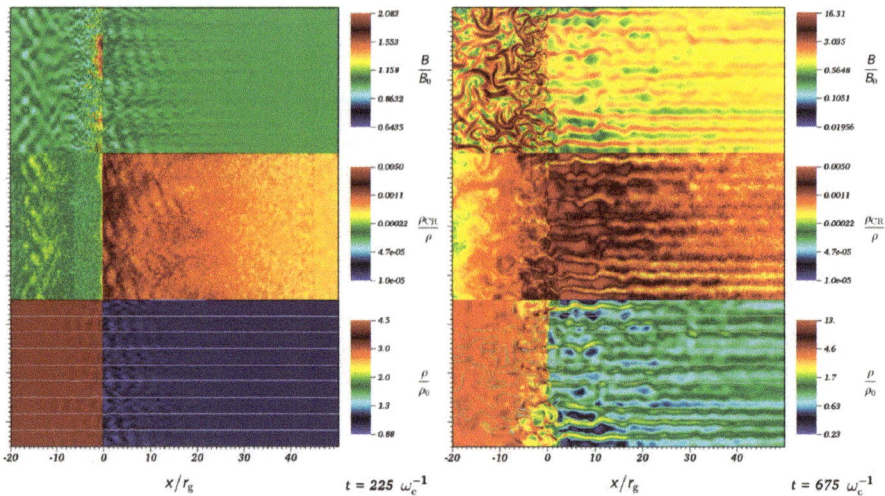

Figure 6. Color maps of the magnetic field amplitude (**top panels**), NTP density (middle panels) and plasma density with the magnetic field stream lines (**bottom panels**) in the early stages of a parallel shock simulation by [183] with a hybrid PIC-MHD code. The gas is streaming through the shock from right to left. **Left panels** ($t = 225\ \omega_c^{-1}$; ω_c is the ion gyrofrequency): the upstream medium shows the start of the streaming instability, while the downstream medium shows the onset of turbulence. **Right panels** ($t = 675\ \omega_c^{-1}$): Both the upstream streaming instability and the downstream turbulence are now fully developed. The shock is warping in response to the instabilities. (Figure adapted from Figures 2 and 4 of [183].)

5. Simulations of Kiloparsec-Scale Jets

5.1. Morphology and Dynamics of kpc-Jets

The first numerical simulations of kiloparsec-scale jets aimed at probing the standard model of FR II radio sources [5]. Among them, the numerical simulations of supersonic (non-relativistic) jets propagating through a homogeneous atmosphere performed by Norman and collaborators [1,185] concentrated on the identification of the basic structural elements (supersonic beam, cocoon, terminal shock, bow shock) and the connection with the morphological elements in powerful radio sources (radio jet, radio lobes, hot spots). These early simulations also served to reveal the mechanism of the jet's working surface governing the jet propagation. The review paper by Burns, Norman and Clarke [186] summarizes the status of the numerical simulations of extragalactic radio sources up to 1990, comprising the first simulations of magnetized jets, the first synthetic images of kiloparsec-scale jets based on synchrotron emission, and the first 3D simulations.

The advent of relativistic codes based on HRSC techniques triggered the study of the effects of relativistic flow speeds and/or relativistic internal energies in the morphology and dynamics of jets in 2D [187–193] and 3D [194–196] simulations. These exploratory relativistic simulations soon incorporated the effects of magnetic fields. The first simulations focused on the propagation of relativistic jets with pure poloidal magnetic fields through ambient media with aligned [197,198] and oblique [199,200] magnetic fields to study how the fields affect the bending properties of relativistic jets. Later studies of large-scale RMHD jets explored the dependence of morphological and dynamic properties of jets on the magnetic field configuration ([201–203]: toroidal fields; [204]: toroidal and poloidal fields), and on the ratios of magnetic energy density and thermal pressure, and magnetic energy density and rest-mass energy density.

5.2. Large-Scale Simulations of kpc-Jets and the FR I/FR II Dichotomy

The simulations discussed in the previous section concentrated on the characterization of the morphology and dynamics of jets in terms of the jet-to-ambient density ratio, jet Mach number, relativistic effects, magnetic field configuration and strength, or injection opening angle. However, none of those simulations have considered the evolution of powerful radio sources as a whole.

The long-term propagation of a powerful jet (FR II jet) through the ambient medium generates a characteristic and well understood structure, formed by: (i) a terminal or reverse shock at the head of the jet where the flow decelerates and heats (the *hot-spot*); (ii) a hot and light region shrouding the jet (the *cocoon*), inflated by the shocked jet particles; and (iii) a dense shell of shocked ambient medium. Begelman and Cioffi [205] described the expansion of this structure assuming a constant jet propagation speed, a complete and instantaneous conversion of the injected jet energy into internal energy in the cocoon and a sideways expansion mediated by a strong shock. Guided by observations showing that cocoons of powerful sources with very different physical size have similar axial ratios, several authors [206–208] developed self-similar expansion models by tuning parameters such as the initial jet opening angle and the ambient density gradient, ignored in the Begelman–Cioffi model.

The first (non-relativistic) long-term simulations of jets aimed to decipher the extended emission of powerful radio sources. Several authors [209–211] used postprocessed bremsstrahlung emission maps to explain the X-ray observations showing evidences of interaction between the radio emitting plasma and the X-ray emitting gas in the ambient medium (see Section 2.4). Other papers [172,174,212] concentrated on the interpretation of the emission at radio wavelengths from MHD simulations of radio galaxies including a detailed treatment of the microphysics: relativistic electron transport, diffusive acceleration at shocks and (synchrotron, inverse Compton from CMB photons) radiative and adiabatic cooling. More recent simulations [213–215] focused on the structure of the magnetic field in the radio lobes and the properties of the synchrotron linearly-polarized emission.

Among the first relativistic simulations of powerful radio sources, are those of Scheck and collaborators [216] aimed at looking for clues on the jet composition ($e^{\pm}$, e/p plasmas) in the large-scale morphology of (2D, axisymmetric, purely hydrodynamical) jets, and the first high-resolution (20 zones per jet radius) 3D simulations of relativistic magnetized jets presented by Mignone and collaborators [217].

Many papers based on simulations of powerful radio sources have focused on the consequences of the interaction of the jet with the interstellar and intergalactic media in an on-going task to comprehend the AGN feedback phenomenon at galactic and supra-galactic scales. These papers are described briefly in the next section. First, let us examine the contribution of numerical simulations of jets to the solution of the FR I/FR II dichotomy issue. As stated in Section 2.3, the ultimate process leading to the morphological division between FR I and FR II sources is the deceleration of the flow to transonic speeds in the low power sources at kpc scales. Numerical simulations have explored different mechanisms responsible of this deceleration. One of these mechanisms is the jet mass loading by stellar winds [55]. However, axisymmetric numerical simulations [218] show that only weak ($L_j \approx 10^{41}$–10^{42} erg s^{-1}) FR I jets can be decelerated by mass entrainment rates consistent with present models of stellar mass loss in elliptical galaxies on scales of 1 kpc. Effective deceleration mechanisms for more powerful jets ($L_j \geq 10^{43}$ erg s^{-1}), involve the development of instabilities and/or strong recollimation shocks. Rossi and collaborators [219] discussed the deceleration of FR I jets resulting from the growth of KH helical instabilities developing at the jet surface, whereas Perucho and collaborators [152] studied the properties of the KHI-driven entrainment in the case of sheared jets. However, none of these papers deal with the origin of the instabilities, which were imposed as initial/boundary conditions. Perucho and Martí [220] connected the jet deceleration with the entrainment of colder and denser ambient gas within the jet due to the growth of KHI pinching modes downstream of a strong recollimation shock. Krause and collaborators [221] studied the properties of this recollimation shock (and the resulting flows) as a function of the initial jet opening angle. A recent work [162] (see Section 4.2) proposes the development of the CI [165] in the outer layers of the reconfining flow passing a recollimation shock as the trigger of the turbulent entrainment of the jet and its subsequent deceleration. Relying on 3D, purely hydrodynamical, non-relativistic simulations of supersonic jets, Massaglia and collaborators [222] studied the morphologies of FR I jets. Interestingly, for low enough kinetic powers ($L_j \leq 10^{43}$ erg s^{-1}), the energy of jets propagating down an inhomogeneous ambient medium, instead of being deposited at the terminal shock, is gradually dissipated by turbulence producing the plumes characteristic of FR I objects.

The simulations discussed in the previous paragraph ignore the effects of magnetic fields in the FR division. In the model of Tchekhovskoy and Bromberg [223], jets with powers below some critical value depending on the galaxy core mass and radius and the magnetic field strength can become kink-unstable within the core, stall, and inflate cavities filled with relativistically hot plasma leading to FR I morphologies (see Figure 7). Finally, let us note that a recent work [224] has considered the impact of the environment (galaxy cluster-like, poor group-like) in the jet morphology. Although focused on FR II sources, these and future simulations of this kind could shed light on how the observable properties of the FR I and FR II jets change due to jet–environment interaction.

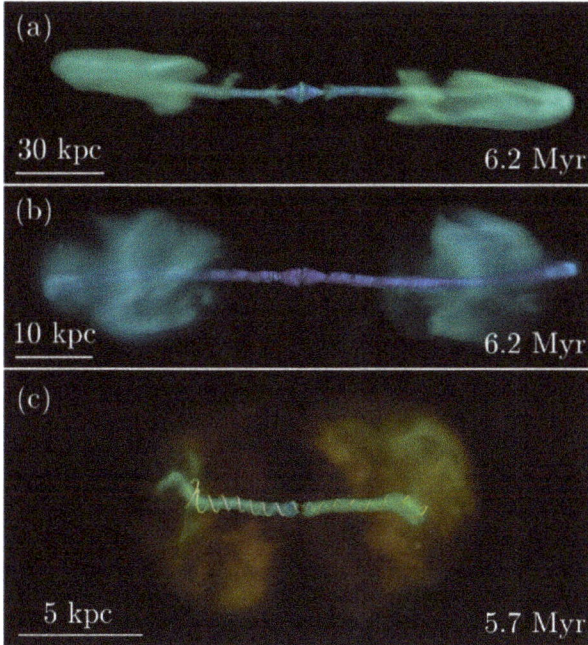

Figure 7. Volume rendering of simulated jets shows the logarithm of density (yellow-green shows high and blue shows low values). White lines show magnetic field lines. (**a**) High-power jets reach 100 kpc distances in 6 Myr. They are mostly straight apart from subtle, large-scale bends seen in FR II sources such as Cygnus A; after running into the ambient medium, the jets end up forming strong backflows, as characteristic of FR II jet sources. (**b**) Same high-power jets as those in panel (**a**) with a break in the ambient density power-law at 10 kpc. Their backflows are less well collimated. (**c**) Low-power jets reach a distance of 5 kpc in 3 Myr. Once there, they succumb to a global kink instability and remain stalled at this distance for 3 Myr. These jets inflate large cavities (shown in yellow) filled with a relativistically-hot plasma, as characteristic of FR I jet sources. (Figure 4 of [223].)

5.3. AGN Feedback

The impact of AGN jets on their galactic scale environment was realized when the first X-ray observations of galaxies and clusters of galaxies showed a clear anticorrelation of the radio- and X-ray emitting plasmas [225]. At the same time, it became apparent that the energy deposition of the jets in the galactic environment could be an important agent in solving the cooling flow problem which manifests itself on a variety of scales, from isolated elliptical galaxies to large clusters of galaxies [226].

As mentioned in the previous section, some works [209–211] aimed at reproducing the X-ray cavities in powerful radio sources. Besides these *active cavities* associated to ongoing jet injection, we also find the so-called *ghost cavities*, which appear disconnected from the active jet-fed cocoon. This second type of cavities are associated with past periods of jet activity and/or low power (FR I) jets. The (axisymmetric) simulations of Churazov and collaborators [227] described the dynamics of buoyant bubbles inflated by an earlier phase of nuclear activity of the galaxy to explain the complex morphology of the X-ray and radio maps in the central 50 kpc region around the galaxy M 87. Quilis and collaborators [228] studied the effect of the energy injection on the cooling rate of a cluster core by means of 3D simulations of the evolution (rise, fall, and mixing) of a hot bubble inflated close to the core center. Their results confirmed this mechanism as a very efficient one for regulating cooling flows. Unlike these simulations, where the lobe inflation was modeled artificially by adding a low-density hot bubble [227] or injecting internal energy [228] within a region close to the gravitational center, Reynolds and collaborators [229] followed the evolution of cavities of powerful jets ($L_j \geq 10^{45}$ erg s^{-1}) across the initial supersonic phase up to their fate as buoyantly rising plumes long after the jet activity has ceased. Let us note by passing that methods similar to that of Quilis and collaborators [228] (i.e., the injection of thermal energy into bubbles placed around the central black hole) have been subsequently implemented into full cosmological simulations [230] and used for example in the Illustris project [231], a series of large-scale hydrodynamical simulations of galaxy formation. Such feedback is a necessary ingredient in cosmological simulations to reproduce the properties of galaxy populations realistically. A more sophisticated AGN jet feedback method [232,233] which deposits mass, momentum and energy has been developed recently.

The early simulations discussed in the previous paragraph focused on the (hydro)dynamics of the cavity/bubble expansion in the cluster medium. However, the results by Bîrzan and collaborators [234] on a systematic study of radio-induced X-ray cavities in 18 systems show that the energy associated with the cavity inflation is about 10–50% the energy necessary to quench the cooling. The conclusion is that a major part of the AGN energy must be transferred to the ICM in other ways not related with the simple hydrodynamic expansion of the bubbles. Options include dissipation of weak shocks and sound waves, turbulence, thermal conduction, shock-heating, heating by cosmic rays, or magnetic fields [235]. Moreover, among the open problems remaining is also to find a robust mechanism by which the (intrinsically directional) AGN jets succeed in heating the ICM isotropically. In the following lines, we concentrate on a few representative papers facing this problem. The 3D-AMR simulations performed by Yang and Reynolds [236] set the state of the art of purely hydrodynamical, non-relativistic simulations including radiative cooling (no viscosity, no heat conduction). The (time-dependent) BH accretion/bipolar jet ejection is modeled via a simplified subgrid model. According to these authors, the isotropization of the heating is the result of a sustained circulation of ambient gas originally displaced by the jets towards the cluster center. Magneto-hydrodynamical simulations performed by the same authors [237] show that anisotropic (along the magnetic field lines) conductive heating is likely significant only for the most massive clusters. Martizzi and collaborators [238] performed (non-magnetized) simulations of jet heating similar to those of Yang and Reynolds and analyzed the dependence of the results on the diffusion of the numerical algorithms, specially the *Riemann solver*. The work reflects the difficulty to achieve a proper numerical convergence for the heating/cooling problem derived from the disparity in scales of the processes involved. Finally, recent papers (e.g., [239,240]; this last reference using a *moving-mesh* code) have simulated the heating by cosmic rays as a mechanism for the AGN energy thermalization in clusters. Supplied by the AGN jets, cosmic rays (likely protons) disperse in the magnetized ICM via streaming, and interact with the ICM via hadronic, Coulomb, and streaming instability heating. Moreover, jets energetically dominated by cosmic rays lose momentum more quickly causing the jet to expand laterally and to displace the ICM close to the center of the clsuter more isotropically.

The papers cited above consider static, idealized cluster cores. Heinz and collaborators [241,242] were the first to take into account the dynamic nature of the cluster gas and detailed cluster physics in their simulations of the interaction of AGN jets with galaxy clusters. The conclusion is that in a cosmologically evolved cluster, the motions of cluster gas effectively distribute the effects of the AGN over a wide angle. Using a more relaxed cluster, Mendygral and collaborators [243] performed MHD simulations confirming the distortion of the morphology of jets and lobes due to the ICM *weather*. Finally, the recent work by Bourne and Sijacki [244] (using a moving-mesh code) shows that the large-scale turbulence generated by orbiting substructures result in line-of-sight velocities and velocity dispersions consistent with the Hitomi observations of the Perseus cluster [245].

Following a different line of work, Perucho and collaborators [246–248] examined the properties of the cluster core heating by relativistic jets. The ultimate motivation behind this study is that modeling the energy input of a powerful jet within a non-relativistic framework leads to inconsistent sets of injection parameters. On one hand, the modeled jets are unrealistically slow (or unphysical). On the other hand, without the contributions of the flow Lorentz factor and the relativistic enthalpy to the inertia, jets are unrealistically massive [249]. The results from 2D axisymmetric simulations [246,247] show that the cavities of powerful relativistic jets are more symmetrical (less elongated) than their non-relativistic counterparts and produce more powerful and long-living bow shocks, and hence are less prone to break into buoyant bubbles, alleviating the problem of the heat isotropization (see Figure 8). Additionally, since the jet particles store much less energy in the form of rest-mass, a larger fraction of energy can be transferred (and, in fact, it is) to the ICM. These basic conclusions seems to survive in 3D simulations [248]. Three-dimensional simulations of lobe inflation in cluster environments by mildly-relativistic magnetized jets below equipartition have been recently performed by English and collaborators [250].

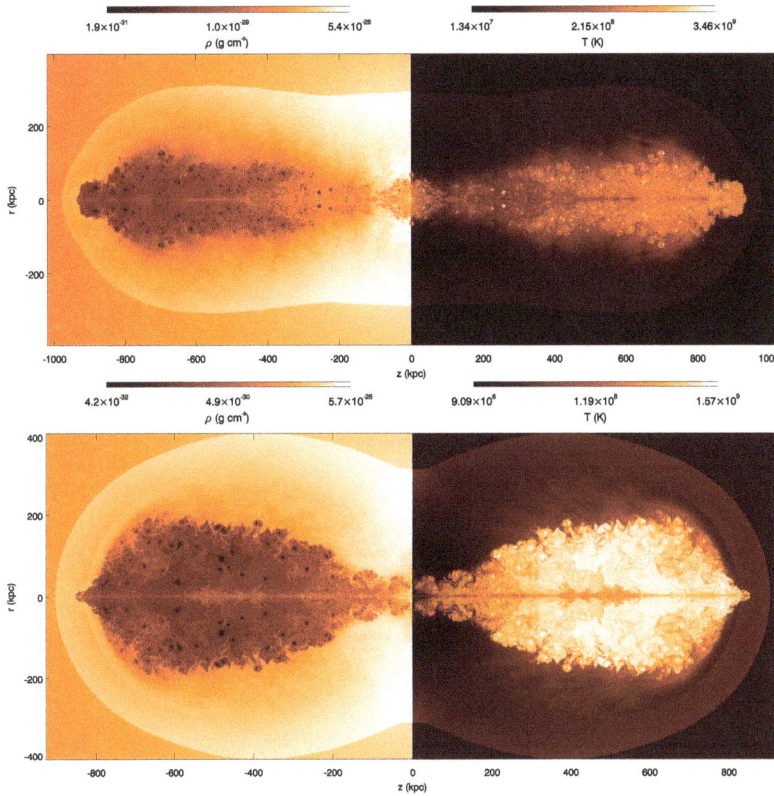

Figure 8. Density (**left**); and temperature (**right**) of two simulations of a powerful jet ($L_j = 10^{46}$ erg/s) propagating through the ICM after jet's switch-off. **Top panel**: non-relativistic, electron-proton jet; **bottom panel**: relativistic leptonic jet. In both cases, the jet density at injection (1 kpc from the cental engine) is 8.3×10^{-29} g cm^{-3}. The jet flow speeds at injection are 0.3 c for the non-relativistic jet, and 0.984 c (Lorentz factor ≈ 5.61) for the relativistic one. To compensate the smaller inertia, the jet radius at injection of the non-relativistic jet has the unrealistic value of 3 kpc (0.1 kpc in the relativistic jet). (Adapted from Figures 7 and 8 of [247].)

As important as unraveling the role of the jet-driven AGN feedback in the solution of the cooling flow process is to determine its influence in the evolution of the host galaxy, in particular through the galaxy's star formation history [251,252]. Gaibler and collaborators [253] performed 3D non-relativistic AMR simulations of the interaction of a powerful AGN jet with the massive gaseous disc of a high-redshift galaxy to asses the impact on the star formation. In [254], special attention is paid to the description of the galaxy ISM described as a two-phase fractal medium with different maximum cloud sizes and volume filling factors. The feedback efficiency, measured by the amount of cloud dispersal generated by the jet–ISM interactions, is investigated through 3D relativistic AMR simulations (see Figure 9).

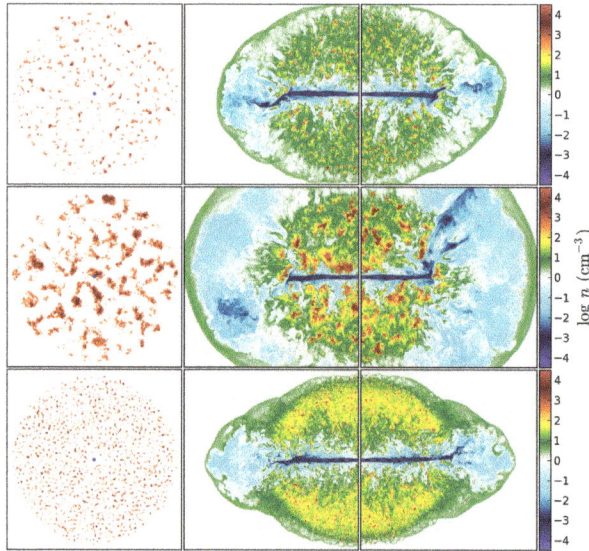

Figure 9. Logarithmic density maps of three simulations of jets propagating through the ISM for different ISM warm gas distributions. Panels are 1 kpc × 1 kpc wide. **Left column:** Face on view of the initial warm gas distribution. **Center and right columns:** Midplane slices at an advanced stage of the simulations for $z = 0$ (reflected about $x = 0$) and $y = 0$, respectively. **Top row:** Very low filling factor run (=0.027); **Middle row:** maximum cloud sizes of 50 pc; **Bottom row:** maximum cloud sizes of 10 pc. (Figure 1 of [254] ©AAS. Reproduced with permission.)

6. Summary and Conclusions

The present review offers a perspective of the contribution of numerical simulations to the understanding of the AGN jet phenomenon along the last four decades of research in the field. Given the disparate scales involved, present simulations of jets concentrate on partial aspects of the problem. These can be broadly divided into the jet formation process (BH accretion, plasma acceleration and collimation), the understanding of the phenomenology of jets at parsec and sub-parsec scales, and the interplay between the jet and the ISM/ICM with important implications in the jet morphology, and the galaxy and cluster evolution (see Section 2). Since these three scenarios also require different physical ingredients, we have reviewed them separately, focusing in their achievements.

Over time, the complexity and physical detail of the simulations have been increasing with the pace of the advance in theoretical models, computational tools and numerical methods. The development of suitable numerical techniques for the (G)RMHD equations (mainly the HRSC methods; Section 1) brought about a revolution in the description of the thermal component of jets (Sections 3.1–3.3, 4.1, 4.2 and 5). Nevertheless, present simulations now demand incorporating microphysical processes (shock acceleration, magnetic reconnection, and radiative processes) in a realistic (i.e., consistent) way. In this regard, the recent attempts to combine (R)MHD and PIC methods into a single hybrid code (Section 4.3) seem very promising.

The AMR finite-volume/finite-difference methods and the more recent moving-mesh techniques (with great impact in the modeling of AGN feedback; Section 5.3) represent elegant and effective ways to improve the numerical resolution in critical regions of the flow, in combination with high-order finite differencing methods. Their use must extend to other areas of the jet modeling.

Finally, as important as the development of suitable numerical techniques is the increase in computing power. In this respect, the use of GPU-accelerated codes (Section 3.2) may represent a significant jump in computing speed.

Funding: This work was supported by the Spanish Ministerio de Economía y Competitividad (grants AYA2015-66899-C2-1-P and AYA2016-77237-C3-3-P) and the Generalitat Valenciana (grant PROMETEOII/2014/069).

Acknowledgments: The author acknowledges the kind invitation of the editors, Athina Meli, Yosuke Mizuno and José Luis Gómez, to contribute to this Special Issue. The author is also indebted to Manel Perucho for a careful reading of the manuscript and to the two anonymous referees whose comments have helped to improve the original version of this review.

Conflicts of Interest: The author declares no conflict of interest.

References

1. Norman, M.L.; Winkler, K.-H.A.; Smarr, L.; Smith, M.D. Structure and dynamics of supersonic jets. *Astron. Astrophys.* **1982**, *113*, 285–302.
2. Scheuer, P.A.G. Models of extragalactic radio sources with a continuous energy supply from a central object. *Mon. Not. R. Astron. Soc.* **1974**, *166*, 513–528. [CrossRef]
3. Blandford, R.D.; Rees, M.J. A 'twin-exhaust' model for double radio sources. *Mon. Not. R. Astron. Soc.* **1974**, *169*, 395–415. [CrossRef]
4. Hawley, J.F.; Fendt, C.; Hardcastle, M.; Nokhrina, E.; Tchekhovskoy, A. Disks and Jets. Gravity, Rotation and Magnetic Fields. *Space Sci. Rev.* **2015**, *191*, 441–469. [CrossRef]
5. Begelman, M.C.; Blandford, R.D.; Rees, M.J. Theory of extragalactic radio sources. *Rev. Mod. Phys.* **1984**, *56*, 255–351. [CrossRef]
6. Leahy, J.P. Interpretation of large scale extragalactic jets. In *Beams and Jets in Astrophysics*; Hughes, P.A., Ed.; Cambridge University Press: Cambridge, UK, 1991; pp. 100–186, ISBN 0-521-33576-0.
7. Böttcher, M.; Harris, D.E.; Krawczynski, H. *Relativistic Jets from Active Galactic Nuclei*; Wiley-VCH: Weinheim, Germany, 2012; ISBN 978-3-527-41037-8.
8. Martí, J.M.; Müller, E. Grid-based Methods in Relativistic Hydrodynamics and Magnetohydrodynamics. *Living Rev. Comput. Astrophys.* **2015**, *1*. [CrossRef] [PubMed]
9. Font, J.A. Numerical Hydrodynamics and Magnetohydrodynamics in General Relativity. *Living Rev. Relativ.* **2008**, *11*. [CrossRef]
10. Lovelace R.V.E. Dynamo model of double radio sources. *Nature* **1976**, *262*, 649–652. [CrossRef]
11. Blandford, R.D. Accretion disc electrodynamics—A model for double radio sources. *Mon. Not. R. Astron. Soc.* **1976**, *176*, 465–481. [CrossRef]
12. Blandford, R.D.; Payne, D.G. Hydromagnetic flows from accretion discs and the production of radio jets. *Mon. Not. R. Astron. Soc.* **1982**, *199*, 883–903. [CrossRef]
13. Li, Z.-Y.; Chiueh, T.; Begelman, M.C. Electromagnetically driven relativistic jets—A class of self-similar solutions. *Astrophys. J.* **1992**, *394*, 459–471. [CrossRef]
14. Vlahakis, N.; Königl, A. Relativistic Magnetohydrodynamics with Application to Gamma-Ray Burst Outflows. I. Theory and Semianalytic Trans-Alfvénic Solutions. *Astrophys. J.* **2003**, *596*, 1080–1103. [CrossRef]
15. Vlahakis, N.; Königl, A. Magnetic Driving of Relativistic Outflows in Active Galactic Nuclei. I. Interpretation of Parsec-Scale Accelerations. *Astrophys. J.* **2004**, *605*, 656–661. [CrossRef]
16. Beskin, V.S.; Nokhrina, E.E. The effective acceleration of plasma outflow in the paraboloidal magnetic field. *Mon. Not. R. Astron. Soc.* **2006**, *367*, 375–386. [CrossRef]
17. Lyubarsky, Y. Asymptotic Structure of Poynting-Dominated Jets. *Astrophys. J.* **2009**, *698*, 1570–1589. [CrossRef]
18. Blandford, R.D.; Znajek, R.L. Electromagnetic extraction of energy from Kerr black holes. *Mon. Not. R. Astron. Soc.* **1977**, *179*, 433–456. [CrossRef]
19. Hirotani, K.; Takahashi, M.; Nitta, S.-Y.; Tomimatsu, A. Accretion in a Kerr black hole magnetosphere—Energy and angular momentum transport between the magnetic field and the matter. *Astrophys. J.* **1992**, *386*, 455–463. [CrossRef]

20. Komissarov, S.S. Central Engines: Acceleration, Collimation and Confinement of Jets. In *Relativistic Jets from Active Galactic Nuclei*; Böttcher, M., Harris, D.E., Krawczynski, H., Eds.; Wiley-VCH: Weinheim, Germany, 2012; pp. 3–16, ISBN 978-3-527-41037-8.

21. Gómez, J.L.; Lobanov, A.P.; Bruni, G.; Kovalev, Y.Y.; Marscher, A.P.; Jorstad, S.G.; Mizuno, Y.; Bach, U.; Sokolovsky, K.V.; Anderson, J.M.; et al. Probing the Innermost Regions of AGN Jets and Their Magnetic Fields with RadioAstron. I. Imaging BL Lacertae at 21 Microarcsecond Resolution. *Astrophys. J.* **2016**, *817*. [CrossRef]

22. Fish, V.L.; Akiyama, K.; Bouman, K.L.; Chael, A.; Johnson, M.; Doeleman, S.; Blackburn, L.; Wardle, J.; Freeman, W. The Event Horizon Telescope Collaboration. Observing—And Imaging—Active Galactic Nuclei with the Event Horizon Telescope. *Galaxies* **2016**, *4*, 54. [CrossRef]

23. Lister, M.L.; Cohen, M.H.; Homan, D.C.; Kadler, M.; Kellermann, K.I.; Kovalev, Y.Y.; Ros, E.; Savolainen, T.; Zensus, J.A. MOJAVE: Monitoring of jets in Active Galactic Nuclei with VLBA experiments. VI. Kinematic analysis of a complete sample of blazar jets. *Astron. J.* **2009**, *138*, 1874–1892. [CrossRef]

24. Blandford, R.D.; Königl, A. Relativistic Jets as Compact Radio Sources. *Astrophys. J.* **1979**, *232*, 34–48. [CrossRef]

25. Rees, M.J. Appearance of Relativistically Expanding Radio Sources. *Nature* **1966**, *211*, 468–470. [CrossRef]

26. Marscher, A.P.; Gear, W.K. Models for high-frequency radio outbursts in extragalactic sources, with application to the early 1983 millimeter-to-infrared flare of 3C 273. *Astrophys. J.* **1985**, *298*, 114–127. [CrossRef]

27. Marscher, A.P.; Jorstad, S.G.; Gómez, J.L.; Aller, M.F.; Teräsranta, H.; Lister, M.L.; Stirling, A.M. Observational evidence for the accretion-disk origin for a radio jet in an active galaxy. *Nature* **2002**, *417*, 625–627. [CrossRef] [PubMed]

28. Marscher, A.P.; Jorstad, S.G.; D'Arcangelo, F.D.; Smith, P.S.; Williams, G.G.; Larionov, V.M.; Oh, H.; Olmstead, A.R.; Aller, M.F.; Aller, H.D.; et al. The inner jet of an active galactic nucleus as revealed by a radio-to-γ-ray outburst. *Nature* **2008**, *452*, 966–969. [CrossRef] [PubMed]

29. Chatterjee, R.; Marscher, A.P.; Jorstad, S.G.; Markowitz, A.; Rivers, E.; Rothschild, R.E.; McHardy, I.M.; Aller, M.F.; Aller, H.D.; Lähteenmäki, A.; et al. Connection between the accretion disk and jet in the radio galaxy 3C 111. *Astrophys. J.* **2011**, *734*. [CrossRef]

30. Cawthorne, T.V.; Jorstad, S.G.; Marscher, A.P. Evidence for Recollimation Shocks in the Core of 1803+784. *Astrophys. J.* **2013**, *772*. [CrossRef]

31. Daly, R.A.; Marscher, A.P. The gasdynamics of compact relativistic jets. *Astrophys. J.* **1988**, *334*, 539–551. [CrossRef]

32. Asada, K.; Nakamura, M. The Structure of the M 87 Jet: A Transition from Parabolic to Conical Streamlines. *Astrophys. J.* **2012**, *745*. [CrossRef]

33. Boccardi, B.; Krichbaum, T.P.; Bach, U.; Mertens, F.; Ros, E.; Alef, W.; Zensus, J.A. The stratified two-sided jet of Cygnus A. Acceleration and collimation. *Astron. Astrophys.* **2016**, *585*. [CrossRef]

34. Nagai, H.; Haga, T.; Giovannini, G.; Doi, A.; Orienti, M.; D'Ammando, F.; Kino, M.; Nakamura, M.; Asada, K.; Hada, K.; Giroletti, M. Limb-brightened Jet of 3C 84 Revealed by the 43 GHz Very-Long-Baseline-Array Observation. *Astrophys. J.* **2014**, *785*. [CrossRef]

35. Mertens, F.; Lobanov, A.P.; Walker, R.C.; Hardee, P.E. Kinematics of the jet in M 87 on scales of 100–1000 Schwarzschild radii. *Astron. Astrophys.* **2016**, *595*. [CrossRef]

36. Lobanov, A.P.; Zensus, J.A. A Cosmic Double Helix in the Archetypical Quasar 3C 273. *Science* **2001**, *294*, 128–131. [CrossRef] [PubMed]

37. Gabuzda, D.C.; Vitrishchak, V.M.; Mahmud, M.; O'Sullivan, S. Radio circular polarization produced in helical magnetic fields in eight active galactic nuclei. *Mon. Not. R. Astron. Soc.* **2008**, *384*, 1003–1014. [CrossRef]

38. Homan, D.C.; Lister, M.L.; Aller, H.D.; Aller, M.F.; Wardle, J.F.C. Full Polarization Spectra of 3C 279. *Astrophys. J.* **2009**, *696*, 328–347. [CrossRef]

39. Potter, W.J.; Cotter, G. New constraints on the structure and dynamics of black hole jets. *Mon. Not. R. Astron. Soc.* **2015**, *453*, 4070–4088. [CrossRef]

40. Böttcher, M.; Reimer, A.; Sweeney, K.; Prakash, A. Leptonic and Hadronic Modeling of Fermi-detected Blazars. *Astrophys. J.* **2013**, *768*. [CrossRef]

41. The MAGIC Collaboration. The blazar TXS 0506+056 associated with a high-energy neutrino: Insights into extragalactic jets and cosmic ray acceleration. *Astrophys. J.* **2018**, *863*. [CrossRef]

42. Laing, R.A. Magnetic fields in extragalactic radio sources. *Astrophys. J.* **1981**, *248*, 87–104. [CrossRef]

43. Asada, K.; Inoue, M.; Uchida, Y.; Kameno, S.; Fujisawa, K.; Iguchi, S.; Mutoh, M. A Helical Magnetic Field in the Jet of 3C 273. *Publ. Astron. Soc. Jpn.* **2002**, *54*, L39–L43. [CrossRef]
44. Hovatta, T.; Lister, M.L.; Aller, M.F.; Aller, H.D.; Homan, D.C.; Kovalev, Y.Y.; Pushkarev, A.B.; Savolainen, T. MOJAVE: Monitoring of Jets in Active Galactic Nuclei with VLBA Experiments. VIII. Faraday Rotation in Parsec-scale AGN Jets. *Astron. J.* **2012**, *144*. [CrossRef]
45. Zamaninasab, M.; Savolainen, T.; Clausen-Brown, E.; Hovatta, T.; Lister, M.L.; Krichbaum, T.P.; Kovalev, Y.Y.; Pushkarev, A.B. Evidence for a large-scale helical magnetic field in the quasar 3C 454.3. *Mon. Not. R. Astron. Soc.* **2013**, *436*, 3351–3356. [CrossRef]
46. Fanaroff, B.L.; Riley, J.M. The morphology of extragalactic radio sources of high and low luminosity. *Mon. Not. R. Astron. Soc.* **1974**, *167*, 31P–36P. [CrossRef]
47. Laing, R.A.; Bridle, A.H. Relativistic models and the jet velocity field in the radio galaxy 3C 31. *Mon. Not. R. Astron. Soc.* **2002**, *336*, 328–352. [CrossRef]
48. Carilli, C.L.; Barthel, P.D. Cygnus A. *Astron. Astrophys. Rev.* **1996**, *7*, 1–54. [CrossRef]
49. Laing, R.A.; Bridle, A.H. Systematic properties of decelerating relativistic jets in low-luminosity radio galaxies. *Mon. Not. R. Astron. Soc.* **2014**, *437*, 3405–3441. [CrossRef]
50. Bridle, A.H.; Hough, D.H.; Lonsdale, C.J.; Burns, J.O.; Laing, R.A. Deep VLA Imaging of Twelve Extended 3CR Sample. *Astron. J.* **1994**, *108*, 766–820. [CrossRef]
51. Ledlow, M.J.; Owen, F.N. 20 CM VLA Survey of Abell Clusters of Galaxies. VI. Radio/Optical Luminosity Functions. *Astron. J.* **1996**, *112*, 9–22. [CrossRef]
52. Baan, W.A. Fluid jets in radio sources. *Astrophys. J.* **1980**, *239*, 433–444. [CrossRef]
53. Bicknell, G.V. A model for the surface brightness of a turbulent low Mach number jet. I—Theoretical development and application to 3C 31. *Astrophys. J.* **1984**, *286*, 68–87. [CrossRef]
54. De Young, D.S. Emission line regions and stellar associations in extended extragalactic radio sources. *Nature* **1981**, *293*, 43–44. [CrossRef]
55. Komissarov, S.S. Mass-Loaded Relativistic Jets. *Mon. Not. R. Astron. Soc.* **1994**, *269*, 394–402. [CrossRef]
56. Fabian, A.C. Observational Evidence of Active Galactic Nuclei Feedback. *Ann. Rev. Astron. Astrophys.* **2012**, *50*, 455–489. [CrossRef]
57. Dunn, R.J.H.; Fabian, A.C. Investigating AGN heating in a sample of nearby clusters. *Mon. Not. R. Astron. Soc.* **2006**, *373*, 959–971. [CrossRef]
58. Rafferty, D.A.; McNamara, B.R.; Nulsen, P.E.J.; Wise, M.W. The Feedback-regulated Growth of Black Holes and Bulges through Gas Accretion and Starbursts in Cluster Central Dominant Galaxies. *Astrophys. J.* **2006**, *652*, 216–231. [CrossRef]
59. Nulsen, P.E.J.; Hambrick, D.C.; MacNamara, B.R.; Rafferty, D.; Bîrzan, L.; Wise, M.W.; David, L.P. The Powerful Outburst in Hercules A. *Astrophys. J.* **2005**, *625*, L9–L12. [CrossRef]
60. Simionescu, A.; Roediger, E.; Nulsen, P.E.J.; Brüggen, M.; Forman, W.R.; Böhringer, H.; Werner, N.; Finoguenov, A. The large-scale shock in the cluster of galaxies Hydra A. *Astron. Astrophys.* **2009**, *495*, 721–732. [CrossRef]
61. McNamara, B.R.; Nulsen, P.E.J.; Wise, M.W.; Rafferty, D.A.; Carilli, C.; Sarazin, C.L.; Blanton, E.L. The heating of gas in a galaxy cluster by X-ray cavities and large-scale shock fronts. *Nature* **2005**, *433*, 45–47. [CrossRef]
62. Gitti. M.; O'Sullivan, E.; Giacintucci, S. Cavities and Shocks in the Galaxy Group HCG 62 as Revealed by Chandra, XMM-Newton, and Giant Metrewave Radio Telescope Data. *Astrophys. J.* **2010**, *714*, 758–771. [CrossRef]
63. Koide, S.; Shibata, K.; Kudoh, T. General Relativistic Magnetohydrodynamic Simulations of Jets from Black Hole Accretions Disks: Two-Component Jets Driven by Nonsteady Accretion of Magnetized Disks. *Astrophys. J.* **1998**, *495*, L63–L66. [CrossRef]
64. Koide, S.; Shibata, K.; Kudoh, T. Relativistic Jet Formation from Black Hole Magnetized Accretion Disks: Method, Tests, and Applications of a General RelativisticMagnetohydrodynamic Numerical Code. *Astrophys. J.* **1999**, *522*, 727–752. [CrossRef]
65. Nishikawa, K.-I.; Richardson, G.; Koide, S.; Shibata, K.; Kudoh, T.; Hardee, P.; Fishman, G.J. A General Relativistic Magnetohydrodynamic Simulation of Jet Formation. *Astrophys. J.* **2005**, *625*, 60–71. [CrossRef]
66. Koide, S.; Meier, D.L.; Shibata, K.; Kudoh, T. General Relativistic Simulations of Early Jet Formation in a Rapidly Rotating Black Hole Magnetosphere. *Astrophys. J.* **2000**, *536*, 668–674. [CrossRef]
67. Koide, S.; Shibata, K.; Kudoh, T.; Meier, D.L. Extraction of Black Hole Rotational Energy by a Magnetic Field and the Formation of Relativistic Jets. *Science* **2002**, *295*, 1688–1691. [CrossRef] [PubMed]

68. Komissarov, S.S. Observations of the Blandford-Znajek process and the magnetohydrodynamic Penrose process in computer simulations of black hole magnetospheres. *Mon. Not. R. Astron. Soc.* **2005**, *359*, 801–808. [CrossRef]

69. Lasota, J.-P.; Gourgoulhon, E.; Abramowicz, M.; Tchekhovskoy, A.; Narayan, R. Extracting black-hole rotational energy: The generalized Penrose process. *Phys. Rev. D* **2014**, *89*, 024041. [CrossRef]

70. Koide, S. Relativistic Outflow Magnetically Driven by Black Hole Rotation. *Astrophys. J.* **2004**, *606*, L45–L48. [CrossRef]

71. Komissarov, S.S. General relativistic magnetohydrodynamic simulations of monopole magnetospheres of black holes. *Mon. Not. R. Astron. Soc.* **2004**, *350*, 1431–1436. [CrossRef]

72. Komissarov, S.S. Direct numerical simulations of the Blandford-Znajek effect. *Mon. Not. R. Astron. Soc.* **2001**, *326*, L41–L44. [CrossRef]

73. Komissarov, S.S. Electrodynamics of black hole magnetospheres. *Mon. Not. R. Astron. Soc.* **2004**, *350*, 427–448. [CrossRef]

74. McKinney, J.C.; Gammie, C.F. A Measurement of the Electromagnetic Luminosity of a Kerr Black Hole. *Astrophys. J.* **2004**, *611*, 977–995. [CrossRef]

75. De Villiers, J.-P.; Hawley, J.F.; Krolik, J.H. Magnetically Driven Accretion Flows in the Kerr Metric. I. Models and Overall Structure. *Astrophys. J.* **2003**, *599*, 1238–1253. [CrossRef]

76. Hirose, S.; Krolik, J.H.; De Villiers, J.-P.; Hawley, J.F. Magnetically Driven Accretion Flows in the Kerr Metric. II. Structure of the Magnetic Field. *Astrophys. J.* **2004**, *606*, 1083–1097. [CrossRef]

77. De Villiers, J.-P.; Hawley, J.F.; Krolik, J.H.; Hirose, S. Magnetically Driven Accretion in the Kerr Metric. III. Unbound Outflows. *Astrophys. J.* **2005**, *620*, 878–888. [CrossRef]

78. McKinney, J.C. General relativistic magnetohydrodynamic simulations of the jet formation and large-scale propagation from black hole accretion systems. *Mon. Not. R. Astron. Soc.* **2006**, *368*, 1561–1582. [CrossRef]

79. Komissarov, S.S.; Barkov, M.V.; Vlahakis, N.; Königl, A. Magnetic acceleration of relativistic active galactic nucleus jets. *Mon. Not. R. Astron. Soc.* **2007**, *380*, 51–70. [CrossRef]

80. McKinney, J.C.; Blandford, R.D. Stability of relativistic jets from rotating, accreting black holes via fully three-dimensional magnetohydrodynamic simulations. *Mon. Not. R. Astron. Soc.* **2009**, *394*, L126–L130. [CrossRef]

81. Tchekhovskoy, A.; Narayan, R.; McKinney, J.C. Efficient generation of jets from magnetically arrested accretion on a rapidly spinning black hole. *Mon. Not. R. Astron. Soc.* **2011**, *418*, L79–L83. [CrossRef]

82. Tchekhovskoy, A.; McKinney, J.C. Prograde and retrograde black holes: whose jet is more powerful? *Mon. Not. R. Astron. Soc.* **2012**, *423*, L55–L59. [CrossRef]

83. McKinney, J.C.; Tchekhovskoy, A.; Blandford, R.D. General relativistic magnetohydrodynamic simulations of magnetically choked accretion flows around black holes. *Mon. Not. R. Astron. Soc.* **2012**, *423*, 3083–3117. [CrossRef]

84. Bisnovatyi-Kogan, G.S.; Ruzmaikin, A.A. The Accretion of Matter by a Collapsing Star in the Presence of a Magnetic Field. *Astrophys. Space Sci.* **1974**, *28*, 45–59. [CrossRef]

85. Narayan, R.; Igumenshchev, I.V.; Abramowicz, M.A. Magnetically Arrested Disk: An Energetically Efficient Accretion Flow. *Publ. Astron. Soc. Jpn.* **2003**, *55*, L69–L72. [CrossRef]

86. Mckinney, J.C.; Tchekhovskoy, A.; Blandford, R.D. Alignment of Magnetized Accretion Disks and Relativistic Jets with Spinning Black Holes. *Science* **2013**, *339*, 49–52. [CrossRef] [PubMed]

87. Liska, M.; Tchekhovskoy, A.; Ingram, A.; van der Klis, M.; Markoff, S. Formation of precessing jets by tilted black hole discs in 3D general relativistic MHD simulations. *Mon. Not. R. Astron. Soc.* **2018**, *474*, L81–L85. [CrossRef]

88. Mizuno, Y.; Hardee, P.; Nishikawa, K.-I. Three-dimensional Relativistic Magnetohydrodynamic Simulations of Magnetized Spine-Sheath Relativistic Jets. *Astrophys. J.* **2007**, *662*, 835–850. [CrossRef]

89. Mizuno, Y.; Lyubarsky, Y.; Nishikawa, K.-I.; Hardee, P.E. Three-Dimensional Relativistic Magnetohydrodynamic Simulations of Current-Driven Instability. I. Instability of a Static Column. *Astrophys. J.* **2009**, *700*, 684–693. [CrossRef]

90. Mizuno, Y.; Hardee, P.; Nishikawa, K.-I. Three-dimensional Relativistic Magnetohydrodynamic Simulations of Current-driven Instability with a Sub-Alfvénic Jet: Temporal Properties. *Astrophys. J.* **2011**, *734*. [CrossRef]

91. Mizuno, Y.; Lyubarsky, Y.; Nishikawa, K.-I.; Hardee, P.E. Three-dimensional Relativistic Magnetohydrodynamic Simulations of Current-driven Instability. III. Rotating Relativistic Jets. *Astrophys. J.* **2012**, *757*. [CrossRef]

92. Mizuno, Y.; Hardee, P. E.; Nishikawa, K.-I. Spatial Growth of the Current-driven Instability in Relativistic Jets. *Astrophys. J.* **2014**, *784*. [CrossRef]

93. Singh, C.B.; Mizuno, Y.; de Gouveia Dal Pino, E.M. Spatial Growth of Current-driven Instability in Relativistic Rotating Jets and the Search for Magnetic Reconnection. *Astrophys. J.* **2016**, *824*. [CrossRef]

94. Porth, O. Three-dimensional structure of relativistic jet formation. *Mon. Not. R. Astron. Soc.* **2013**, *429*, 2482–2492. [CrossRef]

95. Giannios, D.; Uzdensky, D.A.; Begelman, M.C. Fast TeV variability in blazars: jets in a jet. *Mon. Not. R. Astron. Soc.* **2009**, *395*, L29–L33. [CrossRef]

96. Lyubarski, Y. A New Mechanism for Dissipation of Alternating Fields in Poynting-dominated Outflows. *Astrophys. J.* **2010**, *725*, L234–L238. [CrossRef]

97. O'Neill, S.M.; Beckwith, K.; Begelman, M.C. Local simulations of instabilities in relativistic jets—I. Morphology and energetics of the current-driven instability. *Mon. Not. R. Astron. Soc.* **2012**, *422*, 1436–1452. [CrossRef]

98. Dexter, J.; McKinney, J.C.; Markoff, S.; Tchekhovskoy, A. Transient jet formation and state transitions from large-scale magnetic reconnection in black hole accretion discs. *Mon. Not. R. Astron. Soc.* **2014**, *440*, 2185–2190. [CrossRef]

99. Singh, C.B.; Garofalo, D.; de Gouveia Dal Pino, E.M. Magnetic reconnection and Blandford-Znajek process around rotating black holes. *Mon. Not. R. Astron. Soc.* **2018**, *478*, 5404–5409. [CrossRef]

100. Quian, Q.; Fendt, C.; Vourellis, C. Jet Launching in Resistive GR-MHD Black Hole-Accretion Disk Systems. *Astrophys. J.* **2018**, *859*. [CrossRef]

101. Mościbrodzka, M. Modeling Polarized Emission from Black Hole Jets: Application to M 87 Core Jet. *Galaxies* **2017**, *5*, 54. [CrossRef]

102. Gammie, C.F.; McKinney, J.C.; Tóth, G. HARM: A Numerical Scheme for General Relativistic Magnetohydrodynamics. *Astrophys. J.* **2003**, *589*, 444–457. [CrossRef]

103. Gold, R.; McKinney, J.C.; Johnson, M.D.; Doeleman, S.S. Probing the Magnetic Field Structure in Sgr A* on Black Hole Horizon Scales with Polarized Radiative Transfer Simulations. *Astrophys. J.* **2017**, *837*. [CrossRef]

104. O' Riordan, M.; Pe'er, A.; McKinney, J.C. Observational Signatures of Mass-loading in Jets Launched by Rotating Black Holes. *Astrophys. J.* **2018**, *853*. [CrossRef]

105. Broderick, A.E.; Tchekhovskoy, A. Horizon-scale Lepton Acceleration in Jets: Explaining the Compact Radio Emission in M 87. *Astrophys. J.* **2015**, *809*. [CrossRef]

106. Mościbrodzka, M.; Falcke, H.; Shiokawa, H. General relativistic magnetohydrodynamical simulations of the jet in M 87. *Astron. Astrophys.* **2016**, *586*. [CrossRef]

107. Wilson, M.J. Steady relativistic fluid jets. *Mon. Not. R. Astron. Soc.* **1987**, *226*, 447–454. [CrossRef]

108. Dubal, M.R.; Pantano, O. The steady-state structure of relativistic magnetic jets. *Mon. Not. R. Astron. Soc.* **1993**, *261*, 203–221. [CrossRef]

109. Gómez, J.L.; Martí, J.M.; Marscher, A.P.; Ibáñez, J.M.; Marcaide, J.M. Parsec-Scale Synchrotron Emission from Hydrodynamic Relativistic Jets in Active Galactic Nuclei. *Astrophys. J.* **1995**, *449*, L19–L21. [CrossRef]

110. Gómez, J.L.; Martí, J.M.; Marscher, A.P.; Ibáñez, J.M.; Alberdi, A. Hydrodynamical Models of Superluminal Sources. *Astrophys. J.* **1997**, *482*, L33–L36. [CrossRef]

111. Komissarov, S.S.; Falle, S.A.E.G. Simulations of Superluminal Radio Sources. *Mon. Not. R. Astron. Soc.* **1997**, *288*, 833–848. [CrossRef]

112. Mioduszewski, A.J.; Hughes, P.A.; Duncan, G.C. Simulated VLBI Images from Relativistic Hydrodynamic Jet Models. *Astrophys. J.* **1997**, *476*, 649–665. [CrossRef]

113. Gómez, J.L.; Marscher, A.P.; Alberdi, A.; Martí, J.M.; Ibáñez, J.M. Subparsec Polarimetric Radio Observations of 3C 120: A Close-up Look at Superluminal Motion. *Astrophys. J.* **1998**, *499*, 221–226. [CrossRef]

114. Aloy, M.A.; Gómez, J.L.; Ibáñez, J.M.; Martí, J.M.; Müller, E. Radio Emission from Three-dimensional Relativistic Hydrodynamic Jets: Observational Evidence of Jet Stratification. *Astrophys. J.* **2000**, *528*, L85–L88. [CrossRef] [PubMed]

115. Agudo, I.; Gómez, J.L.; Martí, J.M.; Ibáñez, J.M.; Marscher, A.P.; Alberdi, A.; Aloy, M.A.; Hardee, P.E. Jet Stability and the Generation of Superluminal and Stationary Components. *Astrophys. J.* **2001**, *549*, L183–L186. [CrossRef]

116. Aloy, M.A.; Martí, J.M.; Gómez, J.L.; Agudo, I.; Müller, E.; Ibáñez, J.M. Three-dimensional Simulations of Relativistic Precessing Jets Probing the Structure of Superluminal Sources. *Astrophys. J.* **2003**, *585*, L109–L112. [CrossRef]

117. Perucho, M.; Agudo, I.; Gómez, J.L.; Kadler, M.; Ros, E.; Kovalev, Y.Y. On the nature of an ejection event in the jet of 3C 111. *Astron. Astrophys.* **2008**, *489*, L29–L32. [CrossRef]

118. Wehrle, A.E.; Piner, B.G.; Unwin, S.C.; Zook, A.C.; Xu, W.; Marscher, A.P.; Teräsranta, H.; Valtaoja, E. Kinematics of the Parsec-Scale Relativistic Jet in Quasar 3C 279: 1991–1997. *Astrophys. J. Suppl. Ser.* **2001**, *133*, 297–320. [CrossRef]

119. Kadler, M.; Ros, E.; Perucho, M.; Kovalev, Y.Y.; Homan, D.C.; Agudo, I.; Kellermann, K.I.; Aller, M.F.; Aller, H.D.; Lister, M.L.; Zensus, J.A. The Trails of Superluminal Jet Components in 3C 111. *Astrophys. J.* **2008**, *680*, 867–884. [CrossRef]

120. Fromm, C.M.; Perucho, M.; Porth, O.; Younsi, Z.; Ros, E.; Mizuno, Y.; Zensus, J.A.; Rezzolla, L. Jet-torus connection in radio galaxies. Relativistic hydrodynamics and synthetic emission. *Astron. Astrophys.* **2018**, *609*. [CrossRef]

121. Broderick, A.E.; McKinney, J.C. Parsec-scale Faraday Rotation Measures from General Relativistic Magnetohydrodynamic Simulations of Active Galactic Nucleus Jets. *Astrophys. J.* **2010**, *725*, 750–773. [CrossRef]

122. Porth, O.; Fendt, C.; Meliani, Z.; Vaidya, B. Synchrotron Radiation of Self-collimating Relativistic Magnetohydrodynamic Jets. *Astrophys. J.* **2011**, *737*. [CrossRef]

123. Jorstad, S.G.; Marscher, A.P.; Lister, M.L.; Stirling, A.M.; Cawthorne, T.V.; Gear, W.K.; Gómez, J.L.; Stevens, J.A.; Smith, P.S.; Forster, J.R.; Robson, E.I. Polarimetric Observations of 15 Active Galactic Nuclei at High Frequencies: Jet Kinematics from Bimonthly Monitoring with the Very Long Baseline Array. *Astron. J.* **2005**, *130*, 1418–1465. [CrossRef]

124. Lister, M.L.; Aller, M.F.; Aller, H.D.; Homan, D.C.; Kellermann, K.I.; Kovalev, Y.Y.; Pushkarev, A.B.; Richards, J.L.; Ros, E.; Savolainen, T. MOJAVE. X. Parsec-scale Jet Orientation Variations and Superluminal Motion in Active Galactic Nuclei. *Astron. J.* **2013**, *146*. [CrossRef]

125. Cohen, M.H.; Meier, D.L.; Arshakian, T.G.; Homan, D.C.; Hovatta, T.; Kovalev, Y.Y.; Lister, M.L.; Pushkarev, A.B.; Richards, J.L.; Savolainen, T. Studies of the Jet in Bl Lacertae. I. Recollimation Shock and Moving Emission Features. *Astrophys. J.* **2014**, *787*. [CrossRef]

126. Mizuno, Y.; Gómez, J.L.; Nishikawa, K.-I.; Meli, A.; Hardee, P.E.; Rezzolla, L. Recollimation Shocks in Magnetized Relativistic Jets. *Astrophys. J.* **2015**, *809*. [CrossRef]

127. Martí, J.M.; Perucho, M.; Gómez, J.L. The Internal Structure of Overpressured, Magnetized, Relativistic Jets. *Astrophys. J.* **2016**, *831*. [CrossRef]

128. Fuentes, A.; Gómez, J.L.; Martí, J.M.; Perucho, M. Total and Linearly Polarized Synchrotron Emission from Overpressured Magnetized Relativistic Jets. *Astrophys. J.* **2018**, *860*. [CrossRef]

129. Martí, J.M.; Perucho, M.; Gómez, J.L.; Fuentes, A. Recollimation shocks in relativistic jets. *Int. J. Mod. Phys. D* **2018**, *27*. [CrossRef]

130. Komissarov, S.S.; Porth, O.; Lyutikov, M. Stationary relativistic jets. *Comput. Astrophys. Cosmol.* **2015**, *2*. [CrossRef]

131. Porth, O.; Komissarov, S.S. Causality and stability of cosmic jets. *Mon. Not. R. Astron. Soc.* **2015**, *452*, 1089–1104. [CrossRef]

132. Fromm, C.; Porth, O.; Younsi, Z.; Mizuno, Y.; de Laurentis, M.; Olivares, H.; Rezzolla, L. Radiative Signatures of Parsec-Scale Magnetised Jets. *Galaxies* **2017**, *5*, 73. [CrossRef]

133. Shepherd, M.C. Difmap:An Interactive Program for Synthesis Imaging. *ASP Conf. Ser.* **1997**, *125*, 77–84.

134. Högbom, J.A. Aperture Synthesis with a Non-Regular Distribution of Interferometer Baselines. *Astron. Astrophys. Suppl. Ser.* **1974**, *15*, 417–426.

135. Hardee, P.E. The stability of astrophysical jets. *Proc. Int. Astron. Union* **2011**, *275*, 41–49. [CrossRef]

136. Owen, F.N.; Hardee, P.E.; Cornwell, T.J. High-resolution, high dynamic range VLA images of the M 87 jet at 2 centimeters. *Astrophys. J.* **1989**, *340*, 698–707. [CrossRef]

137. Lobanov, A.; Hardee, P.; Eilek, J. Internal structure and dynamics of the kiloparsec-scale jet in M 87. *New Astron. Rev.* **2003**, *47*, 629–632. [CrossRef]

138. Hardee, P.E.; Eilek, J.A. Using Twisted Filaments to Model the Inner Jet in M 87. *Astrophys. J.* **2011**, *735*. [CrossRef]

139. Lobanov, A.P.; Krichbaum, T.P.; Witzel, A.; Kraus, A.; Zensus, J.A.; Britzen, S.; Otterbein, K.; Hummel, C.A.; Johnston, K. VSOP imaging of S5 0836+710: A close-up on plasma instabilities in the jet. *Astron. Astrophys.* **1998**, *340*, L60–L64.

140. Perucho, M.; Lobanov, A.P. Physical properties of the jet in 0836+710 revealed by its transversal structure. *Astron. Astrophys.* **2007**, *469*, L23–L26. [CrossRef]

141. Perucho, M.; Kovalev, Y.; Lobanov, A.P.; Hardee, P.E.; Agudo, I. Anatomy of Helical Extragalactic Jets: The Case of S5 0836+710. *Astrophys. J.* **2012**, *749*. [CrossRef]

142. Perucho, M.; Martí-Vidal, I.; Lobanov, A.P.; Hardee, P.E. S5 0836+710: An FR II jet disrupted by the growth of a helical instability? *Astron. Astrophys.* **2012**, *545*. [CrossRef]

143. Perucho, M.; Lobanov, A.P.; Martí, J.M.; Hardee, P.E. The role of Kelvin-Helmholtz instability in the internal structure of relativistic outflows. The case of the jet in 3C 273. *Astron. Astrophys.* **2006**, *456*, 493–504. [CrossRef]

144. Walker, R.C.; Benson, J.M.; Unwin, S.C.; Lystrup, M.B.; Hunter, T.R.; Pilbratt, G.; Hardee, P.E. The Structure and Motions of the 3C 120 Radio Jet on Scales of 0.6-300 Parsecs. *Astrophys. J.* **2001**, *556*, 756–772. [CrossRef]

145. Walker, R.C.; Hardee, P.E. The implications of helical patterns in 3C 120. *New Astron. Rev.* **2003**, *47*, 645–647. [CrossRef]

146. Hardee, P.E.; Walker, R.C.; Gómez, J.L. Modeling the 3C 120 Radio Jet from 1 to 30 Milliarcseconds. *Astrophys. J.* **2005**, *620*, 646–664. [CrossRef]

147. Cohen, M.H.; Meier, D.L.; Arshakian, T.G.; Clausen-Brown, E.; Homan, D.C.; Hovatta, T.; Kovalev, Y.Y.; Lister, M.L.; Pushkarev, A.B.; Richards, J.L.; Savolainen, T. Studies of the Jet in Bl Lacertae. II. Superluminal Alfvén Waves. *Astrophys. J.* **2015**, *803*. [CrossRef]

148. Perucho, M.; Hanasz, M.; Martí, J.M.; Sol, H. Stability of hydrodynamical relativistic planar jets. I. Linear evolution and saturation of Kelvin-Helmholtz modes. *Astron. Astrophys.* **2004**, *427*, 415–429. [CrossRef]

149. Perucho, M.; Martí, J.M.; Hanasz, M. Stability of hydrodynamical relativistic planar jets. II. Long-term nonlinear evolution. *Astron. Astrophys.* **2004**, *427*, 431–444. [CrossRef]

150. Perucho, M.; Martí, J.M.; Hanasz, M. Nonlinear stability of relativistic sheared planar jets. *Astron. Astrophys.* **2005**, *443*, 863–881. [CrossRef]

151. Perucho, M.; Hanasz, M.; Martí, J.M.; Miralles, J.A. Resonant Kelvin-Helmholtz modes in sheared relativistic flows. *Phys. Rev. E* **2007**, *75*, 056312. [CrossRef]

152. Perucho, M.; Martí, J.M.; Cela, J.M.; Hanasz, M.; de La Cruz, R.; Rubio, F. Stability of three-dimensional relativistic jets: implications for jet collimation. *Astron. Astrophys.* **2010**, *519*. [CrossRef]

153. Radice, D.; Rezzolla, L. THC: A new high-order finite-difference high-resolution shock-capturing code for special-relativistic hydrodynamics. *Astron. Astrophys.* **2102**, *547*. [CrossRef]

154. Radice, D.; Rezzolla, L. Universality and Intermittency in Relativistic Turbulent Flows of a Hot Plasma. *Astrophys. J.* **2013**, *766*. [CrossRef]

155. Zrake, J.; MacFadyen, A.I. Spectral and Intermittency Properties of Relativistic Turbulence. *Astrophys. J.* **2013**, *763*. [CrossRef]

156. Zhang, W.; MacFadyen, A.; Wang, P. Three-Dimensional Relativistic Magnetohydrodynamic Simulations of the Kelvin-Helmholtz Instability: Magnetic Field Amplification by a Turbulent Dynamo. *Astrophys. J.* **2009**, *692*, L40–L44. [CrossRef]

157. Beckwith, K.; Stone, J.M. A Second-order Godunov Method for Multi-dimensional Relativistic Magnetohydrodynamics. *Astrophys. J. Suppl. Ser.* **2011**, *193*. [CrossRef]

158. Zrake, J.; MacFadyen, A.I. Numerical Simulations of Driven Relativistic Magnetohydrodynamic Turbulence. *Astrophys. J.* **2012**, *744*. [CrossRef]

159. Matsumoto, J.; Masada, Y. Two-dimensional Numerical Study for Rayleigh-Taylor and Richtmyer-Meshkov Instabilities in Relativistic Jets. *Astrophys. J.* **2013**, *772*. [CrossRef]

160. Toma, K.; Komissarov, S.S.; Porth, O. Rayleigh-Taylor instability in two-component relativistic jets. *Mon. Not. R. Astron. Soc.* **2017**, *472*, 1253–1258. [CrossRef]

161. Matsumoto, J.; Aloy, M.A.; Perucho, M. Linear theory of the Rayleigh-Taylor instability at a discontinuous surface of a relativistic flow. *Mon. Not. R. Astron. Soc.* **2017**, *472*, 1421–1431. [CrossRef]

162. Gourgouliatos, K.N.; Komissarov, S.S. Reconfinement and loss of stability in jets from active galactic nuclei. *Nat. Astron.* **2018**, *2*, 167–171. [CrossRef]

163. Meliani, Z.; Keppens, R. Transverse stability of relativistic two-component jets. *Astron. Astrophys.* **2007**, *475*, 785–789. [CrossRef]

164. Meliani, Z.; Keppens, R. Decelerating Relativistic Two-Component Jets. *Astrophys. J.* **2009**, *705*, 1594–1606. [CrossRef]

165. Gourgouliatos, K.N.; Komissarov, S.S. Relativistic centrifugal instability. *Mon. Not. R. Astron. Soc.* **2018**, *475*, L125–L129. [CrossRef]

166. Millas, D.; Keppens, R.; Meliani, Z. Rotation and toroidal magnetic field effects on the stability of two-component jets. *Mon. Not. R. Astron. Soc.* **2017**, *470*, 592–605. [CrossRef]

167. Heavens, A.F.; Drury, L.O. Relativistic shocks and particle acceleration. *Mon. Not. R. Astron. Soc.* **1988**, *235*, 997–1009. [CrossRef]

168. Sironi, L.; Petropoulou, M.; Giannios, D. Relativistic jets shine through shocks or magnetic reconnection? *Mon. Not. R. Astron. Soc.* **2015**, *450*, 183–191. [CrossRef]

169. Marscher, A.P. Turbulent, Extreme Multi-zone Model for Simulating Flux and Polarization Variability in Blazars. *Astrophys. J.* **2014**, *780*. [CrossRef]

170. Zakamska, N.L.; Begelman, M.C.; Blandford, R.D. Hot Self-Similar Relativistic Magnetohydrodynamic Flows. *Astrophys. J.* **2008**, *679*, 990–999. [CrossRef]

171. Anantua, R.; Blandford, R.D.; Tchekhovskoy, A. Multiwavelength Observations of Relativistic Jets from General Relativistic Magnetohydrodynamic Simulations. *Galaxies* **2018**, *6*, 31. [CrossRef]

172. Jones, T.W.; Ryu, D.; Engel, A. Simulating Electron Transport and Synchrotron Emission in Radio Galaxies: Shock Acceleration and Synchrotron Aging in Axisymmetric Flows. *Astrophys. J.* **1999**, *512*, 105–124. [CrossRef]

173. Micono, M.; Zurlo, N.; Massaglia, S.; Ferrari, A.; Melrose, D.B. Diffusive shock acceleration in extragalactic jets. *Astron. Astrophys.* **1999**, *349*, 323–333.

174. Tregillis, I.L.; Jones, T.W.; Ryu, D. Simulating Electron Transport and Synchrotron Emission in Radio Galaxies: Shock Acceleration and Synchrotron Aging in Three-dimensional Flows. *Astrophys. J.* **2001**, *557*, 475–491. [CrossRef]

175. Mimica, P.; Aloy, M.A.; Agudo, I.; Martí, J.M.; Gómez, J.L.; Miralles, J.A. Spectral Evolution of Superluminal Components in Parsec-Scale Jets. *Astrophys. J.* **2009**, *696*, 1142–1163. [CrossRef]

176. Fromm, C.M.; Perucho, M.; Mimica, P.; Ros, E. Spectral evolution of flaring blazars from numerical simulations. *Astron. Astrophys.* **2016**, *588*. [CrossRef]

177. Vaidya, B.; Mignone, A.; Bodo, G.; Rossi, P.; Massaglia, S. A Particle Module for the PLUTO Code. II. Hybrid Framework for Modeling Nonthermal Emission from Relativistic Magnetized Flows. *Astrophys. J.* **2018**, *865*. [CrossRef]

178. Sironi, L.; Keshet, U.; Lemoine, M. Relativistic Shocks: Particle Acceleration and Magnetization. *Space Sci. Rev.* **2015**, *191*, 519–544. [CrossRef]

179. Kagan, D.; Sironi, L.; Cerutti, B.; Giannios, D. Relativistic Magnetic Reconnection in Pair Plasmas and Its Astrophysical Applications. *Space Sci. Rev.* **2015**, *191*, 545–573. [CrossRef]

180. Nishikawa, K.-I.; Frederiksen, J.T.; Nordlund, Å.; Mizuno, Y.; Hardee, P.E.; Niemiec, J.; Gómez, J.L.; Pe'er, A.; Duţan, I.; Meli, A.; et al. Evolution of Global Relativistic Jets: Collimations and Expansion with kKHI and the Weibel Instability. *Astrophys. J.* **2016**, *820*. [CrossRef]

181. Nishikawa, K.-I.; Mizuno, Y.; Niemiec, J.; Kobzar, O.; Pohl, M.; Gómez, J.L.; Duţan, I.; Pe'er, A.; Frederiksen, J.; Nordlund, Å.; et al. Microscopic Processes in Global Relativistic Jets Containing Helical Magnetic Fields. *Galaxies* **2016**, *4*, 38. [CrossRef]

182. Bai, X.-N.; Caprioli, D.; Sironi, L.; Spitkovsky, A. Magnetohydrodynamic-particle-in-cell Method for Coupling Cosmic Rays with a Thermal Plasma: Application to Non-relativistic Shocks. *Astrophys. J.* **2015**, *809*. [CrossRef]

183. van Marle, A.J.; Casse, F.; Markowith, A. On magnetic field amplification and particle acceleration near non-relativistic astrophysical shocks: particles in MHD cells simulations. *Mon. Not. R. Astron. Soc.* **2018**, *473*, 3394–3409. [CrossRef]

184. Mignone, A.; Bodo, G.; Vaidya, B.; Mattia, G. A Particle Module for the PLUTO Code. I. An Implementation of the MHD-PIC Equations. *Astrophys. J.* **2018**, *859*. [CrossRef]

185. Smith, M.D.; Norman, M.L.; Winkler, K.-H.A.; Smarr, L. Hotspots in radio galaxies—A comparison with hydrodynamic simulations. *Mon. Not. R. Astron. Soc.* **1985**, *214*, 67–85. [CrossRef]

186. Burns, J.O.; Norman, M.L.; Clarke, D.A. Numerical models of extragalactic radio sources. *Science* **1991**, *253*, 522–530. [CrossRef] [PubMed]

187. van Putten, M.H.P.M. A two-dimensional relativistic (Gamma = 3.25) jet simulation. *Astrophys. J.* **1993**, *408*, L21–L23. [CrossRef]

188. Martí, J.M.; Müller, E.; Ibáñez, J.M. Hydrodynamical simulations of relativistic jets. *Astron. Astrophys.* **1994**, *281*, L9–L12.

189. Duncan, G.C.; Hughes, P.A. Simulations of relativistic extragalactic jets. *Astrophys. J.* **1994**, *436*, L119–L122. [CrossRef]

190. Martí, J.M.; Müller, E.; Font, J.A.; Ibáñez, J.M. Morphology and Dynamics of Highly Supersonic Relativistic Jets. *Astrophys. J.* **1995**, *448*, L105–L108. [CrossRef]

191. Martí, J.M.; Müller, E.; Font, J.A.; Ibáñez, J.M.; Marquina, A. Morphology and Dynamics of Relativistic Jets. *Astrophys. J.* **1997**, *479*, 151–163. [CrossRef]

192. Rosen, A.; Hughes, P.A.; Duncan, G.C.; Hardee, P.E. A Comparison of the Morphology and Stability of Relativistic and Nonrelativistic Jets. *Astrophys. J.* **1999**, *516*, 729–743. [CrossRef]

193. Monceau-Baroux, R.; Keppens, R.; Meliani, Z. The effect of angular opening on the dynamics of relativistic hydro jets. *Astron. Astrophys.* **2012**, *545*. [CrossRef]

194. Aloy, M.A.; Ibáñez, J.M.; Martí, J.M. High-Resolution Three-dimensional Simulations of Relativistic Jets. *Astrophys. J.* **1999**, *523*, L125–L128. [CrossRef]

195. Hughes, P.A.; Miller, M.A.; Duncan, G.C. Three-dimensional Hydrodynamic Simulations of Relativistic Extragalactic Jets. *Astrophys. J.* **2002**, *572*, 713–728. [CrossRef]

196. Choi, E.; Wiita, P.J.; Ryu, D. Hydrodynamic Interactions of Relativistic Extragalactic Jets with Dense Clouds. *Astrophys. J.* **2007**, *655*, 769–780. [CrossRef]

197. Koide, S.; Nishikawa, K.-I.; Mutel, R.L. A Two-dimensional Simulation of Relativistic Magnetized Jet. *Astrophys. J.* **1996**, *463*, L71–L74. [CrossRef]

198. Nishikawa, K.-I.; Koide, S.; Sakai, J.-I.; Christodoulou, D.M.; Sol, H.; Mutel, R.L. Three-Dimensional Magnetohydrodynamic Simulations of Relativistic Jets Injected along a Magnetic Field. *Astrophys. J.* **1997**, *483*, L45–L48. [CrossRef]

199. Koide, S. A Two-dimensional Simulation of a Relativistic Jet Bent by an Oblique Magnetic Field. *Astrophys. J.* **1997**, *478*, 66–69. [CrossRef]

200. Nishikawa, K.-I.; Koide, S.; Sakai, J.-I.; Christodoulou, D.M.; Sol, H.; Mutel, R.L. Three-dimensional Magnetohydrodynamic Simulations of Relativistic Jets Injected into an Oblique Magnetic Field. *Astrophys. J.* **1998**, *498*, 166–169. [CrossRef]

201. Komissarov, S.S. Numerical simulations of relativistic magnetized jets. *Mon. Not. R. Astron. Soc.* **1999**, *308*, 1069–1076. [CrossRef]

202. Mignone, A.; Massaglia, S.; Bodo, G. Relativistic MHD Simulations of Jets with Toroidal Magnetic Fields. *Space Sci. Rev.* **2005**, *121*, 21–31. [CrossRef]

203. Keppens, R.; Meliani, Z.; van der Holst, B.; Casse, F. Extragalactic jets with helical magnetic fields: Relativistic MHD simulations. *Astron. Astrophys.* **2008**, *486*, 663–678. [CrossRef]

204. Leismann, T.; Antón, L.; Aloy, M.A. Relativistic MHD simulations of extragalactic jets. *Astron. Astrophys.* **2005**, *436*, 503–526. [CrossRef]

205. Begelman, M.C.; Cioffi, D.F. Overpressured cocoons in extragalactic radio sources. *Astrophys. J.* **1989**, *345*, L21–L24. [CrossRef]

206. Falle, S.A.E.G. Self-similar jets. *Mon. Not. R. Astron. Soc.* **1991**, *250*, 581–596. [CrossRef]

207. Kaiser, C.R.; Alexander, P. A self-similar model for extragalactic radio sources. *Mon. Not. R. Astron. Soc.* **1997**, *286*, 215–222. [CrossRef]

208. Komissarov, S.S.; Falle, S.A.E.G. The large-scale structure of FR-II radio sources. *Mon. Not. R. Astron. Soc.* **1998**, *297*, 1097–1108. [CrossRef]

209. Clarke, D.A.; Harris, D.E.; Carilli, C.L. Formation of cavities in the X-ray emitting cluster gas of Cygnus A. *Mon. Not. R. Astron. Soc.* **1997**, *284*, 981–993. [CrossRef]

210. Zanni, C.; Bodo, G.; Rossi, P.; Massaglia, S.; Durbala, A.; Ferrari, A. X-ray emission from expanding cocoons. *Astron. Astrophys.* **2003**, *402*, 949–962. [CrossRef]

211. Krause, M. Very light jets II: Bipolar large scale simulations in King atmospheres. *Astron. Astrophys.* **2005**, *431*, 45–64. [CrossRef]

212. Tregillis, I.L.; Jones, T.W.; Ryu, D. Synthetic Observations of Simulated Radio Galaxies. I. Radio and X-Ray Analysis. *Astrophys. J.* **2004**, *601*, 778–797. [CrossRef]

213. Gaibler, V.; Krause, M.; Camenzind, M. Very light magnetized jets on large scales—I. Evolution and magnetic fields. *Mon. Not. R. Astron. Soc.* **2009**, *400*, 1785–1802. [CrossRef]

214. Huarte-Espinosa, M.; Krause, M.; Alexander, P. 3D magnetohydrodynamic simulations of the evolution of magnetic fields in Fanaroff-Riley class II radio sources. *Mon. Not. R. Astron. Soc.* **2011**, *417*, 382–399. [CrossRef]

215. Hardcastle, M.J.; Krause, M.G.H. Numerical modelling of the lobes of radio galaxies in cluster environments—II. Magnetic field configuration and observability. *Mon. Not. R. Astron. Soc.* **2104**, *441*, 1482–1499. [CrossRef]

216. Scheck, L.; Aloy, M.A.; Martí, J.M.; Gómez, J.L.; Müller, E. Does the plasma composition affect the long-term evolution of relativistic jets? *Mon. Not. R. Astron. Soc.* **2002**, *331*, 615–634. [CrossRef]

217. Mignone, A.; Rossi, P.; Bodo, G.; Ferrari, A.; Massaglia, S. High-resolution 3D relativistic MHD simulations of jets. *Mon. Not. R. Astron. Soc.* **2010**, *402*, 7–12. [CrossRef]

218. Perucho, M.; Martí, J.M.; Laing, R.A.; Hardee, P.E. On the deceleration of Fanaroff-Riley Class I jets: Mass loading by stellar winds. *Mon. Not. R. Astron. Soc.* **2014**, *441*, 1488–1503. [CrossRef]

219. Rossi, P.; Mignone, A.; Bodo, G.; Massaglia, S.; Ferrari, A. Formation of dynamical structures in relativistic jets: The FR I case. *Astron. Astrophys.* **2008**, *488*, 795–806. [CrossRef]

220. Perucho, M.; Martí, J.M. A numerical simulation of the evolution and fate of a Fanaroff-Riley type I jet. The case of 3C 31. *Mon. Not. R. Astron. Soc.* **2007**, *382*, 526–542. [CrossRef]

221. Krause, M.; Alexander, P.; Riley, J.; Hopton, D. A new connection between the jet opening angle and the large-scale morphology of extragalactic radio sources. *Mon. Not. R. Astron. Soc.* **2012**, *427*, 3196–3208. [CrossRef]

222. Massaglia, S.; Bodo, G.; Rossi, P.; Capetti, S.; Mignone, A. Making Faranoff-Riley I radio sources. I. Numerical hydrodynamic 3D simulations of low-power jets. *Astron. Astrophys.* **2016**, *596*. [CrossRef]

223. Tchekhovskoy, A.; Bromberg, O. Three-dimensional relativistic MHD simulations of active galactic nuclei jets: Magnetic kink instability and Fanaroff–Riley dichotomy. *Mon. Not. R. Astron. Soc.* **2016**, *461*, L46–L50. [CrossRef]

224. Yates, P.M.; Shabala, S.S.; Krause, M.G.H. Observability of intermittent radio sources in galaxy groups and clusters. *Mon. Not. R. Astron. Soc.* **2018**, *480*, 5286–5306. [CrossRef]

225. Böhringer, H.; Voges, W.; Fabian, A.C.; Edge, A.C.; Neumann, D.M. A ROSAT HRI study of the interaction of the X-ray-emitting gas and radio lobes of NGC 1275. *Mon. Not. R. Astron. Soc.* **1993**, *264*, L25–L28. [CrossRef]

226. Fabian, A.C. Cooling Flows in Clusters of Galaxies. *Ann. Rev. Astron. Astrophys.* **1994**, *32*, 277–318. [CrossRef]

227. Churazov, E.; Brüggen, M.; Kaiser, C.R.; Böhringer, H.; Forman, W. Evolution of Buoyant Bubbles in M87. *Astrophys. J.* **2001**, *554*, 261–273. [CrossRef]

228. Quilis, V.; Bower, R.G.; Balogh, M.L. Bubbles, feedback and the intracluster medium: Three-dimensional hydrodynamic simulations. *Mon. Not. R. Astron. Soc.* **2001**, *328*, 1091–1097. [CrossRef]

229. Reynolds, C.S.; Heinz, S.; Begelman, M.C. The hydrodynamics of dead radio galaxies. *Mon. Not. R. Astron. Soc.* **2002**, *332*, 271–282. [CrossRef]

230. Sijacki, D.; Springel, V.; Di Matteo, T.; Hernquist, L. A unified model for AGN feedback in cosmological simulations of structure formation. *Mon. Not. R. Astron. Soc.* **2007**, *380*, 877–900. [CrossRef]

231. Vogelsberger, M.; Genel, S.; Springel, V.; Torrey, P.; Sijacki, D.; Xu, D.; Snyder, G.; Nelson, D.; Hernquist, L. Introducing the Illustris Project: Simulating the coevolution of dark and visible matter in the Universe. *Mon. Not. R. Astron. Soc.* **2014**, *444*, 1518–1547. [CrossRef]

232. Dubois, Y.; Devriendt, J.; Slyz, A.; Teyssier, R. Self-regulated growth of supermassive black holes by a dual jet-heating active galactic nucleus feedback mechanism: Methods, tests and implications for cosmological simulations. *Mon. Not. R. Astron. Soc.* **2012**, *420*, 2662–2683. [CrossRef]

233. Dubois, Y.; Peirani, S.; Pichon, C.; Devriendt, J.; Gavazzi, R.; Welker, C.; Volonteri, M. The HORIZON-AGN simulation: Morphological diversity of galaxies promoted by AGN feedback. *Mon. Not. R. Astron. Soc.* **2016**, *463*, 3948–3964. [CrossRef]

234. Bîrzan, L.; Rafferty, D.A.; McNamara, B.R.; Wise, M.W.; Nulsen, P.E.J. A Systematic Study of Radio-induced X-Ray Cavities in Clusters, Groups, and Galaxies. *Astrophys. J.* **2004**, *607*, 800–809. [CrossRef]

235. Reynolds, C. S. Jets and AGN feedback. In *Relativistic Jets from Active Galactic Nuclei*; Böttcher, M., Harris, D.E., Krawczynski, H., Eds.; Wiley-VCH: Weinheim, Germany, 2012; pp. 369–394, ISBN 978-3-527-41037-8.

236. Yang, H.-Y.K.; Reynolds, C.S. How AGN Jets Heat the Intracluster Medium—Insights from Hydrodynamic Simulations. *Astrophys. J.* **2016**, *829*. [CrossRef]

237. Yang, H.-Y.K.; Reynolds, C.S. Interplay Among Cooling, AGN Feedback, and Anisotropic Conduction in the Cool Cores of Galaxy Clusters. *Astrophys. J.* **2016**, *818*. [CrossRef]

238. Martizzi, D.; Quataert, E.; Faucher-Ciguère, C.A.; Fielding, D. Simulations of Jet Heating in Galaxy Clusters: Successes and Challenges. *Mon. Not. R. Astron. Soc.* **2019**, *483*, 2465–2486. [CrossRef]

239. Ruszkowski, M.; Yang, H.-Y.K.; Reynolds, C.S. Cosmic-Ray Feedback Heating of the Intracluster Medium. *Astrophys. J.* **2017**, *844*. [CrossRef]

240. Ehlert, K.; Weinberger, R.; Pfrommer, C.; Pakmor, R.; Springel, V. Simulations of the dynamics of magnetized jets and cosmic rays in galaxy clusters. *Mon. Not. R. Astron. Soc.* **2018**, *481*, 2878–2900. [CrossRef]

241. Heinz, S.; Brüggen, M.; Young, A.; Levesque, E. The answer is blowing in the wind: simulating the interaction of jets with dynamic cluster atmospheres. *Mon. Not. R. Astron. Soc.* **2006**, *373*, L65–L69. [CrossRef]

242. Morsony, B.J.; Heinz, S.; Brüggen, M.; Ruszkowski, M. Swimming against the current: Simulations of central AGN evolution in dynamic galaxy clusters. *Mon. Not. R. Astron. Soc.* **2010**, *407*, 1277–1289. [CrossRef]

243. Mendygral, P.J.; Jones, T.W.; Dolag, K. MHD Simulations of Active Galactic Nucleus Jets in a Dynamic Galaxy Cluster Medium. *Astrophys. J.* **2012**, *750*. [CrossRef]

244. Bourne, M.A.; Sijacki, D. AGN jet feedback on a moving mesh: Cocoon inflation, gas flows and turbulence. *Mon. Not. R. Astron. Soc.* **2017**, *472*, 4707–4735. [CrossRef]

245. Hitomi Collaboration. The quiescent intracluster medium in the core of the Perseus cluster. *Nature* **2016**, *535*, 117–121. [CrossRef] [PubMed]

246. Perucho, M.; Quilis, V.; Martí, J.M. Intracluster Medium Reheating by Relativistic Jets. *Astrophys. J.* **2011**, *743*. [CrossRef]

247. Perucho, M.; Martí, J.M.; Quilis, V.; Ricciardelli, E. Large-scale jets from active galactic nuclei as a source of intracluster medium heating: Cavities and shocks. *Mon. Not. R. Astron. Soc.* **2014**, *445*, 1462–1481. [CrossRef]

248. Perucho, M.; Martí, J.M.; Quilis, V. Long-term FRII jet evolution: Clues from three-dimensional simulations. *Mon. Not. R. Astron. Soc.* **2018**. [CrossRef]

249. Perucho, M.; Martí, J.M.; Quilis, V.; Borja-Lloret, M. Radio mode feedback: Does relativity matter? *Mon. Not. R. Astron. Soc.* **2017**, *471*, L120–L124. [CrossRef]

250. English, W.; Hardcastle, M.J.; Krause, M.G.H. Numerical modelling of the lobes of radio galaxies in cluster environments—III. Powerful relativistic and non-relativistic jets. *Mon. Not. R. Astron. Soc.* **2016**, *461*, 2025–2043. [CrossRef]

251. Gaibler, V. Positive and negative feedback by AGN jets in high-redshift galaxies. *Astron. Nachr.* **2014**, *335*, 531–536. [CrossRef]

252. Wagner, A.Y.; Bicknell, G.V.; Umemura, M. Galaxy-scale AGN feedback - theory. *Astron. Nachr.* **2016**, *337*, 167–174. [CrossRef]

253. Gaibler, V.; Khochfar, S.; Krause, M.; Silk, J. Jet-induced star formation in gas-rich galaxies. *Mon. Not. R. Astron. Soc.* **2012**, *425*, 438–449. [CrossRef]

254. Wagner, A.Y.; Bicknell, G.V.; Umemura, M. Driving Outflows with Relativistic Jets and the Dependence of Active Galactic Nucleus Feedback Efficiency on Interstellar Medium Inhomogeneity. *Astrophys. J.* **2012**, *757*. [CrossRef]

galaxies

MDPI

Review

MHD Accretion Disk Winds: The Key to AGN Phenomenology?

Demosthenes Kazanas

NASA/Goddard Space Flight Center, 8800 Greenbelt Rd, Greenbelt, MD 20771, USA; demos.kazanas@nasa.gov;
Tel.: +1-301-286-7680

Received: 8 November 2018; Accepted: 4 January 2019; Published: 10 January 2019

Abstract: Accretion disks are the structures which mediate the conversion of the kinetic energy of plasma accreting onto a compact object (assumed here to be a black hole) into the observed radiation, in the process of removing the plasma's angular momentum so that it can accrete onto the black hole. There has been mounting evidence that these structures are accompanied by winds whose extent spans a large number of decades in radius. Most importantly, it was found that in order to satisfy the winds' observational constraints, their mass flux must increase with the distance from the accreting object; therefore, the mass accretion rate on the disk must decrease with the distance from the gravitating object, with most mass available for accretion expelled before reaching the gravitating object's vicinity. This reduction in mass flux with radius leads to accretion disk properties that can account naturally for the AGN relative luminosities of their Optical-UV and X-ray components in terms of a single parameter, the dimensionless mass accretion rate. Because this critical parameter is the dimensionless mass accretion rate, it is argued that these models are applicable to accreting black holes across the mass scale, from galactic to extragalactic.

Keywords: accretion disks; MHD winds; accreting black holes

1. Introduction-Accretion Disk Phenomenology

Accretion disks are the generic structures associated with compact objects (considered to be black holes in this note) powered by matter accretion onto them. Their formation is a consequence of the fact that the specific angular momentum of the accreting matter at the outer boundary of the flow is larger than its Keplerian value on the accreting body vicinity. The role of the accretion disk is to rid this excess angular momentum of the disk plasma and allow it to accrete onto the gravitating compact object. This process is effected by viscous stresses which at the same time cause the heating of the accreting matter and emission of radiation with rather specific spectral characteristics.

The present article is not a review of accretion disks (the interested reader can consult several such reviews e.g., [1]); instead, it aims to present alternatives to the more conventional accretion disk views which are driven by accumulating phenomenology of the spectroscopic properties of the winds that are ubiquitous in accretion powered compact objects. The hope is that these alternative views will lead to novel, fruitful insights on the structure of these systems. As such, this note focuses on only certain specific issues while ignoring many others along with some of the important works on the subject. Because it argues for the scale invariance of the accretion disk winds, it includes in the discussion properties of winds and accretion disks onto black holes of both Galactic X-ray binaries (XRB) and active galactic nuclei (AGN) to support the mass invariance by as large of mass range as possible. At the end, it is argued that, in their broader sense, the global spectral properties of accreting black holes can be accounted for in terms of a small number of parameters, if one is willing to accept certain facts associated with the general properties of their observed outflows obtained through X-ray spectroscopy.

Much of the work on accretion powered sources has been based on and influenced by the seminal work on the subject, namely that of Shakura and Sunyaev (hereafter SS73) [2]. This work assumes the disks to be steady–state, roughly Keplerian ($v_r \ll v_\phi$), thin ($h \ll R$) and in hydrostatic equilibrium in the vertical direction. The disk temperature is obtained by solving their thermal balance (local energy dissipation equals the energy radiated; see Section 2.1). The dissipation is effected by the viscous stresses, assumed to be proportional to the local gas pressure P ($t_{r\phi} \simeq \alpha P$); $\alpha \sim 0.1$ is an unknown parameter to be determined by observation or simulation. The same viscous stresses serve also to transfer outward the excess angular momentum of the disk plasma, allowing its further accretion.

The usual assumption in most treatments is that all energy involved in the transfer of angular momentum is dissipated locally. However, one should note that since angular momentum cannot be destroyed and since essentially all the disk kinetic energy is stored in its angular motion, none of that energy could be dissipated. However, viscous stresses can dissipate circulation and presumably it is the dissipation of circulation that powers the observed radiation of accretion disks. Because in axisymmetry circulation has the same form as the non-dissipative angular momentum, one can speculate that the observed presence of dissipation in accreting objects involves non-axisymmetric fluid modes.

A further simplifying assumption generally made in modeling accretion disks is that the dissipated energy is thermalized, i.e., that the disk particles achieve a thermal distribution of temperature determined by energy balance, with the emitted radiation being of black body form. This then can determine the disk temperature radial profile: In steady-state disks (mass flux $\dot{M}$ independent of time and radius) the local energy dissipation per unit area is proportional to $3GM\dot{M}/8R^3$; setting equal to σT^4 (black body emission), implies a disk temperature dependence $T \propto R^{-3/4}$, with a multi-color flux density $F_\nu \propto \nu^{1/3}$ photons cm^{-2} s^{-1} (R denotes the cylindrical radial coordinate along the disk equatorial plane and r the spherical coordinate).

The innermost disk temperature can be estimated assuming that the emitting surface is roughly πr_{ISCO}^2 (ISCO is short for the innermost stable circular orbit; for a Schwarzschild black hole $r_{ISCO} \simeq 3R_S \simeq 10^6 M_0$ cm, where $R_S = 2GM/c^2$ is the Schwarzschild radius and $M_n = M/10^n M_\odot$ is the black hole mass in 10^n solar masses. An estimate of the disk highest temperature, then, requires a value of the disk luminosity. Because the maximum luminosity of an accretion powered source is generally considered to be the object's Eddington luminosity, $L_{Edd} \simeq 1.3 \times 10^{38} M_0$ erg/s, the corresponding maximum disk temperature is $T \simeq 10^{7.5} M_0^{-1/4}$ K $= 10^{5.5} M_8^{-1/4}$ K. The fact that quasi-thermal, multi-colo features at the corresponding temperatures were observed at approximately these energies in galactic XRBs and AGNs have established the SS73 disk as a ubiquitous structure in both these classes of sources.

However, besides these quasi-thermal, multi-color spectral components it was established, that a large fraction of these objects' luminosity is shared by a spectral component of power-law form that extends to energies $E \simeq 100$ keV, clearly inconsistent with that of thermal radiation by an accretion disk. In analogy with the Sun, this component was then attributed to the presence of a hot ($T \sim 10^8$–10^9 K) corona, overlaying the accretion disk [3]. Again, in analogy with the solar corona, this was proposed to be powered by magnetic fields that thread the accretion disk and dissipate part of its energy in this hot corona, which then Compton-scatters the disk thermal radiation to produce the observed high energy photons.

The advent of observations and data accumulation, then, established that the ratio of luminosities of these major accretion disk spectral components is not random but varies in a systematic way, generally with the contribution of the quasi-thermal disk component increasing with the objects' bolometric luminosity (a proxy for the accretion rate $\dot{M}$) and that of the harder, power-law one waning. Thus, for AGN (where the multi-color component referred to as the Big Blue Bump (BBB) is in the UV/Optical band, well separated from the power-law X-ra component) it was found that the logarithmic flux slope between the UV (2500 Å) and 2 keV X-ray fluxes, the so-called α_{OX} parameter, increases (in absolute value; it becomes more negative) with increasing source luminosity [4].

A similar correlation appears to be present in XRBs between their multi-color disk component and their power-law X-rays (see [5] for a review). The reason for the observed correlation (Figure 4 of [4]), is not clear (we will present an account in Section 4). Another open issue is the geometry of the X-ray emitting plasma relative to the BBB component. Earlier models based on emission by a magnetically powered corona, assumed that the X-ray emitting plasma was overlying the BBB emitting thin accretion disk. However, issues concerning the cooling of the corona electrons [6], and the corresponding variability due to reprocessing of the X-rays on the thin disk [7], appear to require refinement of this geometric arrangement.

Accretion disk theory got a big boost in the 90's with two significant developments: *(i.)* Balbus and Hawley [8] showed that a fluid rotating with angular velocity Ω and threaded by poloidal magnetic field is stable if $d\Omega^2/dr > 0$ (in disagreement with Rayleigh criterion for unmagnetized flows that demands $d(\Omega^2 r^4)/dr > 0$). According to this criterion, Keplerian accretion disks are unstable, with the magnetic field acting as the agent that helps mediate the transfer of angular momentum necessary for accretion to take place (for a pedagogical discussion of this instability and why it does not reduce to the Rayleigh criterion as the magnetic field goes to zero see [9,10]). *(ii.)* Narayan and Yi [11,12] produced models of accretion disks that are optically thin and geometrically thick ($h \simeq R$), supported at each radius by the pressure of ions, referred to as Advection-Dominated Accretion Flows (ADAF). The rotational velocity of these disks is sub-Keplerian with $v_r \simeq v_\phi \lesssim v_K$.

The reason the ADAF disks are thick is that the proton cooling time (assuming they achieve at some radius r their virial temperature, i.e., $kT_p \simeq GMm_p/r$, heated both by the dissipation of the azimuthal motions and the pdV work of accretion) through Coulomb collisions with the cooler electrons, is longer than the local viscous time scale, which for $h \simeq R$ is only α times longer than the free fall time $t_{ff} \simeq R/v_r$. Such thick flows, are therefore possible only for $t_{cool} > t_{visc}$. As shown in Section 2.2, the condition $t_{cool} > t_{visc}$ reduces to a condition involving just the normalized accretion rate (see Section 2) $\dot{m} < \alpha^2$. It is interesting that this condition involves neither the mass of the object M nor the radius of the flow x (see however [12] for a weak $x-$dependence), implying that such flows are self-similar and scale free if expressed in normalized parameters. Flows with $\dot{m} < \alpha^2$, if they start hot (i.e., with virialized protons), they will remain so all the way to the horizon of the accreting body with their azimuthal and radial velocities a fraction of the Keplerian one ($v_\phi \simeq v_r \sim 0.7v_K$). Because their disk height $h \simeq R$, they resemble spherical accretion; also, because their cooling time is longer than the advection time onto the accreting object, their radiative efficiency is reduced by a factor $t_{visc}/t_{cool} \sim \dot{m}$ and the accretion luminosity L is no longer simply proportional to $\dot{m}$ but to $\dot{m}^2$. As such they are referred to as either ADAF or RIAF (Radiatively Inefficient Accretion Flows).

The great advantage offered by the ADAF paradigm is that it provides the hot electrons demanded by the hard X-ray observations of (galactic and extragalactic) black holes as a result of the general accretion flow dynamics, rather than as a corona, unrestricted by the dynamics of accretion and introduced so that it would accommodate the observations. The notion of ADAF, then, led to a hybrid picture for the spectral decomposition of galactic binary X-ray sources [13] (see Section 2.2).

Besides the above features, ADAF have an additional distinct property: Their Bernoulli integral, *Be*, the sum of their kinetic, thermal and potential energies per unit mass is positive, a fact noted in the original references on the subject [11,12]; therefore, these flows provide for the potential presence of outflows that are apparently ubiquitous in accretion powered sources [14] (this is not the case in SS73 disks whose internal thermal energy is radiated promptly away resulting in $Be < 0$). The reason for the positivity of *Be* was elucidated in a (very important in this author's opinion) paper by Blandford and Begelman [15]: The viscous stresses that transfer outward the fluid's angular momentum, so that it is allowed to sink deeper into the gravitational potential, transfer also mechanical energy (see Section 2.1); while in an SS73 disk this energy is radiated away on time scales shorter than its viscous time scale, in an ADAF it remains stored in the fluid for longer time, that being reason behind the positivity of *Be*.

The solution of the conundrum, offered in [15], is that the excess energy and angular momentum are expelled off the disk in the form of a wind providing thus a combination of advection-dominated inflow - outflow solutions (ADIOS). This leaves the disk with less matter to accrete but with matter that it is now bound gravitationally, i.e., with $Be < 0$. As a result, the accretion rate $\dot{m}$ is no longer constant but depends on the (normalized) radius $x = r/R_S$, i.e., $\dot{m} = \dot{m}(x)$ (R_S is the black hole Schwarzschild radius). The authors of [15] did provide simple models of radius dependent winds with mass flux of the form $\dot{m}(x) \propto x^p$ and $1 > p > 0$. The positive value of p implies that these flows eject to a wind most of the mass available for accretion at their outer edge, thereby the name ADIOS.

The structure of this article is as follows: In Section 2 we give a brief review of the structure of accretion disks and the modifications and effects introduced by the notion of ADAF and ADIOS. In Section 3 we outline the X-ray spectroscopic observations and indicate that they support winds and accretion disks of the form of ADIOS; we present and discuss the results of photoionization calculations of ADIOS-like MHD winds, to show they are consistent with the X-ray absorber observations. In Section 4 we indicate how these winds can reproduce the trend in α_{OX} on luminosity in AGN and conclude in Section 5 with a brief summary of these results and conclusions.

2. The Physics of Accretion Disks

2.1. The Structure of Accretion Disks

The structure of accretion disks is given by the transfer of angular momentum, hydrostatic equilibrium in the direction perpendicular to the disk plane and the mass and energy flux conservations. Finally, the disk spectrum is computed assuming that all energy released locally is dissipated and emitted in black body form.

i. Hydrostatic equilibrium: This assumption of a thin disk implies

$$\frac{dP}{dz} = -\rho \frac{GM}{r^2} \frac{z}{r} \quad \text{or} \quad \frac{P}{h} \simeq \rho \frac{GM}{r^2} \frac{h}{r} \tag{1}$$

upon setting $\Delta P \approx P$ and $\Delta z \approx h \approx z$

$$\frac{P}{\rho} \simeq c_s^2 \simeq \frac{GM}{r} \frac{h^2}{r^2} = V_K^2 \frac{h^2}{r^2} \simeq \Omega^2 h^2 \tag{2}$$

where Ω, V_K are the disk Keplerian frequency and velocity; then the disk height read $h \simeq c_s/\Omega$, where $c_s \simeq (P/\rho)^{1/2}$ is the sound speed in the disk.

ii. Angular Momentum Conservation: If $\dot{J} = \dot{M}(GMr)^{1/2}$ is the rate at which angular momentum is transported inward at radius r by the accretion of matter at accretion rate $\dot{M}$, and $\dot{J}_I = \dot{M}(GMr_I)^{1/2}$ the rate at which angular momentum accreted onto the black hole at its innermost, stable, circular orbit radius (ISCO) $r_I = 3R_S$ (R_S is the black hole Schwarzschild radius), their difference implies the presence of a torque $\mathcal{T} = t_{r\phi}(2h \cdot 2\pi r) \times r = \dot{J} - \dot{J}_I$, which transfers their difference outward. In this expression $t_{r\phi}$ is the viscous stress (i.e., the force per unit area in the ϕ−direction) and $2h$ the total thickness of the disk. It is generally assumed [2] that the viscous stress is proportional to the local pressure P, so that $t_{r\phi} = \alpha P$ with $\alpha < 1$.

Angular momentum conservation then leads to the expression for the stress $t_{r\phi}$

$$2ht_{r\phi} = \frac{\dot{M}}{2\pi r^2}(GMr)^{1/2}\left[1 - \zeta\left(\frac{r_I}{r}\right)^{1/2}\right] = \frac{\dot{M}}{2\pi r^2}(GMr)^{1/2}J(r) \tag{3}$$

The term in square brackets indicates the fact that matter freely falls onto the black hole for radii $r < r_I$ with $\zeta = 1$ if the torques are also zero at the same point.

With the above expression for the stresses, one can then compute the heat generation rate *per disk unit area*, $2Q$ (Q is the emission from each side of the disk), considering that the energy so generated *per unit volume* is $\dot{e} = 2t_{r\phi}\sigma_{r\phi}$, on substituting $2h\,t_{r\phi}$ from Equation (3) above, with $\sigma_{r\phi} = (3/4)\Omega = (3/4)(GM/r^3)^{1/2}$. So,

$$2h\dot{e} = (2ht_{r\phi})(2\sigma_{r\phi}) = 2Q = \frac{3\dot{M}}{4\pi r^2}\frac{GM}{r}J(r) \tag{4}$$

Therefore, the heat generated from a radius $r_1(\gg r_I)$ to infinity is

$$2Q_{tot}(r_1) = \int_{r_1}^{\infty} 2h\dot{e}\, 2\pi r\, dr \simeq \frac{3}{2}\dot{M}\frac{GM}{r_1} \tag{5}$$

One should note that gravitational energy V from infinity to r_1 is released at a rate $GM\dot{M}/r_1$, but because of the virial theorem, $2T + V = 0$, only half of this can be converted to heat, with the rest remaining as orbital energy ($T = V/2$). Thus, the rate at which gravitational energy is converted into heat is $GM\dot{M}/2r_1$.

The difference between the above expression and that of Equation (5) is provided by the viscous stresses which transport outward not only angular momentum but also energy at a rate [16]

$$\dot{E} = \Omega T = \Omega 2\pi r^2\, 2ht_{r\phi} = \frac{GM\dot{M}}{r_1}J(r_1) \simeq \frac{GM\dot{M}}{r_1} \quad (r \gg r_I) \tag{6}$$

As noted in [15], this is an important issue because in cases that the energy transferred by viscous stresses cannot be radiated away on a viscous time scale (e.g., ADAF), it leads to positive Bernoulli integral. Flows in 2D (i.e., r, ϕ) with $Be > 0$ would not be then possible to accrete. To remedy this situation, [15] proposed that the excess energy and angular momentum escape in the θ-direction in the form of a wind, thereby allowing matter in the disk to flow to the compact object. Because the disk span a large number of decades, the authors of [15] provided simple wind models with wind mass flux $\dot{M}_w$ depending on the radius e.g., $\dot{M} = \dot{M}(r) \propto r^p$, $0 < p < 1$, i.e., with mass flux increasing with radius; this should not be surprising, considering that so does the disk specific angular momentum. Therefore, as matter accretes through the disk, it also "peels-off" in a wind away from the disk plane, carrying away the excess disk angular momentum and torque - transferred energy from its inner sections, setting the disk Be to a negative value, while at the same time reducing the mass flux remaining in the disk to be accreted onto the compact object (the notion of a decreasing mass flux in the disk goes against the overwhelming majority of works on the subject). The outward increasing mass flux of these winds has significant observational consequences that appear consistent with observations [17,18]. We propose that this last fact is responsible for much of the AGN and XRB phenomenology, as it will be discussed below.

2.2. General Accretion Disk Scalings

It is instructive to present the accretion disk equations in dimensionless form, as this makes apparent the dependence of their properties relative to their natural units, most notably their accretion rate in terms of the Eddington accretion rate and the Schwarzschild radius. Normalizing the disk radius r by R_S, i.e., setting $r = x\, R_S$, and its accretion rate by the Eddington accretion rate, i.e., setting $\dot{M} = \dot{m} \times \dot{M}_{\rm Edd}$, with $\dot{M}_{\rm Edd} = L_{\rm Edd}/c^2 = 2\pi\, m_p\, cR_S/\sigma_T \propto M$, the hydrostatic equilibrium and angular momentum transfer equations (Equations (2) and (3)) can be both solved for the disk pressure P to obtain (bearing in mind the prescription $t_{r\phi} = \alpha P$)

$$P = \rho\frac{GM}{r^2}\frac{h^2}{r} = \frac{m_p c^2}{2}n(x)x^{-1}\left(\frac{h}{r}\right)^2 \tag{7}$$

$$P = \frac{\dot{M}}{4\pi h\alpha}\frac{(GMr)^{1/2}}{r^2}J(r) = \frac{m_p c^2}{\sigma_T R_S}\dot{m}(x)x^{-5/2}\left(\frac{h}{r}\right)^{-1}\frac{J(x)}{2\sqrt{2}\alpha} \tag{8}$$

with $J(x) = 1 - (3/x)^{1/2}$, and following the arguments given in the introduction we assume that the accretion rate $\dot{m}$ is also a function of the radius x.

From the above equations one can obtain an expression for the disk density $n(x)$

$$n(x) = \frac{\dot{m}(x)}{\sigma_T R_S} x^{-3/2} \left(\frac{r}{h}\right)^3 \frac{J(x)}{2^{1/2}\alpha} \tag{9}$$

and for the energy emitted per unit disk area (from one of its sides)

$$Q = \frac{3}{8\pi} \frac{GM\dot{M}(r)}{r^3} J(r) = \frac{3}{4} \frac{m_p c^3}{\sigma_T R_S} x^{-3} \dot{m}(x) J(x) \tag{10}$$

The total luminosity can be obtained by integrating the above expression over the surface of the disk $2\pi r\, dr$ from r_I to infinity, i.e., (one must use $2Q$ to take into account both sides of the disk)

$$L = \int_{r_I}^{\infty} (2Q)\, 2\pi r\, dr = \frac{GM\dot{M}}{r_I} \left(\frac{3}{2} - \zeta\right) = \frac{\pi m_p c^3}{\sigma_T} \frac{R_S \dot{m}(x)}{x_I} \left(\frac{3}{2} - \zeta\right) \tag{11}$$

It is generally assumed that the pressure P is given either by the sum of gas and radiation pressures, each becoming dominant at different radii of the disk and for different values of the accretion rate $\dot{m}(x)$. Each such approximation leads to different run of the disk parameters with the radius x (see [19] for a detailed study). At present we will include also magnetic pressure. For simplicity, herein we will assume only gas and magnetic pressures P_g, P_B, with the magnetic pressure being dominant (see also [20]) and the stresses being again proportional to $\alpha(P_g + P_B)$. The presence of three components of the magnetic stresses makes the problem necessarily more complicated; however, we will consider here only their contribution to the vertical disk structure. The inclusion of magnetic field contribution in the disk vertical structure divorces the disk height from the local plasma temperature, thereby allowing transitions between hot and cool states as implied by observations and detailed in the next subsection.

Even at this simplified approach, the system likely entails far more detail than presented herein (e.g., field annihilation on an equatorial current sheet). Our simplified model averages all that over the disk height. Then the total pressure, in conjunction with the hydrostatic equilibrium expression, reads

$$P = \frac{B_\phi^2(x)}{4\pi} + n(x)m_p c_s^2 = \rho(x)\frac{GM}{r^3}h(x)^2 \tag{12}$$

where $c_s(x)$ is the sound speed of the gas. Considering that $B_\phi^2/4\pi\rho(x) = V_A^2$ with V_A the Alfvén velocity, and that $GM/r = V_K^2$, the disk Keplerian velocity, the hydrostatic equilibrium condition reads

$$\frac{h^2}{r^2} = \frac{B_\phi^2}{4\pi\rho V_K^2}[1 + \beta(x)] = \frac{V_A^2}{V_K^2}[1 + \beta(x)] \tag{13}$$

where $\beta(x)$ is the usual gas-to-magnetic pressures parameter of the plasma.

From mass conservation (bearing in mind that, as discussed above the accretion rate depends on the radius r),

$$\dot{M}(x) = 2\pi h r\, n(x) m_p V_r \tag{14}$$

and employing Equation (9), we obtain an expression for the radial flow velocity V_r

$$\frac{V_r}{c} = \frac{1}{x^{1/2}} \left(\frac{h}{r}\right)^2 \frac{\alpha}{J(x)} \quad \text{or} \quad \frac{V_r}{V_K} \simeq \left(\frac{h}{r}\right)^2 \frac{\alpha}{J(x)} \tag{15}$$

Considering that the viscous time scale

$$t_{visc} \simeq \frac{r^2}{\nu} \simeq \frac{r}{V_r} \simeq \frac{R_S}{c} x^{3/2} \left(\frac{r}{h}\right)^2 \frac{J(x)}{\alpha} \tag{16}$$

(with ν the coefficient of viscosity), the above expression implies $\nu \simeq r V_r \simeq h V_A [1 + \beta(x)]^{1/2}$, which if V_A is replaced by the disk thermal velocity we obtained the standard result of $\nu \simeq h c_s$.

Finally, assuming proportionality of the magnetic field with the pressure P one can employ the equation of hydrostatic equilibrium and the expression of Equation (9), to obtain the scaling of the magnetic field B_ϕ, i.e.,

$$\frac{B_\phi^2}{4\pi} \simeq \frac{n(x) m_p c^2}{2} \frac{1}{x} = \frac{m_p c^2}{\sigma_T R_S} \frac{\dot{m}(x)}{2} x^{-5/2} \left(\frac{r}{h}\right) \frac{J(x)}{\alpha} \tag{17}$$

and from that an expression for the Alfvén velocity; then one can easily see that the scaling of the rate at which magnetic flux is annihilated at the disk plane, $B_\phi^2 V_A$, is very similar to that of Equation (10), suggesting that magnetic field annihilation could be the major contributor of accretion disk dissipation.

2.3. On ADAF and the Black Hole States

The introduction of Advection-Dominated Accretion Flows [11,12] brought a totally different perspective in accretion flow physics from that of the standard Shakura-Sunyaev [2]. Accretion flows are now allowed to be thick ($h \simeq r$), with their ion temperatures $T_i \sim m_p(GM/r)$ [1]. The radial force balance is now effected not only by the centrifugal force $\Omega^2 r$, but also by the gradient of the radial pressure, and the value of Ω is below its Keplerian value Ω_K. At the same time, the gas is also heated, in addition to viscous stress heating, also by the radial compression of the flow; this alone leads to a temperature $T \propto n(x)^{(\gamma-1)}$, which for adiabatic index $\gamma = 5/3$ and density $n(x) \propto x^{-3/2}$, leads to $T \propto 1/r$ and combined with viscous heating implies $\Omega^2 \propto (5\gamma - 3)/r^3 \to 0$ as $\gamma \to 5/3$, i.e., it precludes rotation of the flow [15]. This issue is circumvented (see [11]) by including the magnetic field contribution to the plasma pressure, which reduces the effective γ below the value 5/3.

As noted in [11], the presence of an ADAF, i.e., a flow with $h \simeq r$ implies that the plasma remains hot, close to its virial temperature on time scales longer than its accretion time scale. Therefore, demanding its cooling time through Coulomb collisions, $t_{cool} \sim 1/n(x)\sigma_T c$ be longer than its viscous (or flow) time scale of Equation (16) implies

$$\frac{R_S}{c} x^{3/2} \left(\frac{r}{h}\right)^2 \frac{J(x)}{\alpha} < \frac{R_S}{c} \frac{1}{\dot{m}(x)} x^{3/2} \left(\frac{h}{r}\right)^3 \frac{\alpha}{J(x)} \quad \text{or} \tag{18}$$

$$\dot{m}(x) < \alpha^2 \tag{19}$$

assuming in the last step that $h \simeq r$ and $J(x) \simeq 1$, i.e., $x \gg 3$. Thus, the presence or not of an ADAF depends on a *single*, *global* condition on the *normalized* accretion $\dot{m}$ and it is independent of the flow radius x or the mass of the system M. This provides a huge economy of assumptions in interpreting the spectra of accreting black holes, of wide applicability because of the mass independence of this criterion. However, while these flows are mass scale invariant they are of reduced radiative efficiency: Because the viscous time scale is $\dot{m}(<1)$ times shorter than the radiative time scale, the power of viscous dissipation, released to heating the protons (same as in the SS73 disks), is advected into the black hole before it can be radiated away; the radiated power then $\dot{m}$ time smaller than that produced

---

[1] The electron temperatures are also hot, but their temperature is determined from the balance between Coulomb heating by the much hotter ions and the cooling processes that produce the observed X-ray emission; the cooling of electrons with $T_e \sim 10^9$ K becomes faster than their heating by protons of any temperature and limits their temperatures at the smallest flow radii to roughly this value.

by viscous dissipation. So for ADAF the expression for the luminosity of Equation (11) should be $L \propto R_S \dot{m}(x)^2 \propto \dot{m}(x)^2 M$, rather than $L \propto \dot{m}(x) M$ that is the case with the standard disks.

The possibility offered by ADAF, namely to include the high energy ($E > 2$ keV) emission of AGN and XRB as an integral part of the global accretion dynamics, has led to models that combine an SS73 disk at large radii (to produce the quasi-thermal component generally present in the spectra) with a transition of the flow to an ADAF at a radius $r < r_{tr}$ that accounts for their high energy X-rays. This way, one could reproduce the entire spectrum with a single accretion flow [13]. To account for the increasing dominance of the quasi-thermal disk component with the source luminosity, observed in individual XRBs on increase of their bolometric luminosity and in AGN statistically (by the variation of their α_{OX} index with luminosity [4]), they proposed that the transition radius, r_{tr}, decreases with increasing luminosity. While appealing at first sight, these models vitiate the ADAF dependence of the single global $\dot{m}$. It is also hard to see how a cool, thin outer disk will convert into an ADAF at $r < r_{tr}$; finally, it is also not obvious why r_{tr} would decrease with source luminosity. Some of these issues will be discussed in Section 4 with consideration of radius dependent accretion rates.

3. Accretion Disks Winds: An AGN Rosetta Stone?

The launch of HST, Chandra and XMM-Newton, ushered a new era in the study of accreting black holes and especially AGN. These observatories discovered that $\simeq 50\%$ of AGN exhibit blue-shifted absorption features in the UV and X-ray spectra [14], presumably a manifestation of the presence of winds that are ubiquitous in accreting black holes. Of particular interest are the absorption features of their X-ray spectra, because they span a large range in velocity ($v \sim 0.5 c$ to $v \simeq 300$–500 km/s) and in ionization states of the plasma (from Fe XXVI to Fe II, much larger than their UV counterparts; the power of X-ray spectroscopy lies in that in the span of 1.5 decades in energy it encompasses ions that span 5 decades in ionization parameter ξ). These absorption features are thought to be the result of a wind photoionized by the AGN continuum and are broadly referred to as Warm Absorbers (WA), because of the inferred temperature of the corresponding plasma.

3.1. X-Ray Absorbers: Their Phenomenology and Implications

Considering that the observed winds are photoionized by the radiation of a point-like X-ray source and that we observe the radiation along a pencil beam from the X-ray source to us, the absorption spectra should probe the continuum of velocities, $v(r)$, columns, $N_H(r)$, and densities, $n(r)$, of the outflowing plasmas and the continuum of ionization parameter $\xi = L/n(r)r^2 = L/N_H(r)r$ along the observer's line of sight (LoS). Therefore, measurement of the absorption depth i.e., equivalent width (EW), (and also the velocity width if possible) of a transition, known to exist at a given range in ξ, provides the local wind column, N_H (and velocity), while its ξ–value (for the observed N_H and L) provides a measure of the distance of the given ion from the ionizing source, along the observer's LoS . Observations of N_H, ξ for multiple ions can then be inverted to probe the wind density (and velocity) structure along our LoS.

This was in essence the approach of Behar and collaborators, ([21], hereafter HBK07) and ([22] hereafter B09). In these two important papers, HBK07 and B09 introduced what they called the Absorption Measure Distribution $AMD \equiv dN_H/d\log\xi$, namely the Hydrogen equivalent column of an ion, N_H, per $\log\xi$. Assuming a power-law functional dependence of N_H on ξ, namely $N_H \propto \xi^\alpha$, they produced a global fit of the detected, distinct ionic species of as many different elements as the observations allowed, to obtain a value for α. (Figure 1(Left)). It is important to note that *for a spatially smooth outflow*, this relation implies a power-law density dependence on the wind along the observer's LoS, i.e., $n(r) \propto r^{-s}$, with s and α related via $s = (2\alpha + 1)/(\alpha + 1)$. The values of α were found to be $\alpha \simeq 0 - 0.3$, implying $s \simeq 1 - 1.24$ and $N_H \propto r^{-s+1} \propto r^0 - r^{0.24}$, a very weak r-dependence.

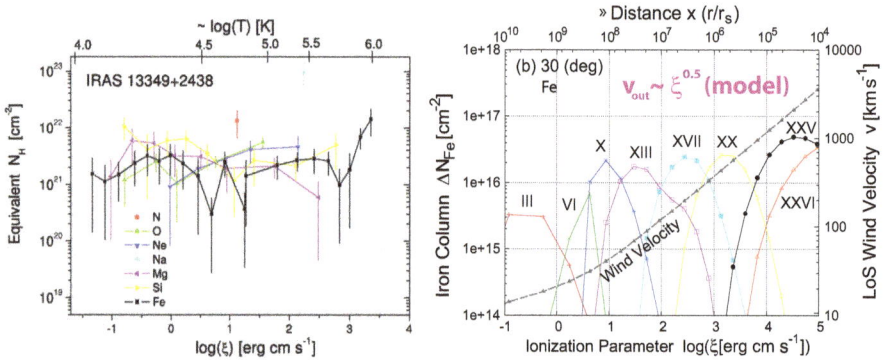

Figure 1. (**Left**) The AMD of IRAS 13349+2438 (from [21]) determined independently from the transitions of each element and ionization state shown. The weak dependence of N_H on ξ is apparent. (**Right**) The ionization structure of Fe for a model with $s = 1$ and $\theta = 30$ deg. The ionization state of Fe decreases with increasing distance; however, the maximum N_H of each ion remains roughly constant (from [17]). Also shown is the wind velocity with distance, indicating that lower ionization ions have lower velocities.

The broad range in ξ (~ 5 decades) and the almost 'flat' $N_H - \xi$ relation observed, preclude that the WA outflows be driven by radiation pressure or X-ray heating; such winds, because they are launched from regions of limited size, R_l, they produce, at distances a few (~ 10) times R_l constant mass flux and velocity, resulting in density $n(r) \propto 1/r^2$ and $N_H \propto r^{-1}$, in gross disagreement with observation; in their accelerating phase, their density dependence is a steeper function of r $n(r) \propto r^{-3}$, implying that their ξ should decrease with r and so should their velocity, $v(r)$, while their column should be larger at smaller r. All these dependencies disagree with observation.

To interpret the AMD relation and the WA observations of HBK07 and B09, Fukumura, Kazanas and collaborators [17,18] (FKCB10a,b) concluded that these winds must be two dimensional (2D), launched across the entire disk domain. Motivated by these considerations they computed the photoionization of the self-similar 2D MHD winds of Contopoulos & Lovelace ([23] hereafter CL94). These are 2D winds, generalizations of those of Blandford & Payne [24]. Because of their broad radial extent, they offer the possibility of a wide range of ξ; indeed the winds of CL94 provide radial density profiles of the form $n(r) \sim r^{-s}$ ($s \simeq 1$), $v(r) \propto r^{-1/2}$, consistent with the values inferred from the AMD analysis.

The winds of [24] imply radial density profiles $n(r) \propto r^{-3/2}$ which are too steep to be consistent with the WA data; however, the broad range of their launching radii has prompted several authors [25,26] to employ them in modeling the AGN UV and optical line profiles.

3.2. The Wind Scaling Relations

The results of FKCB10a, though motivated by specific AGN, are quite general: The MHD Accretion disk winds, whose local velocities scale with the Keplerian (see CL94), are self-similar if the radius r is normalized to the BH Schwarzschild radius R_S ($x = r/R_S$, $R_S = 3\,\mathcal{M}$ km, with $\mathcal{M} \equiv (M/M_\odot)$), their velocity to c ($v(x)/c = x^{-1/2}$) and the mass flux rate $\dot{M}$ to the Eddington rate, i.e., $\dot{m} = \dot{M}/\dot{M}_E$, $\dot{M}_E \equiv L_E/c^2$ ($L_E \simeq 1.3 \times 10^{38}\mathcal{M}$ erg s^{-1} is the Eddington luminosity). So mass conservation in physical and dimensionless units reads respectively

$$\dot{M}(r,\theta) \sim n(r,\theta)\,r^2\,v(r) \quad \text{or} \quad \dot{m}(x)\,\mathcal{M} \propto n(x)\mathcal{N}(\theta)\,x^2\mathcal{M}^2\,x^{-1/2}\,. \tag{20}$$

with $\mathcal{N}(\theta) \simeq e^{5(\pi/2-\theta)}$ (see Figure 2(Left) of FKCB10a). The density normalization is such that (integrated over θ) $n(x \simeq 1)\sigma_T R_S \simeq \tau(x \simeq 1) \sim 3\,10^5 \mathcal{M}\sigma_T\,n(1) \simeq \dot{m}_0$, with $\dot{m}_0$ the (normalized) mass flux at the smallest wind radius ($x \gtrsim 1$). For a density profile such as that obtained from fits to the AGN WA, namely for $n(x) \propto x^{-s}$ we obtain,

$$\dot{m}(x) \simeq \dot{m}_0 x^{-s+3/2} \quad \text{and} \quad N_H(x) \propto \mathcal{N}(\theta) \dot{m}_0 x^{-s+1} \tag{21}$$

The relations of Equation (21) then suggest that for $s \simeq 1$ the wind mass flux increases with radius like $x^{1/2}$, while the column is roughly constant independent of the distance, in agreement with the properties of ADIOS [15].

The wind ionization parameter ξ, can also be cast in dimensionless units: If η, ($\simeq 10\%$) is the radiative efficiency of the accretion process, then the luminosity L can be written as $L \simeq 10^{38} \eta \, \dot{m}_a \, \mathcal{M}$ erg/s ($\dot{m}_a$ is the dimensionless accretion rate onto the BH to produce the luminosity L), yielding the for ξ

$$\xi(x) \simeq \frac{L}{n(r)r^2} \simeq \frac{\eta \dot{m}_a}{N_H(x)x} \simeq 10^8 \frac{\eta}{f_w \mathcal{N}(\theta)} \frac{\mathcal{S}(\nu_{\text{ion}})}{x^{-s+2}} \quad \text{erg cm s}^{-1} \tag{22}$$

where $f_w = \dot{m}_0/\dot{m}_a$ (~ 1) is the ratio of mass flux in wind and in accretion at the smallest radii and $\mathcal{S}(\nu_{\text{ion}})$ is a factor that determines the ionizing fraction of the bolometric luminosity produced by accretion onto the BH, related to the AGN α_{OX}.

With respect to Equations (21) and (22), that provide the expression for the winds' mass flux, column and ionization, one should note that: (a) They are independent of the BH mass $\mathcal{M}$ and as such they could apply to an accreting black hole of any mass [27]. (b) The wind column N_H depends mainly on the dimensionless mass flux at the smallest radii $\dot{m}_0$ and, depending on the value of s, it has a weak dependence on the (dimensionless) distance from the black hole x; however, it has a very strong dependence on the observer inclination angle θ giving the wind the toroidal appearance implied by the AGN unification scheme. (c) The wind ionization parameter ξ is independent of both $\dot{m}$ and M and depends only on the (dimensionless) distance from the black hole x and the observer inclination angle θ. However, the wind's ionization structure and appearance (i.e., the velocities of its absorption features and the angular dependence of their column) depend crucially on the fraction of ionizing photons in the spectrum $\mathcal{S}(\nu_{\text{ion}})$ (α_{OX} serves as its proxy in AGN) [18] and $\mathcal{N}(\theta)$.

Our basic premise is that these winds are launched across the entire disk extent with velocities $v(x) \propto x^{-1/2}$, in all accreting BH; as a result, the scalings discussed above should also correlate with the velocities of observed absorption features. In AGN with high X-ray content (Seyferts) the wind's inner region is fully ionized out to $x \gtrsim 10^3$, resulting in the occurrence of the highest ξ ($\sim 10^4, 10^5$) absorption transitions (Fe XXV, Fe XXVI) at $v \sim cx^{-1/2} \sim 10,000$ km/s; in the lower X-ray content BAL QSOs, the fully ionized region is much smaller ($x \sim 10 - 30$) and the corresponding velocities much higher ($v \sim 0.3c$; APM 08279+5255) [18,28]; finally, in galactic XRBs with $\mathcal{S}(\nu_{\text{ion}}) \simeq 1$ the wind is ionized out to $x \gtrsim 10^5$ and the absorber velocities are $v \sim 300 - 1000$ km/s [27], in agreement with our scalings.

We have discussed in this section the mass scale invariance of the MHD winds and the dependence of their obscuration (i.e., $N_H(x, \theta), \xi(x, \theta)$) and their absorber velocities on sources' content of ionizing radiation and the wind mass flux. To close the gap in AGN (or XRB) phenomenology, one is left with finding a relation between the AGN (dimensionless) accretion rate $\dot{m}$ and the X-ray content, or broadly, the relative strengths of the BBB and the X-ray emission. This is discussed in the next section.

4. Accretion Disk Spectral Energy Distributions (SEDs)

It was shown above that the disk wind density profiles implied by the X-ray absorber observations are of the form $n(x) \propto x^{-s}$, with $s \simeq 1$. However (see Equation (21)), for winds with $s < 3/2$ *the wind mass flux increases with radius*, implying that *the disk accretion rate must decreases toward the BH!* ; as a result, only a small fraction of the available mass accretes onto the black hole [15]. Most importantly, if below some radius, x_{tr}, the local accretion disk rate drops below the critical rate $\dot{m} \simeq \alpha^2$ (α is the disk viscosity parameter), then (see Equation (19)) the disk can potentially make a transition to an

ADAF flow [11], in fact an ADIOS; this hot ($T \simeq 10^9$ K) segment occupies the AGN innermost region and constitutes the site of X-ray emission, in agreement with the X-ray variability properties and the microlensing observations.

However, the condition of Equation (19) for the presence of an ADAF is necessary but not sufficient. One must still address the issue of heating up the "cold" protons of the outer part of the disk to their virial value, at $r < r_{tr}$. In the absence of a more detailed theory, we propose here that their heating takes place at the equatorial current sheet of the accretion disk. The current sheet dissipation can provide energies for protons (and electrons) close to their virial one; the ability to retain these energies sufficiently long to affect the disk structure depends on the disk average density at this specific radius, i.e., on the local value of $\dot{m}$. Therefore, the picture of black hole states envisioned in [13] can be realized for accretion disks with variable (in r) accretion rates.

In Figure 2(Middle,Right) below we show the SEDs of two AGN: the Broad Line Seyfert 1 (BLS1) galaxy NGC 5548 [29] and that of the Narrow-Line Seyfert 1 (NLS1) PG 1244+026 [30]. These exemplify the drastically different SED distributions of these two Seyfert galaxies, even though they both belong to the same broad AGN category. Their most obvious difference is that of the relative importance of their BBB (Optical-UV) and X-ray emissions. The bolometric luminosity of NGC 5548 is dominated by the X-ray emission, while that of the PG 1244+026 by the UV one. Considering that it is broadly assumed that the BBB emission comes from a disk that terminates in the ISCO, one would expect that emission to be always the dominant one.

We propose that this distinctive difference is related to the radius–dependent mass accretion rate of their corresponding disks, effected by the presence of MHD winds which remove progressively most of the mass available for accretion as the disk plasma sinks toward the black hole. As we noted earlier when, at some transition radius r_{tr}, the local accretion rate becomes smaller than α^2, the disk transits to an ADAF (or rather an ADIOS) state for radii $r < r_{tr}$, while maintaining the standard SS73 disk form at $r > r_{tr}$. Considering that at its $r < r_{tr}$ segment, where the X-rays are produced, the luminosity has smaller efficiency with $\dot{m}$, i.e., $L_X \propto \dot{m}(r)^2$ vs. $L_{UV} \propto \dot{m}(r)$ for $r > t_{tr}$, but it is produced at a deeper gravitational potential, their relative ratios will depend on the transition radius r_{tr}. Furthermore, as the global mass flux, i.e., that provided at the outer edge of the accretion flow, $\dot{m}(x)$, and the ensuing luminosity increase, the transition radius r_{tr} will shift to smaller r, with the importance of the thermal component increasing, as suggested by [13] in their attempt to account for the XRB spectral evolution.

With the above qualifications and assumptions, we can express quantitatively the relative importance of the BBB and X-ray components in the AGN spectra. We assume, to be specific, that the index $s \simeq 1$, so that x_{tr} is given by the expression

$$\dot{m}(x_{tr}) = \dot{m}_0 \, x_{tr}^{1/2} \simeq \alpha^2 \tag{23}$$

Then from Equation (11) we obtain the following expressions of $L(x > x_{tr})$ and $L(x < x_{tr})$

$$L(x > x_{tr}) \quad \propto \quad \int_{x_{tr}}^{\infty} \frac{\dot{m}(x)}{x} J(x) d\ln x \sim \frac{\dot{m}_0}{x_{tr}^{1/2}} \left[1 - \frac{\zeta}{2} \left(\frac{x_I}{x_{tr}} \right)^{1/2} \right] \tag{24}$$

$$L(x < x_{tr}) \quad \propto \quad \int_{x_I}^{x_{tr}} \frac{\dot{m}(x)^2}{x} J(x) d\ln x \sim \dot{m}_0^2 \left[\ln \left(\frac{x_{tr}}{x_I} \right) - 2\zeta + 2\zeta \left(\frac{x_I}{x_{tr}} \right)^{1/2} \right] \tag{25}$$

One should note that the ratio R of these two luminosities is proportional to $1/\alpha^2$:

$$R = \frac{L(x > x_{tr})}{L(x < x_{tr})} = \frac{1}{\dot{m}_0 \, x_{tr}^{1/2}} \left(\frac{P1}{P2} \right) = \frac{1}{\alpha^2} \left(\frac{P1}{P2} \right) \tag{26}$$

with the last equality because of Equation (23) and with $P1$, $P2$ the values of the square brackets in Equations (24) and (25) respectively.

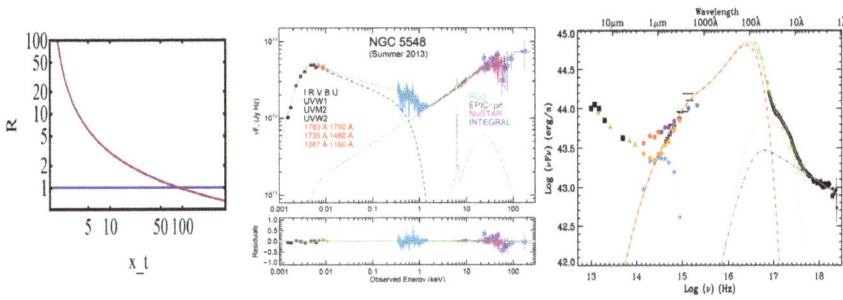

Figure 2. (**Left**) The ratio R of the thermal (BBB) to the X-ray bolometric luminosities as a function of the transition radius x_t from the SS73 to the ADAF disk regime. (**Middle**) The νF_ν spectrum of NGC 5548 (from [29]) indicating a slight dominance of the X-ray luminosity, implying for this case that $x_t \gtrsim 100$. (**Right**) Same as (**Middle**) but for the NLS1 PG 1124+026 (from [30]); in this case the BBB component is dominant implying $x_t \lesssim 5$.

Figure 2(Left) depicts the ratio R as a function of the transition radius x_{tr}, for $x_I = 1.5, \zeta = 1$ and $\alpha = 1/3$.[2] Of interest in this figure is the value of $x_{tr} \simeq 100$ for which $R \simeq 1$, because this sets the size of the disk at which the cool/thin disk quasi-thermal luminosity matches that of the harder radiation produced in the advection-dominated accretion flow section. This will then correspond to the spectrum of NGC 5548 of Figure 2(Middle). On the other hand, the spectrum of PG 1124+026, implies $x_{tr} \simeq 3 - 5$, consistent with the larger $\dot{m}$ associated with NLS1 galaxies. The spectra are also steeper, considering that the transfer of proton energy to electrons is not very efficient in the narrow range of radii that the ADAF flow is allowed in this AGN.

5. Discussion, Conclusions

We described above that the multidimensional phenomenology associated with the appearance and spectral distribution of AGN (and XRB) can be reduced to a small number of dimensionless global parameters of these systems, most notably the dimensionless accretion rate $\dot{m}$ and their disk inclination angle θ. The underlying black hole mass, while important in providing a scale of the absolute luminosity for a given $\dot{m}$, it is also important in determining the temperature of the thermal BBB component of the accretion disk; however, this effect is only weakly dependent on this mass ($\propto M^{-1/4}$), it is well understood and largely under control. Apart from this component, the rest of the accretion disk and wind properties (ionization, column density, velocity) appear to be mass independent. While many details remain elusive, the radial dependence of the disk accretion rate, implied by the apparent increase in the wind mass flux with radius, is the crucial novel notion in this analysis. It is this notion that allows for a coherent picture of their combined X-ray–UV–Optical spectra, with the relative importance and spatial location of these components strongly dependent on the value of $\dot{m}$ as described and in good agreement with observation. Finally, the angular dependence of the MHD winds, implies a strong dependence of their properties on the disk inclination angle, a feature generally known as AGN Unification. While some reasons for the inferred increase in the disk wind mass flux with distance are given above, the detailed underlying physics of this most important issue, namely the connection the disk structure with the wind properties is still not entirely clear. We hope that this note will motivate a more focused activity on this issue. Perhaps the generation of magnetic flux near the disk inner edge by the Cosmic Battery effect [32–34] plays a role in this respect.

---

[2] While this value may appear too large, one should note that [20], argue that for disks at which the pressure is dominated by magnetic fields $\alpha = 1/\sqrt{3}$, while [31] in their study of shear-box simulations with a net magnetic flux find $\alpha \simeq 1$ for disks with gas to (poloidal) magnetic pressure, greater than 0.01.

We have refrained in this note from discussing the line emission in AGN. This constitutes an entire subject onto itself. We believe that it is relevant to the issues discussed herein because the plasma responsible for the absorption will also produce line emission. Actually, the X-ray absorbers that motivated our work probe only a small fraction of the entire wind phase space, namely the small pencil beam from the X-ray source along the observer's LoS; the AGN line emission provides information about the entire phase space of these winds; it remains to be seen whether it is consistent with the ideas promoted herein. There is clearly more work to be done; we believe that our work points to the correct direction.

Finally, the issue of MHD winds discussed in the previous sections, while focused on radio quiet AGN, is also relevant to the physics of radio loud ones, in particular blazars. The reason for that is that these winds, of columns that depend mainly on $\dot{m}$, reprocess and isotropize the disk radiation so that it can be reprocessed by the relativistic jet propagating along the disk axis as "External Inverse Compton" γ-rays. In a recent publication [35] we showed how the blazar phenomenology, known as the "blazar sequence" can be reproduced by variation of a single parameter, that again being $\dot{m}$, indicating an underlying economy of parameters across the entire field of AGN physics.

Funding: This research received no external funding.

Acknowledgments: The author would like to acknowledge financial support by NASA's ADAP program and GI grants from Chandra and Fermi. He would also like to acknowledge numerous discussions and interactions with his collaborators, K. Fukumura, E. Behar, I. Contopoulos, C. Shrader and F. Tombesi.

Conflicts of Interest: The authors declare no conflict of interest.

References

1. Abramowicz, M.A.; Fragile, P.C. Foundations of Black Hole Accretion Disk Theory. *Living Rev. Relativ.* **2013**, *16*, 1–88. [CrossRef] [PubMed]
2. Shakura, N.; Sunyaev, R. Black Holes in Binary Systems. Observational Appearance. *Astron. Astrophys.* **1973**, *24*, 337–355.
3. Galeev, A.A.; Rosner, R.; Vaiana, G.S. Structured Coronae of Accretion Disks. *Astrophys. J.* **1979**, *229*, 318–326. [CrossRef]
4. Steffen, A.T.; Strateva, I.; Brandt, W.N.; Alexander, D.M.; Koekemoer, A.M.; Lehmer, B.D.; Schneider, D.P.; Vignali, C. The X-Ray-to-Optical Properties of Optically Selected AGN over Wide Luminosity and Redshift Ranges. *Astron. J.* **2006**, *131*, 2826–2842. [CrossRef]
5. Done, C.; Gierlinski, M.; Kubota, A. Modelling the Behaviour of Accretion Flows in X-ray Binaries. *Astron. Astrophys. Rev.* **2007**, *15*, 1–66. [CrossRef]
6. Haardt, F.; Maraschi, L. X-Ray Spectra from Two-Phase Accretion Disks. *Astrophys. J.* **1993**, *413*, 507–517. [CrossRef]
7. Berkeley, A.J.; Kazanas, D.; Ozik, J. Modeling the X-Ray-Ultraviolet Correlations in NGC 7469. *Astrophys. J.* **2000**, *535*, 712–720. [CrossRef]
8. Balbus, S.; Hawley, J. A Powerful Local Shear Instability in Weakly Magnetized Disks. *Astrophys. J.* **1991**, *376*, 214–233. [CrossRef]
9. Christodoulou, D.; Contopoulos, I.; Kazanas, D. Interchange Method in Incompressible Magnetized Couette Flow: Structural and Magnetorotational Instabilities. *Astrophys. J.* **1996**, *462*, 865–873. [CrossRef]
10. Christodoulou, D.; Contopoulos, I.; Kazanas, D. Interchange Method in Compressible Magnetized Couette Flow: Magnetorotational and Magnetoconvective Instabilities. *Astrophys. J.* **2003**, *462*, 373–383. [CrossRef]
11. Narayan, R.; Yi, I. Advection-dominated accretion: A self-similar solution. *Astrophys. J.* **1994**, *428*, L13–L16. [CrossRef]
12. Narayan, R.; Yi, I. Advection-dominated accretion: Self-similarity and bipolar outflows. *Astrophys. J.* **1995**, *444*, 231–243. [CrossRef]
13. Esin, A.A.; McClintock, J.E.; Narayan, R. Advection-Dominated Accretion and the Spectral States of Black Hole X-Ray Binaries: Application to Nova Muscae 1991. *Astrophys. J.* **1997**, *489*, 865–889. [CrossRef]

14. Crenshaw, D.M.; Kraemer, S.B.; George, I.M. Mass Loss from the Nuclei of Active Galaxies. *Ann. Rev. Astron. Astrophys.* **2003**, *41*, 117–167. [CrossRef]

15. Blandford, R.D.; Begelman, M.C. On the Fate of Gas Accreting at a Low Rate on to a Black Hole. *Mon. Not. R. Astron. Soc.* **1999**, *211*, P1–P5. [CrossRef]

16. Novikov, I.D.; Thorn, K.S. Astrophysical Black Holes. In *Black Holes*; by DeWitt, C., DeWitt, B.S.; Eds.; Gordon and Breach: New York, NY, USA, 1972; pp. 343–450.

17. Fukumura, K.; Kazanas, D.; Contopoulos, I. Behar, E. MHD Accretion Disk Winds as X-ray Absorbers in Active Galactic Nuclei. *Astrophys. J.* **2010**, *715*, 636–650. [CrossRef]

18. Fukumura, K.; Kazanas, D.; Contopoulos, I. Behar, E. Modeling High-velocity QSO Absorbers with Photoionized MHD Disk Winds. *Astrophys. J.* **2010**, *723*, L228–L232. [CrossRef]

19. Svensson, R.; Zdziarski, A.A. Black Hole Accretion Disks with Coronae. *Astrophys. J.* **1994**, *436*, 599–606. [CrossRef]

20. Pariev, V.I.; Blackman, E.G.; Boldyrev, S.A. Extending the Shakura-Sunyaev Approach to a Strongly Magnetized Accretion Disk Model. *Astron. Astrophys.* **2003**, *407*, 403–421. [CrossRef]

21. Holczer, T.; Behar, E.; Kaspi, S. Absorption Measure Distribution of the Outflow in IRAS 13349+2438: Direct Observation of Thermal Instability? *Astrophys. J.* **2007**, *663*, 799–807. [CrossRef]

22. Behar, E. Density Profiles in Seyfert Outflows. *Astrophys. J.* **2009**, *703*, 1346–1351. [CrossRef]

23. Contopoulos, J.; Lovelace, R.V.E. Magnetically driven jets and winds: Exact solutions. *Astrophys. J.* **1994**, *429*, 139–152. [CrossRef]

24. Blandford, R.D.; Payne, D.G. Hydromagnetic flows from accretion discs and the production of radio jets. *Mon. Not. R. Astron. Soc.* **1982**, *199*, 883–903. [CrossRef]

25. Emmering, R.T.; Blandford, R.D.; Shlosman, I. Magnetic acceleration of broad emission-line clouds in active galactic nuclei. *Astrophys. J.* **1992**, *385*, 460–477. [CrossRef]

26. Bottorff, M.C.; Korista, K.T. Shlosman, I.; Blandford, R.D. Dynamics of Broad Emission-Line Region in NGC 5548: Hydromagnetic Wind Model versus Observations. *Astrophys. J.* **1997**, *479*, 200–221. [CrossRef]

27. Fukumura, K.; Kazanas, D.; Shrader, C. R.; Tombesi, F.; Behar, E.; Contopoulos, I. Magnetic origin of black hole winds across the mass scale. *Nat. Astron.* **2017**, *1*, 62–72. [CrossRef]

28. Chartas, G.; Brandt, W.N.; Gallagher, S.C.; Garmire, G.P. Chandra Detects Relativistic Broad Absorption Lines from APM 08279+5255. *Astrophys. J.* **2002**, *579*, 169–175 [CrossRef]

29. Mehdipour, M.; Kaastra, J.S. Anatomy of the AGN in NGC 5548: I. A global model for the broadband spectral energy distribution. *Astron. Astrophys.* **2015**, *575*, A22–A39. [CrossRef]

30. Jin, C.; Done, C.; Middleton, M.; Ward, M. A long XMM-Newton observation of an extreme narrow-line Seyfert 1: PG 1244+026. *Mon. Not. R. Astron. Soc.* **2013**, *436*, 3173–3185. [CrossRef]

31. Bai, X.-N.; Stone, J.M. Local Study of Accretion Disks with a Strong Vertical Magnetic Field: Magnetorotational Instability and Disk Outflow. *Astrophys. J.* **2013**, *767*, 30–48. [CrossRef]

32. Contopoulos, I.; Kazanas, D. A Cosmic Battery. *Astrophys. J.* **1999**, *508*, 859–863. [CrossRef]

33. Contopoulos, I.; Kazanas, D.; Christodoulou, D. M.: The Cosmic Battery Revisited. *Astrophys. J.* **2006**, *652* 1451–1456. [CrossRef]

34. Christodoulou, D.M.; Contopoulos, I.; Kazanas, D. Simulations of the Poynting-Robertson Cosmic Battery in Resistive Accretion Disks. *Astrophys. J.* **2008**, *674*, 388–407. [CrossRef]

35. Boula, S.; Kazanas, D.; Mastichiadis, A. Accretion Disk MHD Winds and Blazar Classification. *Mon. Not. R. Astron. Soc.* **2019**, *482*, L80–L84. [CrossRef]

MDPI

Article

Physics of "Cold" Disk Accretion onto Black Holes Driven by Magnetized Winds

Sergey Bogovalov

National Research Nuclear University (MEPHI), Kashirskoje Shosse, 31, 115409 Moscow, Russia; ss433@mail.ru;
Tel.: +7-495-788-5699

Received: 30 October 2018; Accepted: 3 January 2019; Published: 14 January 2019

Abstract: Disk accretion onto black holes is accompanied by collimated outflows (jets). In active galactic nuclei (AGN), the kinetic energy flux of the jet (jet power or kinetic luminosity) may exceed the bolometric luminosity of the disk by a few orders of magnitude. This may be explained in the framework of the so called "cold" disk accretion. In this regime of accretion, the disk is radiatively inefficient because practically all the energy released at the accretion is carried out by the magnetized wind. This wind also provides efficient loss of the angular momentum by the matter in the disk. In this review, the physics of the accretion driven by the wind is considered from first principles. It is shown that the magnetized wind can efficiently carry out angular momentum and energy of the matter of the disk. The conditions when this process dominates conventional loss of the angular momentum due to turbulent viscosity are discussed. The "cold" accretion occurs when the viscous stresses in the disk can be neglected in comparison with impact of the wind on the accretion. Two problems crucial for survival of the model of "cold" accretion are considered. The first one is existence of the magnetohydrodynamical solutions for disk accretion purely due to the angular momentum loss by the wind. Another problem is the ability of the model to reproduce observations which demonstrate existence of the sources with kinetic power of jets 2–3 orders of magnitude exceeding the bolometric luminosity of disks. The solutions of the problem in similar prescriptions and numerical solutions without such an assumption are discussed. Calculations of the "unavoidable" radiation from the "cold" disk and the ratio of the jet power of the SMBH to the bolometric luminosity of the accretion disk around a super massive black hole are given in the framework of the Shakura and Sunyaev paradigm of an optically thick α-disk. The exploration of the Fundamental Plane of Black Holes allows us to obtain semi empirical equations that determine the bolometric luminosity and the ratio of the luminosities as functions of the black hole mass and accretion rate.

Keywords: MHD–accretion; accretion discs–jets; AGN

1. Introduction

Classical works on the physics of accretion [1–3] laid the foundations of the theory of disk accretion onto relativistic objects, neutron stars and black holes. In the model of Shakura and Sunyaev [3] (hereafter, SS model), every particle loses angular momentum due to viscous stresses arising in a turbulent plasma. In the geometrically thin and optically thick accretion disks, all the gravitational energy released at the accretion is carried out by radiation.

The bolometric luminosity of a disk accreting onto a nonrotating black hole can be represented as $L_{bol} = \eta \dot{M} c^2$, where $\dot{M}$ is the rate of accretion, c is the speed of light and η is the efficiency of transformation of the rest of the mass into the radiation during the accretion onto a Schwarzschild black hole $\eta \approx 0.1$. It is also convenient to work with the dimensionless mass of the black hole $m = M/M_\odot$, where $M_\odot$ is the solar mass, and dimensionless luminosity expressed in units of the Eddington luminosity $L_{Edd} = 4\pi G m_p M c/\sigma_T$, where G is the gravitational constant, m_p is the proton mass, and σ_T is the Thomson cross-section. At Eddington luminosity, the force arising in Thomson scattering of photons of radiation on an electron equals the force of gravitational attraction of the proton compensating the electric charge of the electron. Correspondingly, the Eddington accretion rate is introduced as $\dot{M}_{Edd} = L_{Edd}/\eta c^2$. Dimensionless accretion rate $\dot{m} = \dot{M}/\dot{M}_{Edd}$. In the standard SS disk, accretion $L_{bol}/L_{Edd} = \dot{m}$.

Observations of AGNs revealed rather dramatic deviations of the theory from reality. The galactic center of our galaxy Sgr A* is especially interesting in this regard. Sgr A* is surprisingly faint despite the rich gas reservoir in its immediate surroundings that should provide a high accretion rate. The accretion onto the SMBH in the Galactic center should be $\dot{M} \sim 10^{-6}$ M$_\odot$/year [4] at Bondi radius. For the mass of SMBH, $M = 4 \times 10^6$ $\dot{M}_{Edd} = 0.072$ M$_\odot$/year giving the accretion rate $\dot{m} \sim 10^{-5}$. However, the bolometric luminosity is no more than $\sim 10^{36}$ ergs^{-1}. This corresponds to $L_{bol}/L_{Edd} = 2.5 \times 10^{-9}$, which is 3–4 orders of magnitude below the value that should be expected in the standard SS disk. This is not an isolated case.

The faintness of Sgr A* led to the development of theoretical models with radiatively inefficient accretion flows (RIAFs). One of the models is the advection dominated accretion flow (ADAF) [5], in which the low luminosity is explained by the combination of a high ratio of radial to tangential gas velocities, and the decoupling of hot protons and cold electrons in low density gas. However, this solution has numerous problems both in the assumptions used and in comparing with the observations. For example, the presence of the magnetic field in the accreted material violates one of the basic assumptions of ADIOS that radiation efficiency of the disk is low [6]. The detection of linear polarization and the low electron densities estimated from the Faraday rotation measure rules out the large accretion rate of the standard ADAF model. This led to the development of convection dominated accretion flow models (CDAFs) [7,8], which favor lower accretion rates and shallower density profiles. The last set of models are models with substantial mass loss like advection-dominated inflow–outflow solutions (ADIOS) [9–11] or jet models [12,13].

While astrophysicists tried to explain the low luminosity of disks, another spectacular property was discovered. Accretion at a low rate is accompanied by an impressive phenomena, which was not expected in the standard models. X-ray binaries and AGN produce jets, well collimated flows of plasma, propagating on a large distance from the source. It has been found that power of the jets from AGN is often much greater than the bolometric luminosity of the disk. For example, the famous galaxy M87 is a characteristic example of an AGN with a very large kinetic luminosity $\sim 10^{44}$ erg/s [14,15] in comparison with the bolometric luminosity of the disk not exceeding 10^{42} erg/s [16]. The example of M87 is also not an isolated case.

It is necessary to keep in mind that estimation of the kinetic and bolometric luminosities of the jets is not a simple task for observers. Starting with the paper [17], jet power in radio galaxies and quasars were estimated using energetics and lifetimes of extended double radio sources. The ratio of kinetic-to-bolometric luminosity can be estimated also from radio and X-ray data. The works [18–24] argued that the radio and X-ray luminosities are likely to be related to the kinetic and bolometric luminosities, respectively. Exploration of these methods shows that, in a large fraction of AGNs, the jet kinetic luminosity exceeds the bolometric luminosity [25–31].

Other estimates follows from gamma-ray astronomy. The jet power in 191 quasars detected by the Fermi Large Area Telescope (LAT) in gamma rays, systematically exceeds the bolometric luminosity [32].

Indirect evidence of high kinetic luminosity of an outflow exceeding the bolometric luminosity is provided by observations of the Galactic Center in TeV gamma-rays [33]. To explain the observed diffuse flux of the VHE gamma-rays from the Galactic Center region, the production rate of protons accelerated up to 1 PeV should be $\sim 10^{38}$ erg/s. Assuming that the accelerator of protons is powered by the kinetic energy of the outflow (a wind or jets) from the SMBH in the Galactic Center (Sgr A*), even in the case of 100% conversion of the bulk kinetic energy to non thermal particles, the kinetic luminosity of the outflow would be two orders of magnitude larger than the bolometric luminosity of Sgr A*, which is estimated to be close to 10^{36} erg/s [34].

Sometimes, the jet power exceeds the Eddington limit. Observations of the very powerful and bright in gamma-rays AGN 3C 454.3 during the outbursts of this object show that the apparent luminosity in GeV gamma-rays could exceed 10^{50} erg/s [35–38]. The mass of the black hole in this AGN is estimated in the region $(0.5–4) \times 10^9$ M$\odot$. Thus, the Eddington luminosity is in the range of $(0.6–5) \times 10^{47}$ erg/s. Because of the Doppler boosting effect, the intrinsic gamma-ray luminosity of this source is much smaller, by several orders of magnitude, than the apparent luminosity. However, the estimates of the jet kinetic luminosity in any realistic scenario give a value exceeding the Eddington luminosity [39]. In general, the estimates of the bolometric and kinetic luminosities are model dependent [40]. Nevertheless, it is unlikely that the estimated values differ from the actual values more than an order of magnitude. Therefore, it is difficult to avoid a conclusion that some AGNs demonstrate extremely high kinetic luminosities of jets which are not only above the bolometric luminosity, but in some cases can exceed the Eddington luminosity of the central SMBH.

Compilation of data about X-ray binaries and AGN results in the following general picture. At high accretion rates, the accretion occurs in accordance with the SS model. The disk is bright and there is no (or there is only weak) evidence of jets [41]. At accretion rate $\dot{m} < 10^{-2} - 10^{-1}$, the disk becomes radiatively inefficient and accretion is accompanied by powerful jets. This picture is a challenge for astrophysicists. It is necessary to answer two key questions. The first one is why the accretion process is radiatively inefficient at low accretion rates and why jets with the kinetic luminosity exceeding the bolometric luminosity are produced in this regime of accretion and are not produced at high accretion rates (or are produced with low efficiency).

The problem of low radiative efficiency of disks is conventionally explained by accretion in ADAF mode. In the standard SS model, the accreting material is cooled efficiently. All the energy released through viscosity is radiated. The accreted gas is much cooler than the local virial temperature. The orbiting material has a vertical thickness much smaller than the radius. However, if the cooling is not able to keep up with the heating, then a part of the released energy will have to be advected with the accreted gas. The gas has a higher temperature, but lower luminosity than in the SS disks. The analysis of this kind of flow resulted in a model of geometrically thick but optically thin disks with suppressed bolometric luminosity of the disk [5]. In these disks, the height of the disk is of the order of the radius while the radial velocity of the accreted matter is higher than in the SS model. The density of matter in the disk appears much lower than in the SS disk. If the free–free processes dominate in the emissivity of the disk, this results in strong reduction of radiation from the disk. Only a small fraction of the released gravitational energy goes into radiation. ADAF is a very inefficient regime of accretion regarding transformation of the gravitational energy of the accreted material into radiation or energy of jets. The major part of the energy is advected into the black hole and goes into increasing of its mass.

However, ADAF does not solve the problem with energetics of jets. This model needs an additional source of energy to energize them. A rotating (Kerr) black hole can supply the jets by the energy of rotation. A Kerr black hole placed in an external magnetic field makes it co-rotate, producing an effect similar to the rotation of the pulsar magnetosphere. This results in an extraction of rotational energy and angular momentum from the black hole which is carried out by an electron–positron wind. This is the so called Blandford and Znajek effect [42]. Estimates made by Blandford and Znajek have shown that the energy of the outflow is small compared with the radiation from conventional SS disks. However, this is valid only if we consider moderate rotation and conventional values of the magnetic field according to the SS model. A new model of a Magnetically Arrested Disk (MAD) has been introduced in the work [43]. This model is based on the assumption that the interstellar magnetic field $\sim \mu$G is dragged to the center by the converging accreting plasma like it was shown in [44] to the level where the magnetic field disrupts the disk. The value of the magnetic field in this case essentially exceeds the conventional magnetic field in SS disks. Numerical simulations in fully 3D geometry show that the energy flux in the outflow from the black hole can achieve a value of the order $3\dot{M}c^2$, provided that the black hole rotates close to the maximal possible angular momentum [45].

Thus, one of the possible models able to explain the energetics of accretion and outflow from AGN needs the following set of assumptions. It is necessary to assume that accretion occurs in the ADAF regime, that the main source of energy of jets is the energy of rotation of the black hole rotating close to the maximal limit and finally that accretion results in the formation of the Magnetically Arrested Disk (MAD) in its inner part. In this review, an alternative approach to the problem of disk accretion is discussed. In the alternative approach, the only available energy of AGNs including jets is the gravitational energy released in the accretion.

In conventional theories of disk accretion, turbulent viscous stresses provide loss of the angular momentum of the accreted matter. However, the angular momentum and rotational energy can be lost due to another mechanism. Wind from a rotating magnetized object can carry away its angular momentum and rotational energy. Starting with the classical work of Parker [46], it was clear that all main sequence stars, including the Sun, eject matter in the form of winds. Schatzman [47] proposed that as the winds contain a frozen-in magnetic field that goes back to the star, the angular momentum loss is leveraged many times. The importance of stellar winds in extracting angular momentum from main sequence stars was recognized by Mestel [48,49]. Pulsars also lose their energy of rotation due to a similar mechanism. The wind of electron–positron plasma produced in the pulsar magnetosphere carries out all the rotational energy of the pulsar. It is important to pay attention that the loss of the angular momentum is accompanied by a corresponding loss of the energy of rotation without essential heating of the stars and pulsars. In the case of accretion disks, we are sure that they produce outflows in the form of magnetized winds and jets. It is natural to assume that, in addition to the loss of the angular momentum due to the viscosity, the matter in the disk loses its angular momentum due to the magnetized wind. This idea was first formulated by Blandford and Payne [50]. Pelletier and Pudritz [51] pointed out that the loss of the angular momentum due to the wind can dominate over the loss of the angular momentum due to the viscosity under rather conventional conditions. Later, this idea has been explored in many works of the Grenoble group [52–55], which called this type of flow around black holes Magnetized Accretion–Ejection structures (MAES). Several other authors over the years explored similar approaches in different physical contexts [56–60]. In the last works of the Grenoble group [61,62], the radiation from the Jet Emitting Disks (JED) is discussed in the context of X-ray binaries. Starting with our first work [63] devoted to the same problem, we focused on the fact that these disks can be the key to the solution of the problem of high ratio of the kinetic luminosity of the jets over the bolometric luminosity of the disks. Actually, Ferreira and Pelletier [53,55] noted much earlier that, under conventional conditions, the jet can carry out almost all the angular momentum and energy from the accretion disk. Due to this, the radiation from the disk appears suppressed and we arrive to another model of a radiatively inefficient disk. However, unlike ADAF, in this case, the system black hole and the disk is a very efficient system. It transforms almost all the energy released at the accretion in to the kinetic energy of jets. In this case, we have $L_{kin}/L_{Edd} \approx \dot{m}$. No additional sources of energy are necessary.

The review is based mainly on the results obtained in the National Research Nuclear university (MEPHI), although a lot of results have been obtained earlier by the Grenoble Group (see Refs. [52–55]). We acknowledge this in the appropriate places.

2. Magnetic Field of the Disk and Wind

Assumed Structure of the Magnetic Field in the Disk

In the case of ideal plasma (viscosity and electric resistivity are neglected), the magnetic field is determined by advection of the field lines by the accreted plasma to the center. This process was firstly considered in [64]. However, the matter in the disk is evidently not ideal. Turbulence produces rather strong turbulent viscosity and electric resistivity. Therefore, the processes of diffusion of matter across the magnetic field lines also take place in the disk. Moreover, the process of diffusion appears so strong that prevents accumulation of the magnetic flux at the central part of the accretion disk what makes the operation of the Blandford and Znajek effect problematic [65,66]. Some ideas about how to avoid this problem are discussed in [67,68].

The majority of the works devoted to the process of disk accretion consider the disk as a thin layer of plasma penetrated by the field lines of one polarity. The value of the magnetic field and magnetic field pressure is defined by two processes: by the process of advection of the field lines to the disk center and by the process of diffusion of the field lines in the opposite direction. Equilibrium of these processes provides a steady state structure of the magnetic field. This approach was used by the Grenoble group starting with work [52]. However, there is a general understanding that the process of disk dynamo plays a major role in the production of the magnetic field inside the disk [69–73]. In addition, a quite specific dynamo mechanism can operate in the disks [74,75], although they remain debatable [76].

The dynamo mechanism results in formation of small scale loops of the magnetic field which basically defines the viscosity of the matter of the disk. Numerical simulations show that the loops emerge on the surface of the magnetic field in accordance with the predictions made in [69], and expand into surrounding space at the differential rotation of the field line foot points in the disk [77,78]. Pressure of plasma and centrifugal forces leads to the opening of the field lines and to the formation of magnetized wind along open field lines. The schematic structure of the magnetic field lines inside and outside the disk is shown in Figure 1. It is reasonable to consider the disk with the wind in the quasi steady state. In this state, the average value $< B^2 >$ does not vary with time, while the time derivative $\frac{\partial \mathbf{B}}{\partial t} \neq 0$. The average pressure of the magnetic field does not change with time, but the polarity of $\mathbf{B}$ varies with time, so that the average over time value of $\mathbf{B}$ is equal to zero. This is valid for the magnetic field inside the disk and for the magnetic field in the wind. The field lines of the wind of the opposite polarities are separated by current sheets. We assume that, like in the case of the Sun, the process of field annihilation due to field line reconnection in the current sheets takes place inside the disk and corona, while, in the wind, the activity of the current sheets in the wind is suppressed. Therefore, the dynamics of plasma in the current sheets can be considered in the ideal MHD approximation. All the dissipative processes connected with the final electric conductivity and viscosity are neglected. The current sheets in this approximation are MHD discontinuities with zero thickness. Observations of a fine structure of the magnetic field of the solar wind support this assumption. The small scale magnetic field of the coronal streamers produces multiple current sheets in the interplanetary space that are observed even at the Earth orbit [79,80].

In this picture, the magnetic field in the wind changes with time. Although the magnetic pressure can be constant in the steady state flow, the magnetic field permanently changes polarity because a new magnetic field emerges from the interior of the disk and advection replaces the field lines of one polarity with field lines of the opposite polarity. Nevertheless, the problem of the wind outflow with such a magnetic field can be reduced to the problem of the wind outflow in the unipolar magnetic field like it was done in the work [50].

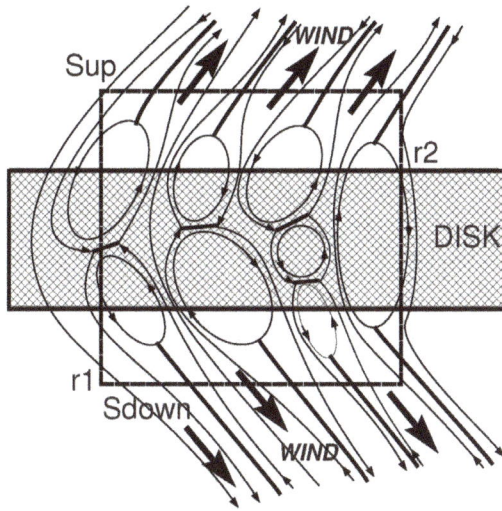

Figure 1. The structure of the magnetic field in the accretion disc and in the out flowing wind. The disc is shadowed. The magnetic field lines in the disc are distributed chaotically. At the base of the wind from the disc, all the magnetic field lines are opened. Their polarities are random. Therefore, the total magnetic flux leaving one side of the disc equals zero. The box drawn in dashed thick lines is the region of integration of conservation laws connecting the properties of the disc and the wind.

3. Basic Equations of the Accretion and Outflow

The flow in the disk has huge hydrodynamic and magnetic Reynolds numbers. This means that the viscosity and electrical conductivity are determined by turbulence. The collisional viscosity and electrical conductivity can be neglected. Therefore, we will start with the ideal plasma approximation. The condition of ideality has the form

$$\mathbf{E} + \frac{1}{c}\mathbf{v} \times \mathbf{B} = 0, \tag{1}$$

where $\mathbf{E}$ is the electric field, $\mathbf{B}$—magnetic field and $\mathbf{v}$ is the velocity of the plasma.

The dynamics of plasma is defined by the momentum conservation equation having the following form in the tensor representation [81]

$$\frac{\partial \rho v_i}{\partial t} + \frac{\partial \sigma_{ik}}{\partial x_k} = -\rho GM \frac{R_i}{R^3}, \tag{2}$$

where

$$\sigma_{ik} = \rho v_i v_k + p \delta_{ik} - \frac{1}{4\pi}\left(B_i B_k - \frac{1}{2}B^2 \delta_{ik}\right). \tag{3}$$

The second equation expresses the energy conservation in the form

$$\frac{\partial W}{\partial t} + \frac{\partial q_k}{\partial x_k} = 0, \tag{4}$$

$W = \rho e + \frac{B^2}{8\pi}$ is the sum of the thermal and magnetic field energy densities, and e is the thermal energy per particle. We consider here only nonrelativistic flows. The term related to the electric field is omitted in W. The flux of the energy density equals

$$q_i = \rho v_i \left(w + \frac{v^2}{2} - \frac{GM}{R} \right) + \frac{c}{4\pi} [E \times B]_i + Q_i,$$ (5)

where Q_i is the density of the energy flux of radiation.

In Cartesian coordinates, the angular momentum vector is introduced as $l_i = \varepsilon_{imp} x_m \rho v_p$ [82], where ε_{imp} is the unit antisymmetric tensor. Application of this transformation to Equation (2) gives

$$\frac{\partial l_i}{\partial t} + \frac{\partial m_{ik}}{\partial x_k} = 0,$$ (6)

where $m_{ik} = \varepsilon_{imp} x_m \sigma_{pk}$. Projection of this equation on the z axis gives

$$\frac{\partial l_z}{\partial t} + \frac{\partial m_{zk}}{\partial x_k} = 0,$$ (7)

where $l_z = \rho v_\varphi r$ and

$$m_{zk} = \rho v_k r v_\varphi - \frac{1}{4\pi} r B_k B_\varphi.$$ (8)

The conservation equations for the matter and the magnetic fluxes are

$$\frac{\partial \rho}{\partial t} + \frac{\partial \rho v_k}{\partial x_k} = 0,$$ (9)

and

$$\frac{\partial B_k}{\partial x_k} = 0.$$ (10)

These equations are supplemented by the equation of induction

$$\frac{\partial \mathbf{B}}{\partial t} + \text{curl} \mathbf{E} = 0.$$ (11)

4. Conservation Laws for the Disk

In this section, we obtain vertically integrated equations conventionally explored in the theory of accretion disks. We study a steady state axisymmetric accretion and outflow. Because of turbulent motion inside the disk, all variables vary in time on small time scales. Below, we consider equations for ensemble-averaged variables. The ensamble is the large number of identical accretion disks. In the steady state flow, these variables are constant in time. Averaging is expressed by the brackets $< ... >$. Application of the averaging operation to the terms in the equations with time derivatives makes them equal to zero.

Let us consider a control volume in the form of ring with rectangular cross section in the poloidal plane as it is shown in Figure 1. The cross section is shown with a thick dashed line. The control volume includes a fragment of the disk and corona. The upper and lower boundaries of the volume are located at the base of the wind above the disc. All the magnetic field lines of the wind are rooted here.

Integration of the equations over the control volume gives us the equations in integral form. Conservation of the angular momentum, energy and mass are as follows:

$$\oint_S < m_{zk} > dS_k = 0,$$ (12)

$$\oint_S < q_k > dS_k = 0, \tag{13}$$

$$\oint_S < \rho v_k > dS_k = 0. \tag{14}$$

Integration here is performed along a closed surface surrounding the volume. The surface consists of the sides S_1 at radius r_1, side S_2 at radius r_2 and upper and down sides S_{up} and S_d. Integration gives

$$
\begin{aligned}
&- \int_{-h}^{h} r^2 < \rho v_\varphi v_r > dz \Big|_{r1} + \int_{-h}^{h} r^2 < \rho v_\varphi v_r > dz \Big|_{r2} \\
&+ \frac{1}{4\pi} \int_{-h/2}^{h} r^2 < B_r B_\varphi > dz \Big|_{r1} - \frac{1}{4\pi} \int r^2 < B_r B_\varphi > dz \Big|_{r2} \\
&+ 2 \int_{r1}^{r2} \left(r < \rho v_\varphi v_z > - \frac{1}{4\pi} r < B_\varphi B_z > \right)_{Sup} r\, dr = 0.
\end{aligned}
\tag{15}
$$

Integration across the disc is performed in the interval on z from $-h$ to h correspondingly at the radiuses r1 and r2. Integrations along S_{up} and S_d are equal to each other because the vector $d\mathbf{S}$ and the component of the velocity v_z change sign simultaneously. Therefore, we simply double the integration along the surface S_{up}. This equation can be rewritten in the form

$$
\begin{aligned}
&- \int_{-h/2}^{h} r^2 < \rho v_r > < v_\varphi > dz \Big|_{r1} + \int_{-h/2}^{h} r^2 < \rho v_r > < v_\varphi > dz \Big|_{r2} \\
&- \int_{-h}^{h} r^2 (< \delta \rho v_r \delta v_\varphi > - \frac{1}{4\pi} < B_r B_\varphi >) dz \Big|_{r1} + \int r^2 (< \delta \rho v_r \delta v_\varphi > - \frac{1}{4\pi} < B_r B_\varphi >) dz \Big|_{r2} \\
&+ 2 \int_{r1}^{r2} \left(r < \rho v_\varphi v_z > - \frac{1}{4\pi} r < B_\varphi B_z > \right)_{Sup} r\, dr = 0,
\end{aligned}
\tag{16}
$$

where symbol δ means deviation of the value from average. The term

$$t_{\varphi r} = -(< \delta(\rho v_r) \delta v_\varphi > - \frac{1}{4\pi} < B_r B_\varphi >) \tag{17}$$

is the efficient viscosity of the matter caused by the turbulent motion and magnetic fields inside the disk.

Equation (16) is reduced to the differential form as follows:

$$
\begin{aligned}
&\frac{\partial}{r \partial r} \left(r^2 (\int_{-h}^{h} V_k < \rho v_r > - t_{r\varphi}) dz \right) |_{disc} \\
&+ 2r \left(< \rho v_\varphi v_z > - \frac{1}{4\pi} < B_\varphi B_z > \right) |_{wind} = 0.
\end{aligned}
\tag{18}
$$

The subscripts $|_{disc}$ and $|_{winds}$ denote the variables describing the disc and the wind at the base (at the surface S_{up}). According to this equation, the angular momentum of the disc is carried out by the out flowing plasma and by the magnetic stresses in the outflow. Introducing the accretion rate in the disk as

$$\dot{M} = -2\pi r \int_{-h}^{h} < \rho v_r > dz, \tag{19}$$

we finally obtain for conservation of the angular momentum of the disk the following equation

$$\frac{\partial}{r\partial r}\left(rV_k\dot{M} + 4\pi r^2 t_{r\varphi}h\right)\big|_{disc}$$

$$-4\pi r\left(V_k < \rho v_z > -\frac{1}{4\pi} < B_\varphi B_z >\right)\big|_{wind} = 0. \tag{20}$$

Everywhere below, we accept that the azimuthal velocity of the plasma in the disc $< v_\varphi >= V_k$, where $V_k = \sqrt{GM/r}$ is the Kepler velocity and the disk is geometrically thin with aspect ratio $h/r << 1$. Similar manipulations with the energy conservation equation give the following equation

$$\frac{\partial}{r\partial r}\int_{-h/2}^{h}r\left(<\rho v_r w > +\frac{< v_r B^2 >}{4\pi} + \frac{<\rho v_r v^2 >}{2} - <\rho v_r > \frac{GM}{r} - \frac{< B_r(\mathbf{vB}) >}{4\pi} + < Q_r >\right)dz|_{disc}$$

$$+2\left(<\rho v_z\left(\frac{w+v^2}{2} - \frac{GM}{R}\right) > +\frac{c}{4\pi} < [E\times B]_z > +Q_z\right)|_{wind} = 0. \tag{21}$$

In this equation, the term $< \rho v_r \frac{v^2}{2} >$ can be expanded as

$$<\rho v_r v^2 >=<\rho v_r V_k^2 > + <\rho v_r 2V_k\delta v_\varphi > + <\rho v_r\delta v_\varphi^2 > + <\rho v_r^3 > + <\rho v_r v_z^2 >, \tag{22}$$

taking into account that $V_k \gg \delta v_\varphi$, $V_k \gg \delta v_z$ and $V_k \gg \delta v_r$, we remain with

$$<\rho v_r v^2 >\approx<\rho v_r > V_k^2 + 2V_k <\rho v_r\delta v_\varphi > . \tag{23}$$

Starting at Ref. [3], it is assumed that the energy density of the chaotic magnetic field in the disk is of the order of density of the turbulent energy, $\rho v_t^2/2 \sim B^2/4\pi$. Both of them are much less than $\rho V_k^2/2$. These terms are also omitted in Equation (21). The term $< B_r(\mathbf{vB}) >$ is expanded into

$$< B_r(\mathbf{vB}) >=< B_r^2 v_r > + < B_r B_\varphi > V_k + < v_z B_z B_r > . \tag{24}$$

The term $< B_r B_\varphi > V_k$ is much larger than the remaining. We neglect them. For the same reasons,

$$<\rho v_z(\frac{w+v^2}{2}) >\approx<\rho v_z > \frac{V_k^2}{2}. \tag{25}$$

The z component of the Poynting flux

$$< [E\times B]_z >=< E_r B_\varphi > - < E_\varphi B_r >, \tag{26}$$

where $E_r = \frac{1}{c}(v_z B_\varphi - v_\varphi B_z)$ and $E_\varphi = \frac{1}{c}(v_r B_z - v_z B_r)$. Keeping in Equation (26) the largest term, we obtain that

$$< [E\times B]_z >\approx -\frac{V_k}{c} < B_z B_\varphi > . \tag{27}$$

After substitution of all these equations into Equation (21) taking into account Equations (30) and (19), we obtain the equation expressing energy conservation in the disk

$$\frac{1}{r}\frac{\partial}{\partial r}\left(\dot{M}(\frac{V_k^2}{2} - \frac{GM}{r}) + 4\pi rV_k t_{r\varphi}h\right)_{disc}$$

$$-4\pi\left(<\rho v_z > \left(\frac{V_k^2}{2} - \frac{GM}{R}\right) - \frac{V_k}{4\pi} < B_\varphi B_z > +Q_z\right)_{wind} = 0. \tag{28}$$

The last equation is the matter conservation

$$\frac{\partial}{r\partial r}\left(r\int_{-h/2}^{h}\rho v_r\,dz\right)_{disc} + (2\rho v_z)_{wind} = 0. \tag{29}$$

Exploring definition (19), the last equation takes the form

$$\frac{\partial \dot{M}}{\partial r} - 4\pi r\rho v_z = 0. \tag{30}$$

Subtraction from Equation (28) of Equation (20) multiplied with the Keplerian angular velocity $\Omega_k = V_k/r$ results in the energy flux of radiation from one side of the surface of the disk being

$$Q = Q_z = t_{r\varphi}rh\frac{\partial \Omega_k}{\partial r} \tag{31}$$

like in the classical accretion disk of Shakura and Sunyaev [3]. Below, instead of Q_z, we will use Q.

5. The Problem of the Wind Outflow

5.1. Invariance Principle

Below, we consider only axisymmetric flows when all the averaged variables do not depend on the azimuthal angle. In this case, the problem of the wind outflow can be essentially simplified. However, the problem of the plasma outflow in the poloidal magnetic field which changes polarity remains too complicated. This problem can be simplified by reduction to the same problem in the unipolar magnetic field.

It follows from equations of ideal MHD that the dynamics of plasma in ideal MHD is invariant in relation to a reversal of the direction of the magnetic field lines in an arbitrary flux tube. This property of ideal MHD flows was used for the solution of the problem of plasma outflow from pulsars [83]. This property was called the invariance principle.

The plasma flow in the nonrelativistic limit is described by the set of ideal MHD Equations (1), (2), (4), (7) and (9)–(11).

Let us assume that we have some solution which is described by the functions $\mathbf{B}(r,t)$, $\rho(r,t)$, $\mathbf{V}(r,t)$ and $P(r,t)$. It is easy to show that changing of polarity of the magnetic field (and corresponding electric field) in an arbitrary flux tube does not change dynamics of plasma.

Let us introduce a scalar function $\eta(r,t)$ with the property that $\eta = 1$ everywhere except inside the chosen flux tube where $\eta = -1$. This function satisfies the following two conditions:

$$\mathbf{B} \cdot \nabla \eta = 0, \tag{32}$$

and

$$\frac{\partial \eta}{\partial t} + \mathbf{V} \cdot \nabla \eta = 0. \tag{33}$$

The second equation is the consequence of Equations (1), (11) and (32). The value of η is advected together with the plasma.

Then, the solution $\eta\mathbf{B}(r,t)$, $\rho(r,t)$, $\mathbf{V}(r,t)$ and $P(r,t)$ also satisfies the system of Equations (1), (2), (4), (7) and (9)–(11). Indeed, the tensor of the momentum flux density (3) is bi quadratic in relation to the magnetic field. It does not change because $\eta^2 = 1$. This means that the forces affecting the plasma do not change with this transformation.

Let us consider how this principle operates in the most simplified model of the wind outflow from the surface of a star. The distribution of the normal component of the poloidal magnetic field is axisymmetric. In that case, we get the model with the axisymmetric wind. According to the invariance principle, the change of the direction of magnetic field lines in some flux tube does not affect the dynamics of the plasma. Let us assume that we obtain a solution for the axisymmetric rotator, as shown schematically in Figure 2a. Then, a reversal of the sign of some magnetic field lines in an arbitrary poloidal flux tube gives us a solution which is not axisymmetric and non stationary, as shown in Figure 2b. This is a solution for the plasma outflow from a rotator with axisymmetric B^2 at the surface but with a magnetic spot of the opposite polarity on the upper hemisphere. Figure 2b shows the cross-section of such a magnetic field by the poloidal plane. The stream lines are the same as for the axisymmetric case. However, the poloidal magnetic field changes sign in the magnetic spot corresponding to the flux tube of the opposite polarity. The path of the field line in this flux tube in 3D is shown by a dashed line. These spots propagate in the poloidal plane with the velocity of the plasma and hence the pattern is non stationary. It is clear that the number of such magnetic spots and their position at the base surface can be arbitrary.

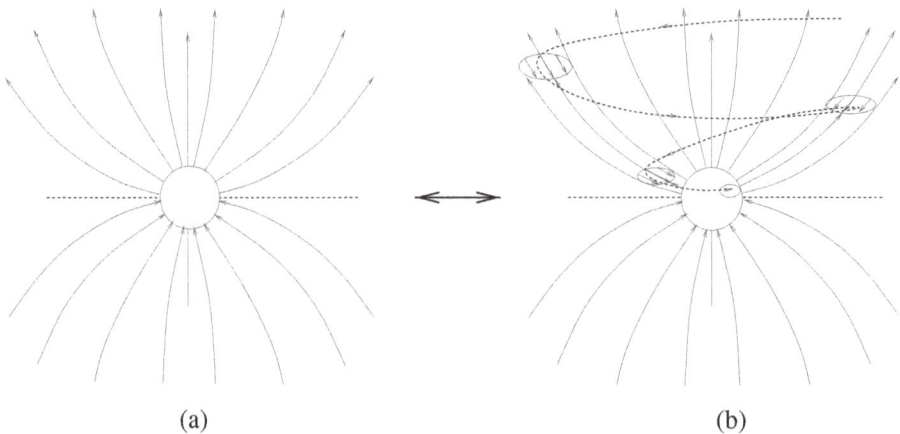

(a)   (b)

Figure 2. Plasma flow from an axisymmetric rotator with an initially split-monopole magnetic field, as in (**a**). Reversing the direction of the poloidal magnetic field in an arbitrary flux tube does not change the dynamics of the problem while we obtain the configuration shown in (**b**) which describes a non stationary and nonaxisymmetric plasma flow from a rotator with a magnetic spot of opposite polarity on the base surface. The distribution of such spots can be arbitrary.

Now, let us return to the accretion disk. In our model, the accretion disk can be considered as a layer at the equatorial plane with thickness $2h$. All magnetic field lines are open and chaotically change polarity in the wind. The total magnetic flux penetrating in to the disk equals zero. This means that the disk brings to the black hole zero total magnetic flux. The formation of such a wind has been investigated numerically in [78]. This naturally solves the old problem which was pointed out in the first works on the accretion of the magnetized plasma [44,64]. The magnetic flux of one polarity can be accumulated at the center preventing the accretion. There is no chance to annihilate this flux due to magnetic field reconnection. In our case, the situation changes dramatically. The magnetic flux can fully annihilate near the black hole, where the geometrical scales become much smaller in comparison with scales in the disk because the plasma is filled by a large amount of current sheets separating flux tubes of opposite polarities. Reconnection of the field lines near the black hole horizon can be accompanied by the sporadic ejection of mass of plasma.

The invariance principle allows us to essentially simplify the solution of the problem of the wind outflow from the disk. According to this principle, we can replace the direction of all the field lines making them unidirectional in every hemisphere. The dynamics of the wind does not change. After that, we arrive to the wind outflow in the unidirectional magnetic field like in the pioneering work by

Blandford and Payne [50]. This procedure is shown in Figure 3 and corresponds to the transition from the upper to the lower panel.

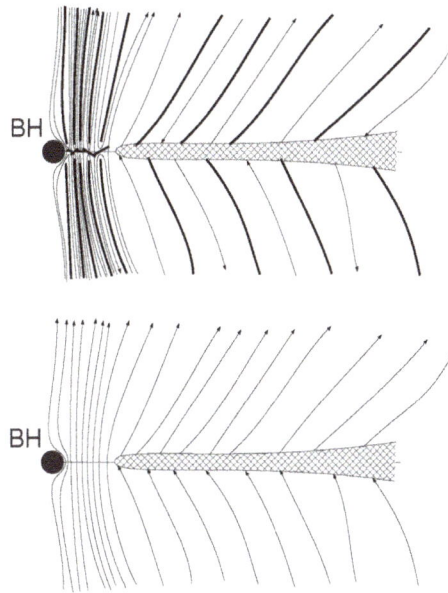

Figure 3. The **upper** panel shows the structure of the magnetic field outside the disk. The magnetic field lines (thin lines) leaving the disk from both sides chaotically change polarity. The flux tubes of opposite polarities are separated by the current sheets (thick lines). This system of field lines is compressed at the black hole horizon where an efficient reconnection takes place annihilating the magnetic field. The replacement of the magnetic field lines leaving the disk by the similar lines of one polarity does not change the dynamics of the wind and satisfy Maxwell's equations. This allows us to use the model shown in the **lower** panel where the magnetic field is unipolar.

5.2. The Role of the Azimuthal Electric Field in the Wind

In the limit of axisymmetric flow in the unipolar magnetic field, the azimuthal component of Equation (11) and the frozen-in condition (1) for the same component give a couple of equations

$$\frac{\partial}{\partial r}(rE_\varphi) = 0,\tag{34}$$

and

$$E_\varphi + \frac{1}{c}(v_z B_r - v_r B_z) = 0.\tag{35}$$

The solution of Equation (34) results in

$$E_\varphi = \frac{A}{r},\tag{36}$$

where A is some constant. This solution diverges at $r \to 0$. It was pointed out in [84] that E_φ is not equal to zero at the accretion of an ideal plasma onto a gravitating center. Indeed, as it follows from Equation (35), E_φ at the base of the wind is equal to $\frac{1}{c}v_r B_z$ provided that $v_z \to 0$ at the center of the disc. At accretion $v_r \neq 0$, there is thus $E_\varphi \neq 0$ as well. Thus, in the region of the accretion flow, E_φ can not be neglected because it is connected directly with the radial velocity of the plasma in the disc.

As it follows from Equation (35) in this case, the velocity has a poloidal component $v_\perp$ orthogonal to the poloidal magnetic field line. The plasma in the wind is advected to the center together with the matter in the disk. Nevertheless, this component of the electric field can be neglected when we consider the dynamics of the wind under the condition $v_\perp \ll v$, where v is the full velocity of the plasma. If we take into account that $v \sim V_k$ and that $v_\perp = c \frac{E_\varphi}{B_p}$ the toroidal electric field can be neglected under the condition

$$\frac{B_{p0}r_0^2}{B_p r^2} \frac{v_r(0)r}{V_k r_0} \ll 1. \tag{37}$$

The value $B_p r^2$ roughly equals the full flux of the magnetic field through the surface limited by the field line. It can not change strongly along the field line, $\frac{B_{p0}r_0^2}{B_p r^2} \sim 1$. Therefore, inequality (37) is inevitably violated at the distance $r \gg r_0 \frac{V_k}{v_r(0)}$. However, we do not care about validity of Equation (37) in all of space.

In order to take into account the impact of the wind on the process of accretion, it is necessary to calculate the product $B_z B_\varphi$ at the base of the flow. The flow of the wind does not depend on the conditions down stream, the so called fast mode surface where the velocity of plasma equals the fast mode magnetosonic velocity. To be more exact, this should be called the fast mode separatrix surface [85]. However, the difference between them is not important for us here. Thus, the product $B_z B_\varphi$ at the base of the wind is determined by the flow in the zone limited by the fast mode surface. Therefore, the toroidal electric field in the wind can be neglected if

$$\frac{v_r(0)R_F}{V_k r_0} \ll 1, \tag{38}$$

where R_F is the radius of the fast mode surface.

This condition can be understood from another point of view. The equilibrium state in the zone limited by the fast mode surface is formed during the time an MHD signal spends for travel from the base to the fast mode surface $\sim R_F/V_F$, where V_F is the fast mode velocity that is close to the Alfven velocity at the Alfven surface. At the Alfven surface, the velocity of plasma is $\sim V_k$. The impact of the advection of the matter in the disk on the dynamics of the wind can be neglected if the root of the magnetic field line is displaced over the distance $\Delta r \ll r_0$ during this time. In this case, we arrive at the same Equation (38).

5.3. Along Field Line MHD Equations of the Wind

If the azimuthal electric field can be neglected, $E_\varphi = 0$, then, according to Equation (35), the poloidal velocity is directed along the poloidal magnetic field. In this case, we have

$$l_p = \rho r v_p \left(v_\varphi - \frac{B_p}{4\pi\rho v_p} B_\varphi \right), \tag{39}$$

for the angular momentum flux density along a poloidal filed line and

$$q_p = \rho v_p \left(\frac{v^2}{2} - \frac{GM}{R} - \Omega r \frac{B_p}{4\pi\rho v_p} B_\varphi \right) \tag{40}$$

for the energy density flux along a poloidal field line. If we take into account that the fluxes of the angular momentum $l_p dS$, energy $q_p dS$, matter ρv_p and magnetic field flux $B_p dS$ are conserved as it is demonstrated in Figure 4, it can be obtained that the following two integrals of motion take place along the field lines

$$r v_\varphi - \frac{r B_p B_\varphi}{4\pi\rho v_p} = L, \tag{41}$$

and

$$\frac{v^2}{2} - \frac{GM}{R} - \Omega \frac{rB_pB_\varphi}{4\pi\rho v_p} = H. \tag{42}$$

The first equation from this couple is the conservation of the angular momentum per particle L and the second one is the conservation of the energy per particle H along a field line.

An additional consequence from $E_\varphi = 0$ is that the poloidal electric field E_p is perpendicular to the poloidal magnetic field B_p. In this case, Equation (11) gives that the product $E_p dl$ is conserved along a field line, where dl is the distance between two neighbor field lines. If we take into account that $dS = 2\pi r dl$, we obtain that $E_r = -r\Omega B_z/c$ and $E_z = r\Omega B_r/c$ or $E_p = \frac{r\Omega}{c} B_p$. The frozen-in condition for the poloidal component of the electric field gives in this case that

$$r\Omega B_p + v_p B_\varphi = v_\varphi B_p. \tag{43}$$

Combining Equation (43) with Equation (41) results in

$$rv_\varphi = \frac{L - r^2\Omega \frac{B_p^2}{4\pi\rho v_p^2}}{1 - \frac{B_p^2}{4\pi\rho v_p^2}}. \tag{44}$$

The denominator of this expression goes to zero at the Alfvenic point where $v_p = B_p/\sqrt{4\pi\rho}$. The nominator of this expression must equal to 0 in this point to provide regularity of v_φ. From this condition, we obtain that the momentum per particle equals $L = \Omega r_A^2$, where r_A is the cylindrical radius at the Alfvenic point where the plasma velocity equals the local Afvenic velocity $V_A = \sqrt{B_p/4\pi\rho}$.

Taking into account that $B_p/v_p = B_z/v_z$ in the wind, it is easy to determine that the product

$$rB_zB_\varphi = -4\pi\rho v_z\Omega_k r_0^2(\lambda - 1), \tag{45}$$

at the base of the wind located at the radius r_0. Following [54], $\lambda = r_A^2/r_0^2$.

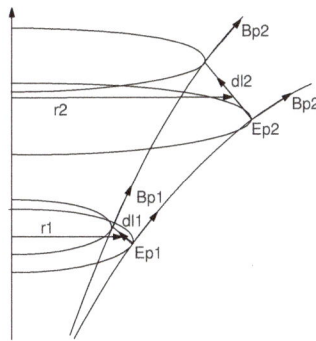

Figure 4. Fluxes of the magnetic field, matter, energy and angular momentum between any two close field lines with the cross section $dS = 2\pi r dl$ are conserved. The condition curl $\mathbf{E} = 0$ gives that the product $E_p dl$, where dl is the distance between these field lines is also conserved.

6. Disk-Wind Connection at the "Cold" Accretion

Taking into account the equation for mass conservation (30), the equations for angular momentum conservation can be rewritten in the form

$$\dot{M}\frac{\Omega_k}{2} + \frac{1}{r}\frac{\partial}{\partial r}\left(4\pi r^2 t_{r\varphi}h\right)|_{disc}$$

$$+r < B_\varphi B_z > |_{wind} = 0. \tag{46}$$

The energy conservation is fulfilled automatically provided that Equation (31) takes place. It follows from Equation (46) that the impact of the wind on the dynamics of the wind is reduced to the value of $B_\varphi B_{z,wind}$. We omit here brackets <> because in the steady state axisymmetric wind $B_\varphi B_{z,wind}$ is constant.

Taking into account Equation (45), Equation (46) can be rewritten in the form

$$\dot{M}\frac{\Omega_k}{2} + \frac{1}{r}\frac{\partial}{\partial r}\left(4\pi r^2 t_{r\varphi}h\right)|_{disc}$$

$$-4\pi\rho v_z \Omega_k r_0^2(\lambda - 1) = 0. \tag{47}$$

This equation clearly demonstrates that every particle of the wind carries out an amount of the angular momentum per particle equal to $\Omega_k r_0^2(\lambda - 1)$. In the case of purely hydrodynamical wind ($\mathbf{B} = 0$), this value equals 0 because $\lambda = 1$. The wind does not carry out any angular momentum from the disk. The magnetic field dramatically changes the situation. Thanks to it, $\lambda > 1$ and every particle of the wind carries out not only its own angular momentum, but also some fraction of the angular momentum of the particles remaining in the disk. A natural question arises. At what conditions does the wind carry out more angular momentum than it is transported outward by the viscous stresses?

As it was pointed out by [51], the momentum loss due to the wind will dominate the losses caused by the viscous stresses provided that

$$4\pi r t_{r\varphi}h \ll r^2 < B_\varphi B_z > |_{wind}. \tag{48}$$

In the opposite case, we have the standard SS version of the disk accretion [3].

The physical sense of inequality (48) becomes clear if we note that according to the assumptions of [3] and recent numerical simulations of the magnetic field generation [72]

$$-t_{r\varphi} \sim \frac{B_{disk}^2}{4\pi}, \tag{49}$$

where the magnetic field B_{disk} is taken inside the disk. We distinguish the magnetic field inside the disk from the magnetic field at the base of the wind. In the works of the Grenoble group [53,54] and other authors [58–60], these values are similar because the magnetic field vertically crosses the disk. In reality, the magnetic field inside the disk can essentially exceed the field at the base of the wind [72]. The regime of "cold" disk accretion occurs when

$$\frac{\theta h}{r} \ll 1, \tag{50}$$

where

$$\theta = \frac{4\pi t_{r\varphi}}{< B_\varphi B_z > |_{wind}} \sim \frac{B_{disk}^2}{B_{wind}^2}. \tag{51}$$

For the geometrically thin disks with $h/r \ll 1$, "cold" disk accretion certainly has room for existence provided that θ is not extremely large.

Dissipative terms defined by $t_{r\varphi}$ can be neglected in Equations (28) and (46) in the regime of cold accretion. The equations for the angular momentum and energy conservation take the form

$$\dot{M}\frac{\Omega_k}{2} + r < B_\varphi B_z > |_{wind} = 0, \tag{52}$$

$$\frac{1}{2}\frac{\partial V_k^2 \dot{M}}{\partial r} + 4\pi r \rho v_z H|_{wind} = 0. \tag{53}$$

The luminosity of the disk is determined by Equation (31).

Our approach to the problem of disk accretion due to the wind methodologically differs from the approach used by the Grenoble group and other researchers. In the works of the Grenoble Group, the accretion and outflow are considered as one self-consistent process using a self similarity prescription. It was named Magnetized Accretion-Ejection structures (MAES) [52]. The advantage of this approach is that, simultaneously with the solution of the problem of accretion, the structure of the disk is determined. However, this is simultaneously a disadvantage because it is necessary to strongly simplify the processes in the disk and neglect for example dynamo processes and remain in the frameworks of self similarity prescriptions. Keeping in mind practically the same physical picture, we follow another methodology. We separate the problem of accretion from the problem of internal structure of the disk. The arguments in favor of this approach are the following.

The rate of the angular momentum loss by the disk due to the wind is determined by the product $B_z B_\varphi$ at the surface of the disk or, equivalently, at the base of the wind. It is known from the theory of MHD winds that, in order to solve the problem of the magnetized wind outflow from the surface of the rotating object (in our case the surface of the disk), it is necessary to specify a certain number of boundary conditions at the base of the wind [83]. At the base of the wind, the magnetic field pressure dominates the gas pressure, $v_s \ll V_A$, where v_s and V_A are the sound and Alfvenic velocities. Therefore, it is natural to assume that $V_A > v_z > v_s$. This means that the mass flux density should be specified as the boundary condition [83]. In addition, we have to specify as boundary conditions the temperature of the plasma, the pressure, the normal component of the magnetic field B_z and the rotational velocity of the object—five parameters in total. It is remarkable that the toroidal component of the magnetic field B_φ is determined from the solution of the problem of the wind outflow. This means that the angular momentum loss of the disk can be determined by solving the problem of the wind outflow only at the specified mass flux density distribution in the wind, temperature of the plasma, pressure, and B_z at the base of the wind.

To determine these parameters, it is necessary to solve the problem of the internal structure of the disk. At present, no convincing solution of this problem is obtained. Therefore, it is reasonable to consider the disk as a layer which provides us, due to some processes, five parameters at the surface, which are the functions of the coordinate at the disk surface. In the case of cold accretion, these parameters reduce to two unknown parameters because temperature and pressure at the base of the wind can be taken equal to 0 because, at the base, the gas pressure is much lower than the magnetic field pressure. This is a so called cold wind. The angular velocity equals the Keplerian velocity of the disk. We remain with two unknown functions: ρv_z and B_z, which can be specified at the base of the wind as boundary conditions for the problem of the wind outflow. Below, we will see that, in the fully self-consistent solution, ρv_z can not be an arbitrary function. This function is defined by the distribution of B_z over the disk surface at the fixed $\dot{M}$ at the inner edge of the disk. Thus, the process of cold accretion actually depends only on $\dot{M}$ at the inner edge of the disk and the distribution of B_z over the surface of the disk.

The magnetic field at the surface of the disk is evidently limited from below. Our numerical modeling of the self-consistent disk outflow shows (see below) that $B_\varphi \sim B_z$ at the base of the wind.

In this case, the disk accretion is possible if the magnetic field at the surface of the disk satisfies the inequality

$$B_z^2|_{wind} > \frac{\dot{M}\Omega_k}{2r}. \tag{54}$$

Basic Properties of the "Cold" Accretion

Equation (47) can be rewritten in the form

$$\frac{\partial}{\partial r}(rV_k\dot{M}) - \frac{\partial\dot{M}}{\partial r}r_A(r)^2\Omega_k(r) = 0, \tag{55}$$

where $r_A(r)$ is the Alfven radius of the force line rooted into disc at the point with radius r. The solution of Equation (55) gives

$$\dot{M} = \dot{M}_{edge}\exp\int_{r_{in}}^{r}\frac{dr}{2r(\lambda(r)-1)}, \tag{56}$$

where $\dot{M}_{edge}$ is the accretion rate at the inner radius of the disc r_{in}.

Equation (53) also can be rewritten in the following form. Using Label (30), it is easy to obtain that

$$\frac{\partial}{\partial r}\frac{\dot{M}V_k^2}{2}\bigg|_{disc} + \frac{\partial\dot{M}}{\partial r}H_{wind} = 0. \tag{57}$$

We are interested in the solutions that allow particles to go to infinity from the disc. The necessary condition for this is $H > 0$. This condition means that the energy per particle is positive. This is necessary (but not sufficient) in order to have positive v^2 at a large distance from the source. The substitution of the explicit dependance (56) for $\dot{M}$ into Equation (57) results in

$$H = (2\lambda - 3)\frac{GM}{2r} = (2\lambda - 3)\frac{V_k^2}{2}. \tag{58}$$

This means that "cold" disk accretion is possible only at the magnetic field which satisfies the condition $\lambda > 3/2$ first obtained in [50]. Equation (58) shows also that the energy of a particle in the wind can essentially exceed the virial energy of the particle in the disk. Potentially, this fact gives us the key to the solution of the problem of large Lorentz factors of the jets from AGNs. Kinetic energy of the particles in the jets evidently strongly exceeds the virial energy of the particles at the last stable orbit of the disk.

7. Self-Consistent Solution of the Problem

To make sure that the model of "cold" accretion can reproduce real processes of the disk accretion onto a black hole, we have to obtain a convincing solution of the problem of self-consistent "cold" accretion and to verify whether this model can give a ratio of the kinetic luminosity of jets over the bolometric luminosity of the disk compatible with observations.

A lot of results in this regard have been obtained by the Grenoble Group. They used a self-similar prescription of the solution and considered the problem of the disk structure, accretion and outflow unified. The self-similarity imposes on the physics of the disk some extra demands, which do not satisfy the full set of equations. Nevertheless, solutions obtained by the Grenoble Group support the assumption that the process of accretion when the majority of the angular momentum of the accreted material is carried out by the wind can be realized indeed [54].

To obtain solutions beyond the self-similarity limitations, we use another approach. As we already discussed, we avoid consideration of the processes inside the disk. The disk is considered as a layer with the zero thickness. The processes inside the disk provide the boundary conditions at the base of the wind mass flux density ρv_z and magnetic field flux B_z. Self-consistency of the outflow and accretion means that the equation for mass conservation (30) and the equation for angular momentum conservation (52) are solved together with the full system of Equations (1), (2), (4), (7) and (9)–(11) defining the flow of the wind from the disk. Below, we present in short the basic properties of solution of the self-consistent problem in the self similar approximation and the numerical solution of the problem without self similarity assumptions.

7.1. Basic Properties of Self Similar Solutions

The most comprehensive study of types of self similar flows has been investigated by the Grenoble group [54]. The self similarity in the form proposed initially by [50] is used. In this kind of self similarity, all the variables depend on the coordinates in the form

$$G(z,r) = r^\delta \tilde{G}\left(\frac{z}{r}\right), \tag{59}$$

where z, r are the cylindrical coordinates, and δ is the self similarity index.

The steady-state equations for an ideal, cold plasma (with pressure $p = 0$) (4), (7), (9)–(11) can be rewritten in vector form as

$$\rho(\mathbf{v}\nabla)\mathbf{v} = -\frac{1}{8\pi}\nabla\,\mathbf{B}^2 + \frac{1}{4\pi}(\mathbf{B}\nabla)\mathbf{B} - \rho\,\frac{GM\mathbf{R}}{R^3}. \tag{60}$$

According to the self similarity assumption, all the variables in these equations can be presented as

$$\begin{aligned}
\mathbf{v}(r,z) &= r^{-\delta_v}\tilde{\mathbf{v}}(z/r), \\
\rho(r,z) &= r^{-\delta_\rho}\tilde{\rho}(z/r), \\
\mathbf{B}(r,z) &= r^{-\delta_B}\tilde{\mathbf{B}}(z/r).
\end{aligned} \tag{61}$$

This representation of the variables says that they are scaled as the power law of r and all functions depend on the angle ζ defined as $\tan\zeta = z/r$.

The superscripts δ_v, δ_ρ, and δ_B are determined from the following conditions. Substituting of Equations (61) into Equation (60) leads to the equations

$$2\delta_B - \delta_\rho = 2\delta_v = 1. \tag{62}$$

It follows from them that

$$\begin{aligned}
\mathbf{v}(r,z) &= r^{-1/2}\,\tilde{\mathbf{v}}(z/r), \\
\rho(r,z) &= r^{-\delta}\tilde{\rho}(z/r), \\
\mathbf{B}(r,z) &= r^{-\frac{(1+\delta)}{2}}\,\tilde{\mathbf{B}}(z/r).
\end{aligned} \tag{63}$$

It is evident that, in the self similar solution, λ is constant for all field lines. According to Equation (56) the accretion rate in the disc varies with r as follows:

$$\dot{M} = \dot{M}_{edge}\left(\frac{r}{r_{in}}\right)^{\frac{1}{2(\lambda-1)}}. \tag{64}$$

This dependence was explored first in [52]. The combination $1/2(\lambda - 1)$ was called the ejection index. Substitution of (64) into (30) taking into account (63) gives the following relationship between λ and δ

$$\delta = \frac{3}{2} - \frac{1}{2(\lambda-1)}, \tag{65}$$

which has been first used in the work [54]. λ can change from $\frac{3}{2}$ up to ∞. It is interesting that at the same time the index of self similarity varies only in the limits between $\frac{1}{2}$ and $\frac{3}{2}$.

It is interesting to compare this solution [63] with the classical solution of Blandford and Payne [50]. Their solution was not self-consistent. Equations (52) and (30) were not used. Therefore, index δ is not connected with λ by Equation (65). The scaling of the magnetic field and density was taken in the form $B \sim r^{-5/4}$, $\rho \sim r^{-3/2}$ arbitrary as one of the possible scalings. The self similar solution for "cold" accretion reproduces the solution of Blandford and Payne with additional connection expressed by Equation (65). The scaling of the magnetic field depends on λ as $B \sim r^{-(5/4 - \frac{1}{2(\lambda-1)})}$ and density as $\rho \sim r^{(-\frac{3}{2} - \frac{1}{2(\lambda-1)})}$. It is interesting to point out that observations show that the density of plasma in real objects is more likely scaled with an index less than 3/2 [86], which can be explained by the fact that $\lambda \neq \infty$ or by deviation of the flow from self similarity.

There is another interesting result obtained in [63]. The solutions at relatively low $\lambda = 25$ in [63] and $\lambda = 30$ in [50] are quite similar. All flow lines diverge from the axis experiencing slight collimation. However, already at $\lambda = 64$, the solution obtained in [63] demonstrates perfect cylindrical collimation of the flow to the rotational axis at a rather large distance from the disk. It would be important to verify this result in the numerical solutions. This is one of the reasons why the numerical solutions of this problem are of special interest for us.

7.2. Numerical Solution of the Self-Consistent Problem

In the numerical self-consistent solution, the equations for the mass and angular momentum conservation ((30), (52)) are solved together with the full system of Equations (1), (2), (4), (7), (9)–(11) determining the flow of the wind from the disk.

The problem of the flow of the wind is solved numerically by the method of relaxation explored practically in all the numerical solutions of the problems of the wind outflow from astrophysical objects. The dimension of the box of the numerical simulation was (1000×800) expressed in gravitational radius $r_g = \frac{2GM}{c^2}$ and is located above the equatorial plane. The disk and plasma are assumed to be cold. The disk is located in the equatorial plane in the interval from 3 to 300 r_g. The verification of the solution and details of the solution of the problem by the relaxation method are presented in work [87]. Here, we focus on how to specify the boundary conditions at the disk in order to satisfy Equations (52) and (30).

The basic steps of the solution of this problem are the following. Firstly, we specify the distribution of B_z over the surface of the disk from Equation (54) as

$$B^2_{z,min} = \frac{1}{2} \frac{\dot{M}\Omega_k}{r}. \tag{66}$$

For the calculations, we take the field three times higher than defined by (66).

After that, we specify the mass flux density $j = \rho v_{z,0}$ at the disk surface. The flow from the disk is sub Alfvenic. The initial velocity of the plasma is below the local Alfvenic velocity. The initial velocity at the disk is taken equal to $v_{z,0} = 0.01 \cdot V_k$. The density in this case is equal to $\rho_0 = j/v_{z,0}$. Equation (30) gives us the distribution of the mass loss rate along the radius of the disk r. However, we do not know what will be the product $B_{z,0}B_{\varphi,0}$ at the disk surface. While B_z is specified at the disk, the component of the magnetic field $B_{\varphi,0}$ is defined from the solution of the problem of the wind outflow in all the computational domain. We stress that the product $B_z B_{\varphi,0}$ is not a simple function of j. It depends on the distribution j over all the disk rather than on the value of j at the same point. Moreover, even if the solution of the problem of the wind outflow is solved, the obtained distribution $B_{z,0}B_{\varphi,0}$ will not satisfy the equation for angular momentum conservation (52). The solution is not self-consistent in this case.

We propose the following iterative procedure [88].

1. The mass flux at the inner edge of the disk $\dot{M}_{edge}$ is specified and kept constant during the modeling;
2. The initial distribution of the mass flux $\dot{M}_n(r)$ as a function of r is assumed. This is the iteration number $n = 0$. The most simple assumption at $n = 0$ is $\dot{M}_n(r) = \dot{M}_{in}$. Thus, $\dot{M}_0(r)$ is constant.

3. The initial distribution of the mass flux from the disk $(\rho v_z)_n$ is specified. At $n = 0$, the distribution of $(\rho v_z)_n$ is not connected with the accretion rate and the solution of the problem of the wind outflow is not self-consistent with the process of accretion.

4. Internal iterations are performed. The objective of the internal iteration is to define the distribution of j which fulfills the angular momentum Equation (52).They are indexed by an additional index k. Therefore, the mass flux from the disk depends on two indexes and takes the form $j_{n,k}$. At every step k, the problem of the wind outflow is solved in all of the computational domains. The steady state solution is obtained and thus we obtain new value of $B_z B_\varphi$ at the disk surface. Of course, this value does not satisfy Equation (52) for the specified $\dot{M}_n$. Here, we make the next step of the internal iterations according to the following procedure. Let us introduce function

$$\Phi_k = -2r^{5/2} B_z B_\varphi \tag{67}$$

specified at the disk surface exploring the steady state solution in the computational domain. The equation for the angular momentum conservation (52) is reduced to

$$\dot{M}\sqrt{GM} = \Phi_k(r). \tag{68}$$

The new mass flux $(\rho v_z)_{n,k+1}$ is defined by the equation

$$j_{n,k+1} = \frac{\dot{M}_n\sqrt{GM}}{(\delta\Phi_k + (1-\delta)\dot{M}_n\sqrt{GM})} j_{n,k}, \tag{69}$$

where δ is the parameter of relaxation. This step means that, if the $\Phi(r)_k < \dot{M}_n(r)\sqrt{GM}$, then the mass flux from the point of the disk with radius r increases and decreases if the opposite inequality takes place. Finally, we find the distribution of the mass flux j_n which satisfies Equation (52). However, it still does not satisfy the conservation of the mass in the disk—Equation (30).

5. In this step, Equation (30) is solved taking the obtained mass distribution j_n and boundary condition $\dot{M} = \dot{M}_{edge}$ at r_{in}. We obtain a new distribution of the accretion rate $\dot{M}_{n+1}$. This is the external iteration. After that, the entire procedure is repeated.

As a result of this entire procedure, the steady state solution of the problem of the wind outflow with boundary conditions, which satisfies Equations (52) and (30), is obtained. As an example of the procedure, we present the results of three consequent external iterations (iterations on n). Figure 5 shows the distribution of $\dot{M}_n$ and $\Phi_{(n,k)}$ at some n and k. This is still not a fully converged solution. Nevertheless, there is a good coincidence of two curves at $r < 200$. At $r > 200$, the solution still has not converged. Here, we demonstrate the largest problem of the proposed method. While the convergence on the external iterations is rather fast, the internal iterations converge slowly. This happens for the following reason. The reaction of the wind on the variation of the mass flux of the wind at the base of the flow is delayed and this delay increases with r.

The toroidal magnetic field at the surface of the disk is defined by all the flow until the Alfvenic surface. According to the theory of steady state magnetized winds, the toroidal magnetic field is defined by regularity conditions at the Alfvenic surface [83,85]. Therefore, after variation of the mass flux at the base of the flow, the information about this event should propagate to the Alfvenic surface and return back to specify a new value of the toroidal magnetic field at the base. This delay can be estimated as the time τ necessary for a signal to travel from the surface of the disk to the Alfvenic surface. The signal propagates with the Alfvenic velocity $V_A = \frac{B}{\sqrt{4\pi\rho}}$. Travel time is estimated as $\tau \sim R_A/V_A$, where R_A is proportional to r. Assuming that V_A scales as $r^{-1/2}$ like the Keplerian velocity, τ scales as $r^{3/2}$. Then, in our case, the relaxation time of the flow at the outer edge of the disk exceeds the relaxation time at the inner edge by a factor of 1000. This explains the extremely slow convergence of the flow to the self-consistent solution.

Nevertheless, already obtained results convince us that the self-consistent solution of the cold accretion exists not only under the self similar assumptions. Apparently, at the magnetic field exceeding B_{min}, it is possible to find a density of the mass flow from the disk that will satisfy Equations (52) and (30) expressing mass and angular momentum conservation. This means that actually the solution of the problem depends only on distribution of the normal component of B_z over the surface of the disk and accretion rate at the inner radius $\dot{M}_{edge}$. The mass flux distribution is defined at the solution of the self-consistent problem.

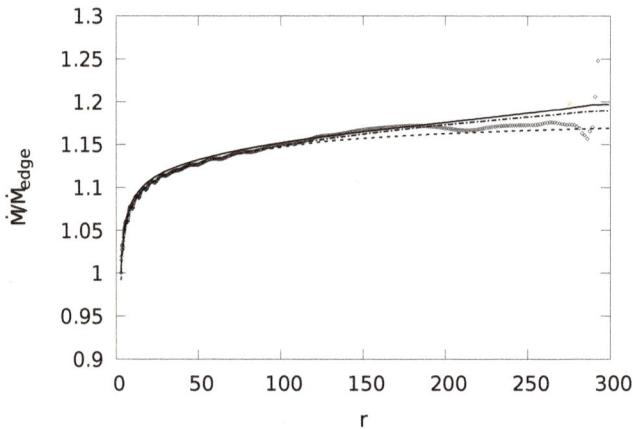

Figure 5. Distributions of $\dot{M}_{n+1}$ (solid line) and $\dot{M}_n$ (dashed-dotted line) and $\dot{M}_{n-1}$ in comparison with Φ_{n+1} (open circles) at some n.

8. Ratio of Kinetic to Bolometric Luminosity of the Disk

One of the main objectives of the model of the cold accretion is the explanation of the high ratio of the kinetic luminosity of jets from AGN over the bolometric luminosity of the disks. Therefore, one of the most important tests for the model is its ability to account for this ratio. In this section, we present an estimation of this ratio. Energy dissipated in the disk is defined by Equation (31). This energy is distributed among nonthermal and thermal radiation emitted by the disk. All together gives the bolometric luminosity of the disk. The rate of dissipation (31) depends on the thickness of the disk. Its calculation is possible only in a specific model of the disk structure. Here, we explore the SS model of geometrically thin but optically thick disk [3]. In this model,

$$t_{r\varphi} = -\alpha \rho v_s^2, \tag{70}$$

where α is the viscosity parameter, and the height of the disk above equatorial plane h equals

$$h = r\frac{v_s}{V_k}. \tag{71}$$

Assumption (70) allows us to estimate the radial velocity of matter in the disk as follows

$$v_r = \frac{\alpha}{\theta} v_s . \tag{72}$$

This velocity essentially exceeds the radial velocity of matter in the SS disk at $\theta \sim 1$ because of the higher efficiency of the angular momentum loss. In accordance with the SS model, we assume that all the dissipated energy goes into heating of the disk and finally is carried out by thermal radiation from the disk surface.

8.1. Thermal Radiation from the Disk during "Cold" Accretion

In [3], three regimes of disk accretion were considered: (a) the radiation pressure exceeding the gas pressure and the Thomson scattering dominating over free–free absorption; (b) the gas pressure dominating over radiation pressure but the Thomson scattering dominating free–free absorption; and (c) the gas pressure dominating over radiation pressure and the opacity of the matter defined by free–free absorption. We consider only the cases when gas pressure dominates over radiation pressure. These are regimes (b) and (c). Below, we will see that, when radiation dominates, accretion proceeds in the Shakura–Sunayev regime.

8.2. Scattering Dominating over Free–Free Absorption

Firstly, we consider the case when Thomson scattering dominates over free–free absorption (Thomson regime). Radiation pressure P_{rad} equals $\varepsilon/3$, where $\varepsilon = bT^4$. The sound velocity is defined as $v_s^2 = kT/m_p$, where m_p is the proton mass. According to [3], the heat conductivity of the disk is defined by the transport of radiation. Then,

$$\varepsilon = \frac{3}{4} \frac{Q\sigma u_0}{c}, \tag{73}$$

where $\sigma = 0.4 \ \mathrm{cm}^2/\mathrm{g}$ is the Thomson opacity, and $u_0 = 2\rho h$. The rate of heating of the disk follows from Equations (31), (51), (52) and equals

$$Q = \frac{3\theta \dot{M} V_k v_s}{16\pi r^2}. \tag{74}$$

We used here that $h = v_s/\Omega$. The solution of these two equations under assumption (70) yields the temperature inside the disk

$$T = \frac{\sqrt{3}}{4\sqrt{\pi}} \left(\frac{\theta^2 \dot{M}^2 V_k \sigma}{b\alpha c r^3} \right)^{\frac{1}{4}}. \tag{75}$$

The sound velocity equals

$$v_s = \frac{3^{1/4}}{2(\pi)^{1/4}} \frac{k^{1/2} V_k^{1/8} (\theta \dot{M})^{1/4} \sigma^{1/8}}{m_p^{1/2} b^{1/8} \alpha^{1/8} c^{1/8} r^{3/8}}, \tag{76}$$

and the density flux of radiation from one side of the disk is expressed as

$$Q = \frac{3^{5/4}}{32\pi^{5/4}} \frac{(\theta \dot{M})^{5/4} V_k^{9/8} k^{1/2} \sigma^{1/8}}{r^{19/8} m_p^{1/2} b^{1/8} \alpha^{1/8} c^{1/8}}. \tag{77}$$

Let us express $\dot{M} = \dot{m} \dot{M}_{Edd}$, the radius r in $r = (3r_g)x$ and the mass M in the solar masses $M = m M_\odot$. In these variables, we obtain

$$Q = 0.77 \times 10^{23} \frac{(\theta \dot{m})^{5/4}}{m^{9/8} x^{47/16} \alpha^{1/8}}, \ \mathrm{erg/s/cm}^2. \tag{78}$$

The integration of this expression over the disk gives the bolometric luminosity of the disk

$$L_{bol} = 0.84 \times 10^{36} \frac{(\theta \dot{m})^{5/4} m^{7/8}}{\alpha^{1/8}}, \ \mathrm{erg/s}. \tag{79}$$

The ratio of L_{bol}/L_{Edd} is

$$\frac{L_{bol}}{L_{Edd}} = 6 \times 10^{-3} \frac{(\theta \dot{m})^{5/4}}{(\alpha m)^{1/8}}. \tag{80}$$

The kinetic luminosity of the jets equals the total energy release of accretion. Therefore,

$$L_{kin} = \frac{\dot{M}c^2}{12} = 1.4 \times 10^{38} m\dot{m}, \text{ erg/s.} \tag{81}$$

Then, the ratio of the kinetic luminosity over the bolometric luminosity equals

$$\frac{L_{kin}}{L_{bol}} = 170\frac{(m\alpha)^{1/8}}{\dot{m}^{1/4}\theta^{5/4}}. \tag{82}$$

The bolometric luminosity can be expressed in the conventional variables:

$$L_{bol} = \frac{4}{5}\theta \dot{M} V_{k0} v_{s0}, \tag{83}$$

where V_{k0} and v_{s0} are the Keplerian and sound velocities at the inner edge of the disk. Taking into account that the kinetic luminosity

$$L_{kin} = \frac{\dot{M}V_{k0}^2}{2}, \tag{84}$$

the condition $L_{bol}/L_{kin} \ll 1$ becomes

$$\frac{8}{5}\frac{\theta v_{s0}}{V_{k0}} = \frac{8}{5}\frac{\theta h}{r} \ll 1, \tag{85}$$

which practically coincides with the condition of applicability of the "cold" disk accretion approximation defined by Equation (50). A similar condition has been obtained earlier in [55]. The condition $L_{kin} \gg L_{bol}$ indicates that accretion occurs in the "cold" regime.

The temperature in the disk

$$T = 2.5 \times 10^7 \frac{\sqrt{\theta\dot{m}}}{\alpha^{1/4}x^{7/8}m^{1/4}}, \text{ K} \tag{86}$$

is less than the temperature in the Shakura–Sunyaev disk [3] disk results in $\theta \sim 1$.

Let us calculate the ratio of radiation pressure over the gas pressure in the disk,

$$\frac{P_{rad}}{P_{gas}} = \frac{3}{32\pi}\frac{\theta\dot{M}\sigma}{rc} = 0.85\frac{\theta\dot{m}}{x}. \tag{87}$$

This means that all our estimates are valid when $0.85\theta\dot{m} < 1$. Other disk parameters are estimated as follows. The density equals

$$\rho = \frac{1}{2\sqrt{3\pi}}\frac{\sqrt{\theta\dot{M}}V_k^{3/4}m_p b^{1/4}c^{1/4}}{r^{5/4}k\alpha^{3/4}\sigma^{1/4}} = 0.6\frac{\sqrt{\theta\dot{m}}}{m^{3/4}x^{13/8}\alpha^{3/4}}, \text{ g/cm}^3. \tag{88}$$

The aspect ratio of the disk is

$$\frac{h}{r} = 3.7 \times 10^{-3}\frac{(\dot{m}\theta)^{1/4}x^{1/16}}{(\alpha m)^{1/8}}. \tag{89}$$

The true optical depth $\tau^* = \sqrt{\sigma \cdot \sigma_{ff}} \cdot u_0$ of the disk is expressed as

$$\tau^* = 51(\theta\dot{m})^{1/8}m^{3/16}x^{5/32}\alpha^{-13/16}, \tag{90}$$

where $\sigma_{ff} = 0.11 \cdot T^{-7/2} n$, cm^2/g is the free–free opacity of the disk. The surface temperature of the disk T_S is defined from the equation $bcT_s^4/4 = Q$ and has the form

$$T_s = 5 \times 10^6 \frac{(\theta \dot{m})^{5/16}}{m^{9/32} x^{47/64} \alpha^{1/32}} \ \text{K}. \tag{91}$$

8.3. Free–Free Absorption Dominating over Scattering

Under the condition

$$4.6 \times 10^{-3} \frac{(\alpha m)^{1/10} x^{23/20}}{(\theta \dot{m})} > 1, \tag{92}$$

free–free absorption exceeds Thomson scattering. Hereafter, we call this regime "free-free". Similar calculations yield the following temperature distribution inside the disk

$$T = 10^7 \frac{(\theta \dot{m})^{6/17}}{x^{12/17} (\alpha \cdot m)^{4/17}} \ \text{K}. \tag{93}$$

The bolometric luminosity of the disk is

$$L_{bol} = 0.6 \times 10^{36} (\theta \dot{m})^{20/17} m^{15/17} \alpha^{-2/17} \ \text{erg/s}, \tag{94}$$

while

$$\frac{L_{bol}}{L_{Edd}} = 4.4 \times 10^{-3} \frac{(\theta \dot{m})^{20/17}}{(\alpha m)^{2/17}}. \tag{95}$$

The ratio of kinetic luminosity over the bolometric luminosity equals

$$\frac{L_{kin}}{L_{bol}} = \frac{228 (\alpha m)^{2/17}}{\dot{m}^{3/17} \theta^{20/17}}. \tag{96}$$

The full optical depth, the density of plasma and the aspect ratio of the disk are given by

$$\tau = 93 (\theta \dot{m})^{4/17} m^{3/17} x^{1/34} \alpha^{-14/17}, \tag{97}$$

$$\rho = \frac{1.2 (\theta \dot{m})^{11/17}}{(\alpha m)^{13/17} \cdot x^{61/34}} \ \text{g/cm}^3, \tag{98}$$

$$\frac{h}{r} = 2.5 \times 10^{-3} x^{5/34} (\theta \dot{m})^{3/17} (\alpha m)^{-2/17}, \tag{99}$$

respectively. The dissipated flux of energy per unit square from one side of the disk is

$$Q = 5.2 \times 10^{22} \frac{\theta^{20/17} \dot{m}^{20/17}}{m^{19/17} x^{97/34} \alpha^{2/17}}. \tag{100}$$

Finally, the surface temperature is equal to

$$T_s = 5.5 \times 10^6 \frac{(\theta \dot{m})^{5/17}}{m^{19/68} x^{97/136} \alpha^{1/34}} \ \text{K}. \tag{101}$$

9. Comparison with the Fundamental Plane of Black Holes

The fundamental plane encapsulates the relationship between the compact radio luminosity, X-ray luminosity, and the black hole mass, and provides a good description of the data over a very large range of black hole masses. There are reasons to believe that the Fundamental Plane (hereafter, FP) of black holes reproduces the actual relationship between the kinetic luminosity of jets and the bolometric luminosity of the disks. In [29], the position of objects of different masses in the coordinates L_{kin}/L_{bol} and L_{bol}/L_{Edd} has been collected in one FP. If this is true, the FP can be used to extract information about the dependence of θ on $\dot{m}$ and m. All data at the FP can be approximated by a power law function of the form

$$\log \frac{L_{kin}}{L_{bol}} = (A - 1) \log(\frac{L_{bol}}{L_{Edd}}) + B \tag{102}$$

with A in the range (0.43–0.47) and B in the range from -0.94 to -1.37. For our estimates, the values $A = 0.457$ and $B = -1.1$ around the average have been chosen.

If the empirical relationship (102) is valid, then

$$\frac{L_{bol}}{L_{Edd}} = 10^{-\frac{B}{A}} \dot{m}^{\frac{1}{A}}. \tag{103}$$

We used here that $L_{kin}/L_{Edd} = \dot{m}$ in the regime of "cold" accretion. This semi empirical relationship is very useful because it connects the bolometric luminosity of the disk with the accretion rate directly. The equations defining L_{bol} and L_{kin}/L_{bol} obtained for $A = 0.457$ and $B = -1.1$ are as follows:

$$L_{bol} = 3.6 \times 10^{40} m\dot{m}^{2.19} \text{ erg/s}, \tag{104}$$

and

$$\frac{L_{kin}}{L_{bol}} = 3.9 \times 10^{-3} \dot{m}^{-1.19}. \tag{105}$$

The empirical relationship (102) allows us to estimate θ. Obviously, a constant θ is not consistent with observations. It must depend on $\dot{m}$ and m. It is natural to assume that θ depends on $\dot{m}$ as a power law

$$\theta = D\dot{m}^{\gamma}. \tag{106}$$

In the Thomson regime,

$$X = \frac{L_{bol}}{L_{Edd}} = 6 \times 10^{-3} \frac{\dot{m}^{5(\gamma+1)/4} D^{5/4}}{(\alpha m)^{1/8}}, \tag{107}$$

and

$$Y = \frac{L_{kin}}{L_{bol}} = \frac{168(\alpha m)^{1/8}}{D^{5/4} \dot{m}^{(5\gamma+1)/4}}. \tag{108}$$

After simple algebraic calculations, we obtain that Y depends on X as in Equation (102) if

$$A = \frac{4}{5(\gamma + 1)}, \tag{109}$$

For $A = 0.457$, the value $\gamma = 3/4 = 0.75$. Then, the value $B = -1.1$ is obtained for $D = 5 \times 10^3 (\alpha m)^{1/10}$. Thus, in the Thomson regime,

$$\theta = 5 \times 10^3 \dot{m}^{3/4} (\alpha m)^{1/10}. \tag{110}$$

Similar calculations in the free–free regime yield

$$\theta = 11.2 \times 10^3 \dot{m}^{0.86} (\alpha m)^{1/10}. \tag{111}$$

The power of $\dot{m}$ is chosen so as to provide uniform dependence of Y on X of the form (102) with constant A in both regimes.

The dependencies (110) and (111) look physically reasonable. They show that, the smaller the accretion rate, the more uniform is the magnetic field across the disk.

A plot of $\theta/(\alpha m)^{1/10}$ is presented in Figure 6. θ corresponding to FP agrees with the assumption of "cold" accretion because this curve is located well below the curve separating the regime of cold accretion from the Shakura–Sunyaev regime. In Figure 6, the dashed-dotted line separates regions of domination of the gas pressure and regions of domination of the radiation pressure as defined by Equation (87). The thin solid line separates the Thomson regime from the free–free regime.

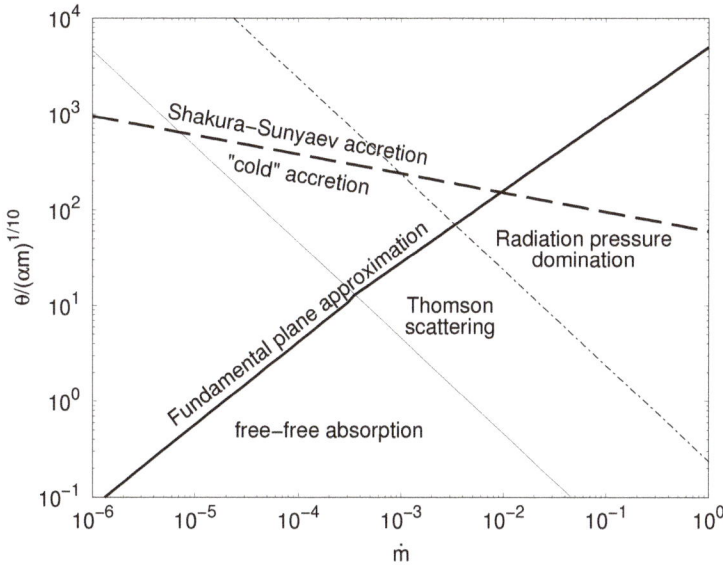

Figure 6. Dependence of $\theta/(\alpha m)^{1/10}$ on $\dot{m}$. The Shakura–Sunyaev accretion regime takes place above the thick dashed line. Below this line, the regime of "cold" accretion takes place. Thomson scattering dominates above the thin solid line, while, below this line, free–free absorption gives the major contribution in the opacity of the medium. The dashed-dotted line (calculated for $m = 10^8$) divides the plane in two parts where radiation pressure (above) and gas pressure (below) dominate.

10. Comparison with Specific Sources

It is interesting to apply the estimated dependencies to specific sources. Below, we consider M87 and the SMBH in the galactic center, Sgr A*. We will see below that both sources are in the free–free regime. Therefore, we used Equation (111) in our estimations. For ease of calculation, we consider the case with $\alpha = 0.1$.

10.1. M 87

For this object, $\dot{m}$ and m can be easily estimated. The kinetic luminosity of this object is $L_{kin} = 10^{44}$ erg/s, which we assume is equal to the total rate of gravitational energy released in the accretion. The mass of the central black hole is equal to $m = 3.5 \times 0^9$ [89]. With the Eddington luminosity equal to $L_{Edd} = 1.4 \times 10^{38} m$ erg/s, we find $\dot{m} = L_{kin}/L_{Edd} = 2 \times 10^{-4}$. From Equations (111) and (96), we obtain that $\theta = 54$ and $L_{kin}/L_{bol} \approx 95$. According to Equation (104), $L_{bol} = 10^{42}$ erg/s in accordance with observations. The optical depth of the disk exceeds $\tau > 10^4$.

10.2. Sgr A*

The mass of SMBH in Sgr A* is equal to $m = 4 \times 10^6$ while the bolometric luminosity is $L_{bol} \sim 10^{36}$ erg/s [34]. The kinetic luminosity of the outflow from the disk around SMBH in Sgr A* is not known. The flux of TeV gamma-rays from the Galactic Center can be explained by very high energy accelerated protons with a luminosity close to 10^{38} erg/s. The kinetic luminosity of the wind has to be higher. Let us estimate $\dot{m}$ from the bolometric luminosity of the disk (see Equation (104)). In this case, $\dot{m} = 8 \times 10^{-6}$ and $\theta = 1.7$ at $\tau \sim 10^3$. The value of $\dot{m}$ agrees with estimates of the accretion rate obtained from Bondi accretion of stellar winds of the order of 10^{21} g/s [90]. From Equation (81), we obtain that

$$L_{kin} = 4.4 \times 10^{39} \text{ erg/s} . \tag{112}$$

The kinetic luminosity of the wind from the Galactic accretion disk exceeds the bolometric luminosity of the disk 4.4×10^3 times. Remarkably, this power is sufficient to explain the flux of PeV protons from the Galactic Center.

11. Conclusions

The main energy release in the AGNs occurs as a flux of kinetic energy of the jets. The question of the energy source of the jets is the main question to which the theory must first answer. There are two answers to this question. In the first case, the main source of energy of the jet is the energy of rotation of the black hole which is transformed into energy of $e^{\pm}$ jets due to the Blandford and Znajek mechanism [42]. Only a small fraction of the gravitational energy of the accreted matter is released in the form of radiation of the disk. The major part of the gravitational energy goes into increasing the mass of the SMBH. This model needs additional assumptions. Accretion has to occur in the ADAF regime to provide radiatively inefficient disks. The speed of rotation of the SMBH must be close to the maximal possible and accretion must occur in the regime of the Magnetically Arrested Disks. In this regime, the magnetic field is so strong that it destroys the disk. Otherwise, it is not possible to provide the necessary energetics of jets.

In another case of "cold" accretion, the only source of energy is the gravitational energy of the accreted matter. The major part of this energy goes into the energy of magnetized wind and a small fraction of the energy is released in the form of radiation from the disk. An attractive feature of this model is a natural explanation for the high kinetic power of the jets compared to the luminosity of the accretion disk. As a bonus, the kinetic energy of particles in a jet can be orders of magnitude greater than the kinetic energy of particles in a Kepler orbit.

"Cold" accretion does not need special conditions or exotic magnetic fields. This regime of accretion is implemented when a magnetized wind expires from the disk. The existence of the winds from the disks is confirmed by numerous observations. "Cold" accretion goes into Shakura–Sunyaev accretion [3] when the loss of the angular momentum due to viscose stresses dominates the loss due to the wind. Nevertheless, "cold" accretion occurs even when the magnetic field inside the accretion disk essentially exceeds the magnetic field at the base of the wind. This is explained by geometry. The angular momentum transport due to viscosity is proportional to the magnetic pressure in the disk times the thickness of the disk h while the flux of the angular momentum from the disk is proportional to the magnetic pressure at the base of the wind times the radius r. The ratio of viscous losses to losses due to the wind is $\sim \theta \cdot \left(\frac{h}{r}\right)$, where θ roughly equals the ratio of magnetic pressures inside and at the surface of the disk. Therefore, the Shakura–Sunyaev regime of accretion is realized when $\theta > \left(\frac{r}{h}\right)$.

Estimates of the ratio of the kinetic luminosity of the jets to the bolometric luminosity of the disk show that the current observations can be explained in the framework of "cold" accretion. Of course, the assumption that θ is constant evidently contradicts observations. Detailed comparison of the theoretical predictions with the Fundamental plane of the black holes shows that θ has to increase two orders of magnitude with $\dot{m}$. This behavior of θ agrees with the results of modeling of the magnetic field distribution in the disk [72]. This estimate allows us to obtain certain conclusions about the realization of the regime of "cold" disk accretion. At small accretion rates $\dot{m} < 10^{-2}$, the estimated value of θ lies in the region well below the line where the Shakura–Sunyaev model is valid. The magnetic pressure inside the disk appears less than the magnetic pressure estimated in the model [3]. It is reasonable to assume that, at relatively low rates of accretion, $\dot{m} < 10^{-2}$, accretion occurs predominantly in the regime of "cold" accretion. At higher values of $\dot{m} > 0.1$, the accretion occurs in the regime of Shakura and Sunyaev. The transition between the two regimes takes place at a value of $\dot{m}$ between 0.01 and 0.1 that agrees well with the transition from very bright to very dim disks around SMBH with powerful outflows deduced in [41]. Remarkably, the rough estimate of the dependence of θ on $\dot{m}$ gives good agreements with observations of two SMBHs, M87 and Sgr A*. It is highly unlikely that such an agreement could be accidental.

Funding: This research was funded by the Russian Science Foundation project N 16-12-10443.

Conflicts of Interest: The authors declare no conflict of interest.

References

1. Shakura, N.I. Disk Model of Gas Accretion on a Relativistic Star in a Close Binary System. *Sov. Astron.* **1972**, *49*, 921.
2. Pringle, J.E.; Rees, M.J. Accretion Disc Models for Compact X-Ray Sources. *Astron. Astrophys.* **1972**, *21*, 1.
3. Shakura, N.I.; Sunyaev, R.A. Black holes in binary systems. Observational appearance. *Astron. Astrophys.* **1973**, *24*, 337–355.
4. Baganoff, F.K.; Maeda, Y.; Morris, M.; Bautz, M.W.; Brandt, W.N.; Cui, W.; Doty, J.P.; Feigelson, E.D.; Garmire, G.P.; Pravdo, S.H.; et al. Chandra X-Ray Spectroscopic Imaging of Sagittarius A* and the Central Parsec of the Galaxy. *Astrophys. J.* **2003**, *591*, 891–915. [CrossRef]
5. Narayan, R.; Yi, I. Advection-dominated Accretion: Underfed Black Holes and Neutron Stars. *Astrophys. J.* **1995**, *452*, 710. [CrossRef]
6. Bisnovatyi-Kogan, G.S.; Lovelace, R.V.E. Influence of Ohmic Heating on Advection-dominated Accretion Flows. *Astrophys. J.* **1997**, *486*, 43–46. [CrossRef]
7. Quataert, E.; Gruzinov, A. Convection-dominated Accretion Flows. *Astrophys. J.* **2000**, *539*, 809–814. [CrossRef]
8. Narayan, R.; Quataert, E.; Igumenshchev, I.V.; Abramowicz, M.A. The Magnetohydrodynamics of Convection-dominated Accretion Flows. *Astrophys. J.* **2002**, *577*, 295–301. [CrossRef]
9. Blandford, R.D.; Begelman, M.C. On the fate of gas accreting at a low rate on to a black hole. *Mon. Not. R. Astron. Soc.* **1999**, *303*, 1–5. [CrossRef]
10. Stone, J.M.; Pringle, J.E.; Begelman, M.C. Hydrodynamical non-radiative accretion flows in two dimensions. *Mon. Not. R. Astron. Soc.* **1999**, *310*, 1002–1016. [CrossRef]
11. Blandford, R.D.; Begelman, M.C. Two-dimensional adiabatic flows on to a black hole—I. Fluid accretion. *Mon. Not. R. Astron. Soc.* **2004**, *349*, 68–86. [CrossRef]
12. Falcke, H.; Markoff, S. The jet model for Sgr A*: Radio and X-ray spectrum. *Astron. Astrophys.* **2000**, *362*, 113–118.
13. Yuan, F.; Markoff, S.; Falcke, H. A Jet-ADAF model for Sgr A*. *Astron. Astrophys.* **2002**, *383*, 854–863. [CrossRef]
14. Bicknell, G.V.; Begelman, M.C. Understanding the Kiloparsec-Scale Structure of M87. *Astrophys. J.* **1996**, *467*, 597. [CrossRef]
15. Reynolds, C.S.; Di Matteo, T.; Fabian, A.C.; Hwang, U.; Canizares, C.R. The 'quiescent' black hole in M87. *Mon. Not. R. Astron. Soc.* **1996**, *283*, 111–116. [CrossRef]
16. Biretta, J.A.; Stern, C.P.; Harris, D.E. The radio to X-ray spectrum of the M87 jet and nucleus. *Astron. J.* **1991**, *101*, 1632–1646. [CrossRef]
17. Rawlings, S.; Saunders, R. Evidence for a common central-engine mechanism in all extragalactic radio sources. *Nature* **1991**, *349*, 138–140. [CrossRef]
18. Heinz, S.; Sunyaev, R.A. The non-linear dependence of flux on black hole mass and accretion rate in core-dominated jets. *Mon. Not. R. Astron. Soc.* **2003**, *343*, 59–64. [CrossRef]
19. Merloni, A.; Heinz, S.; di Matteo, T. A Fundamental Plane of black hole activity. *Mon. Not. R. Astron. Soc.* **2003**, *345*, 1057–1076. [CrossRef]
20. Falcke, H.; Körding, E.; Markoff, S. A scheme to unify low-power accreting black holes. Jet-dominated accretion flows and the radio/X-ray correlation. *Astron. Astrophys.* **2004**, *414*, 895–903. [CrossRef]
21. Körding, E.G.; Fender, R.P.; Migliari, S. Jet-dominated advective systems: Radio and X-ray luminosity dependence on the accretion rate. *Mon. Not. R. Astron. Soc.* **2006**, *369*, 1451–1458. [CrossRef]
22. Merloni, A.; Heinz, S. Measuring the kinetic power of active galactic nuclei in the radio mode. *Mon. Not. R. Astron. Soc.* **2007**, *381*, 589–601. [CrossRef]

23. De Gasperin, F.; Merloni, A.; Sell, P.; Best, P.; Heinz, S.; Kauffmann, G. Testing black hole jet scaling relations in low-luminosity active galactic nuclei. *Mon. Not. R. Astron. Soc.* **2011**, *415*, 2910–2919. [CrossRef]

24. Saikia, P.; Körding, E.; Falcke, H. The Fundamental Plane of black hole activity in the optical band. *Mon. Not. R. Astron. Soc.* **2015**, *450*, 2317–2326. [CrossRef]

25. Körding, E.G.; Jester, S.; Fender, R. Measuring the accretion rate and kinetic luminosity functions of supermassive black holes. *Mon. Not. R. Astron. Soc.* **2008**, *383*, 277–288. [CrossRef]

26. Ma, M.L.; Cao, X.W.; Jiang, D.R.; Gu, M.F. A New Approach for Estimating Kinetic Luminosity of Jet in AGNs. *Chin. J. Astron. Astrophys.* **2008**, *8*, 39–49. [CrossRef]

27. López-Corredoira, M.; Perucho, M. Kinetic power of quasars and statistical excess of MOJAVE superluminal motions. *Astron. Astrophys.* **2012**, *544*, A56. [CrossRef]

28. Daly, R.A. Spin properties of supermassive black holes with powerful outflows. *Mon. Not. R. Astron. Soc.* **2016**, *458*, 24–28. [CrossRef]

29. Daly, R.A.; Stout, D.A.; Mysliwiec, J.N. The Fundamental Plane of Black Hole Activity Represented in Terms of Dimensionless Beam Power and Bolometric Luminosity. *arXiv* **2016**, arXiv:1606.01399.

30. Fernandes, C.A.C.; Jarvis, M.J.; Rawlings, S.; Martínez-Sansigre, A.; Hatziminaoglou, E.; Lacy, M.; Page, M.J.; Stevens, J.A.; Vardoulaki, E. Evidence for a maximum jet efficiency for the most powerful radio galaxies. *Mon. Not. R. Astron. Soc.* **2011**, *411*, 1909–1916. [CrossRef]

31. Punsly, B. High Jet Efficiency and Simulations of Black Hole Magnetospheres. *Astrophys. J.* **2011**, *728*, 17. [CrossRef]

32. Ghisellini, G.; Tavecchio, F.; Maraschi, L.; Celotti, A.; Sbarrato, T. The power of relativistic jets is larger than the luminosity of their accretion disks. *Nature* **2014**, *515*, 376–378. [CrossRef]

33. Abramowski, A.; Aharonian, F.; Benkhali, F.A.; Akhperjanian, A.G.; Angüner, E.O.; Backes, M.; Balzer, A.; Becherini, Y.; Tjus, J.B.; Berge, D.; et al. Acceleration of petaelectronvolt protons in the Galactic Centre. *Nature* **2016**, *531*, 476–479. [CrossRef]

34. Genzel, R.; Eisenhauer, F.; Gillessen, S. The Galactic Center massive black hole and nuclear star cluster. *Rev. Mod. Phys.* **2010**, *82*, 3121–3195. [CrossRef]

35. Striani, E.; Vercellone, S.; Tavani, M.; Vittorini, V.; D'Ammando, F.; Donnarumma, I.; Pacciani, L.; Pucella, G.; Bulgarelli, A.; Trifoglio, M.; et al. The Extraordinary Gamma-ray Flare of the Blazar 3C 454.3. *Astrophys. J.* **2010**, *718*, 455–459. [CrossRef]

36. Ackermann, M.; Ajello, M.; Baldini, L.; Ballet, J.; Barbiellini, G.; Bastieri, D.; Bechtol, K.; Bellazzini, R.; Berenji, B.; Blandford, R.D.; et al. Fermi Gamma-ray Space Telescope Observations of Gamma-ray Outbursts from 3C 454.3 in 2009 December and 2010 April. *Astrophys. J.* **2010**, *721*, 1383–1396. [CrossRef]

37. Vercellone, S.; Striani, E.; Vittorini, V.; Donnarumma, I.; Pacciani, L.; Pucella, G.; Tavani, M.; Raiteri, C.M.; Villata, M.; Romano, P.; et al. The Brightest Gamma-Ray Flaring Blazar in the Sky: AGILE and Multi-wavelength Observations of 3C 454.3 During 2010 November. *Astrophys. J.* **2011**, *736*, L38. [CrossRef]

38. Abdo, A.A.; Ackermann, M.; Ajello, M.; Allafort, A.; Baldini, L.; Ballet, J.; Barbiellini, G.; Bastieri, D.; Bellazzini, R.; Berenji, B.; et al. Fermi Gamma-ray Space Telescope Observations of the Gamma-ray Outburst from 3C454.3 in November 2010. *Astrophys. J.* **2011**, *733*, 26. [CrossRef]

39. Khangulyan, D.V.; Barkov, M.V.; Bosch-Ramon, V.; Aharonian, F.A.; Dorodnitsyn, A.V. Star-Jet Interactions and Gamma-Ray Outbursts from 3C454.3. *Astrophys. J.* **2013**, *774*, 113. [CrossRef]

40. Pjanka, P.; Zdziarski, A.A.; Sikora, M. The power and production efficiency of blazar jets. *Mon. Not. R. Astron. Soc.* **2017**, *465*, 3506–3514. [CrossRef]

41. Churazov, E.; Sazonov, S.; Sunyaev, R.; Forman, W.; Jones, C.; Böhringer, H. Supermassive black holes in elliptical galaxies: Switching from very bright to very dim. *Mon. Not. R. Astron. Soc.* **2005**, *363*, L91–L95. [CrossRef]

42. Blandford, R.D.; Znajek, R.L. Electromagnetic extraction of energy from Kerr black holes. *Mon. Not. R. Astron. Soc.* **1977**, *179*, 433–456. [CrossRef]

43. Narayan, R.; Igumenshchev, I.V.; Abramowicz, M.A. Magnetically Arrested Disk: An Energetically Efficient Accretion Flow. *Publ. Astron. Soc. Jpn.* **2003**, *55*, L69–L72. [CrossRef]

44. Bisnovatyi-Kogan, G.S.; Ruzmaikin, A.A. The accretion of matter by a collapsing star in the presence of a magnetic field. II—self-consistent stationary picture. *Astrophys. Space Sci.* **1976**, *42*, 401–424. [CrossRef]

45. McKinney, J.C.; Tchekhovskoy, A.; Blandford, R.D. General relativistic magnetohydrodynamic simulations of magnetically choked accretion flows around black holes. *Mon. Not. R. Astron. Soc.* **2012**, *423*, 3083–3117. [CrossRef]

46. Parker, E.N. Dynamics of the Interplanetary Gas and Magnetic Fields. *Astrophys. J.* **1958**, *128*, 664. [CrossRef]

47. Schatzman, E. A theory of the role of magnetic activity during star formation. *Ann. d'Astrophys.* **1962**, *25*, 18.

48. Mestel, L. Magnetic braking by a stellar wind—I. *Mon. Not. R. Astron. Soc.* **1968**, *138*, 359. [CrossRef]

49. Mestel, L. Magnetic braking by a stellar wind—II. *Mon. Not. R. Astron. Soc.* **1968**, *140*, 177. [CrossRef]

50. Blandford, R.D.; Payne, D.G. Hydromagnetic flows from accretion discs and the production of radio jets. *Mon. Not. R. Astron. Soc.* **1982**, *199*, 883–903. [CrossRef]

51. Pelletier, G.; Pudritz, R.E. Hydromagnetic disk winds in young stellar objects and active galactic nuclei. *Astrophys. J.* **1992**, *394*, 117–138. [CrossRef]

52. Ferreira, J.; Pelletier, G. Magnetized accretion-ejection structures. 1. General statements. *Astron. Astrophys.* **1993**, *276*, 625.

53. Ferreira, J.; Pelletier, G. Magnetized accretion-ejection structures. III. Stellar and extragalactic jets as weakly dissipative disk outflows. *Astron. Astrophys.* **1995**, *295*, 807.

54. Ferreira, J. Magnetically-driven jets from Keplerian accretion discs. *Astron. Astrophys.* **1997**, *319*, 340–359.

55. Ferreira, J.; Petrucci, P.O.; Henri, G.; Saugé, L.; Pelletier, G. A unified accretion-ejection paradigm for black hole X-ray binaries. I. The dynamical constituents. *Astron. Astrophys.* **2006**, *447*, 813–825. [CrossRef]

56. Lubow, S.H.; Papaloizou, J.C.B.; Pringle, J.E. On the Stability of Magnetic Wind-Driven Accretion Discs. *Mon. Not. R. Astron. Soc.* **1994**, *268*, 1010. [CrossRef]

57. Livio, M.; Pringle, J.E.; King, A.R. The Disk-Jet Connection in Microquasars and Active Galactic Nuclei. *Astrophys. J.* **2003**, *593*, 184–188. [CrossRef]

58. Cao, X.; Spruit, H.C. The Large-scale Magnetic Fields of Thin Accretion Disks. *Astrophys. J.* **2013**, *765*, 149. [CrossRef]

59. Li, S.L.; Begelman, M.C. Thermal Stability of a Thin Disk with Magnetically Driven Winds. *Astrophys. J.* **2014**, *786*, 6. [CrossRef]

60. Li, S.L. The High-efficiency Jets Magnetically Accelerated from a Thin Disk in Powerful Lobe-dominated FRII Radio Galaxies. *Astrophys. J.* **2014**, *788*, 71. [CrossRef]

61. Marcel, G.; Ferreira, J.; Petrucci, P.; Henri, G.; Belmont, R.; Clavel, M.; Malzac, J.; Coriat, M.; Corbel, S.; Rodriguez, J.; et al. A unified accretion-ejection paradigm for black hole X-ray binaries. II. Observational signatures of jet-emitting disks. *arXiv* **2018**, arXiv:1803.04335.

62. Marcel, G.; Ferreira, J.; Petrucci, P.; Belmont, R.; Malzac, J.; Clavel, M.; Henri, G.; Coriat, M.; Corbel, S.; Rodriguez, J.; et al. A unified accretion-ejection paradigm for black hole X-ray binaries. III. Spectral signatures of hybrid disk configurations. *Astron. Astrophys.* **2018**, *617*, A46. [CrossRef]

63. Bogovalov, S.V.; Kelner, S.R. Accretion and Plasma Outflow from Dissipationless Discs. *Int. J. Mod. Phys. D* **2010**, *19*, 339–365. [CrossRef]

64. Bisnovatyi-Kogan, G.S.; Ruzmaikin, A.A. The Accretion of Matter by a Collapsing Star in the Presence of a Magnetic Field. *Astrophys. Space Sci.* **1974**, *28*, 45–59. [CrossRef]

65. Van Ballegooijen, A.A. Magnetic fields in the accretion disks of cataclysmic variables. In *Accretion Disks and Magnetic Fields in Astrophysics*; Belvedere, G., Ed.; Springer: Dordrecht, The Netherlands, 1989; pp. 99–106.

66. Lubow, S.H.; Papaloizou, J.C.B.; Pringle, J.E. Magnetic field dragging in accretion discs. *Mon. Not. R. Astron. Soc.* **1994**, *267*, 235–240. [CrossRef]

67. Bisnovatyi-Kogan, G.S.; Lovelace, R.V.E. Large-Scale B-Field in Stationary Accretion Disks. *Astrophys. J.* **2007**, *667*, L167–L169. [CrossRef]

68. Bisnovatyi-Kogan, G.S.; Lovelace, R.V.E. Vertical Structure of Stationary Accretion Disks with a Large-scale Magnetic Field. *Astrophys. J.* **2012**, *750*, 109. [CrossRef]

69. Galeev, A.A.; Rosner, R.; Vaiana, G.S. Structured coronae of accretion disks. *Astrophys. J.* **1979**, *229*, 318–326. [CrossRef]

70. Begelman, M.C.; Armitage, P.J. A Mechanism for Hysteresis in Black Hole Binary State Transitions. *Astrophys. J.* **2014**, *782*, L18. [CrossRef]

71. Begelman, M.C.; Armitage, P.J.; Reynolds, C.S. Accretion Disk Dynamo as the Trigger for X-Ray Binary State Transitions. *Astrophys. J.* **2015**, *809*, 118. [CrossRef]

72. Salvesen, G.; Simon, J.B.; Armitage, P.J.; Begelman, M.C. Accretion disc dynamo activity in local simulations spanning weak-to-strong net vertical magnetic flux regimes. *Mon. Not. R. Astron. Soc.* **2016**, *457*, 857–874. [CrossRef]

73. Riols, A.; Rincon, F.; Cossu, C.; Lesur, G.; Ogilvie, G.I.; Longaretti, P.Y. Magnetorotational dynamo chimeras. The missing link to turbulent accretion disk dynamo models? *Astron. Astrophys.* **2017**, *598*, A87. [CrossRef]

74. Contopoulos, I.; Kazanas, D. A Cosmic Battery. *Astrophys. J.* **1998**, *508*, 859–863. [CrossRef]

75. Contopoulos, I.; Nathanail, A.; Katsanikas, M. The Cosmic Battery in Astrophysical Accretion Disks. *Astrophys. J.* **2015**, *805*, 105. [CrossRef]

76. Bisnovatyi-Kogan, G.S.; Lovelace, R.V.E.; Belinski, V.A. A Cosmic Battery Reconsidered. *Astrophys. J.* **2002**, *580*, 380–388. [CrossRef]

77. Romanova, M.M.; Ustyugova, G.V.; Koldoba, A.V.; Chechetkin, V.M.; Lovelace, R.V.E. Dynamics of Magnetic Loops in the Coronae of Accretion Disks. *Astrophys. J.* **1998**, *500*, 703–713. [CrossRef]

78. Parfrey, K.; Giannios, D.; Beloborodov, A.M. Black hole jets without large-scale net magnetic flux. *Mon. Not. R. Astron. Soc.* **2015**, *446*, L61–L65. [CrossRef]

79. Crooker, N.U.; Burton, M.E.; Siscoe, G.L.; Kahler, S.W.; Gosling, J.T.; Smith, E.J. Solar wind streamer belt structure. *J. Geophys. Res.* **1996**, *101*, 24331–24342. [CrossRef]

80. Kahler, S.; Crooker, N.U.; Gosling, J.T. Properties of interplanetary magnetic sector boundaries based on electron heat-flux flow directions. *J. Geophys. Res.* **1998**, *103*, 20603–20612. [CrossRef]

81. Landau, L.D.; Lifshitz, E.M. *Electrodynamics of Continuous Media*; Elsevier: Amsterdam, The Netherlands, 1960.

82. Landau, L.D.; Lifshitz, E.M. *The Classical Theory of Fields*; Elsevier: Amsterdam, The Netherlands, 1975.

83. Bogovalov, S.V. Boundary conditions and critical surfaces in astrophysical MHD winds. *Astron. Astrophys.* **1997**, *323*, 634–643.

84. Contopoulos, J. General Axisymmetric Magnetohydrodynamic Flows: Theory and Solutions. *Astrophys. J.* **1996**, *460*, 185. [CrossRef]

85. Bogovalov, S.V. On the Theory of Magnetohydrodynamic Winds from a Magnetosphere of Axisymmetric Rotators. *Mon. Not. R. Astron. Soc.* **1994**, *270*, 721. [CrossRef]

86. Fukumura, K.; Kazanas, D.; Shrader, C.; Behar, E.; Tombesi, F.; Contopoulos, I. Magnetized Disk Winds in NGC 3783. *Astrophys. J.* **2018**, *853*, 40. [CrossRef]

87. Bogovalov, S.V.; Tronin, I.V. Numerical simulations of dissipationless disk accretion. *Astron. Lett.* **2017**, *43*, 595–601. [CrossRef]

88. Bogovalov, S.V.; Tronin, I.V. Toward the self-consistent model of cold disk accretion. *Int. J. Mod. Phys. D* **2018**, *27*, 1844005. [CrossRef]

89. Walsh, J.L.; Barth, A.J.; Ho, L.C.; Sarzi, M. The M87 Black Hole Mass from Gas-dynamical Models of Space Telescope Imaging Spectrograph Observations. *Astrophys. J.* **2013**, *770*, 86. [CrossRef]

90. Di Matteo, T.; Quataert, E.; Allen, S.W.; Narayan, R.; Fabian, A.C. Low-radiative-efficiency accretion in the nuclei of elliptical galaxies. *Mon. Not. R. Astron. Soc.* **2000**, *311*, 507–521. [CrossRef]

galaxies

MDPI

Article

Relativistic Aspects of Accreting Supermassive Black Hole Binaries in Their Natural Habitat: A Review

Roman Gold

Institute for Theoretical Physics, Johann Wolfgang Goethe-Universität Max-von-Laue-Str. 1, 60438 Frankfurt am Main, Germany; gold@itp.uni-frankfurt.de or rgold@perimeterinstitute.ca or rg.roman.gold@gmail.com; Tel.: +49-(0)69-798-47873

Received: 4 December 2018; Accepted: 24 May 2019; Published: 31 May 2019

Abstract: In this review a summary is given on recent theoretical work, on understanding accreting supermassive black hole binaries in the gravitational wave (GW)-driven regime. A particular focus is given to theoretical predictions of properties of disks and jets in these systems during the gravitational wave driven phase. Since a previous review by Schnittman 2013, which focussed on Newtonian aspects of the problem, various relativistic aspects have been studied. In this review we provide an update on these relativistic aspects. Further, a perspective is given on recent observational developments that have seen a surge in the number of proposed supermassive black hole binary candidates. The prospect of bringing theoretical and observational efforts closer together makes this an exciting field of research for years to come.

Keywords: GRMHD; numerical relativity; relativistic astrophysics; jets; high energy astrophysics

1. Introduction

The study of supermassive black hole binary evolution is an old research field dating back at least as far as 1980 [1,2] where key physical processes operating on a diverse range of length scales were identified. In this pioneering work it was readily realized that following a galaxy merger event, the two black holes assumed at the centers of each progenitor galaxy would slowly sink towards the center of the common gravitational potential via a combination of dynamical friction and more generally through N-body interactions with the stellar population. While the arguments draw a rather clear picture free from major ambiguities at large separations, the behavior nearing pc scale separations and lower would later prove to be much more involved.

1.1. kpc → pc: *Galaxy Mergers and Relaxation, Key Physics: Newtonian Self-Gravity*

The Hubble space telescope famously delivered direct images of mergers of galaxies on kpc scales [3] thereby proving that galaxies do collide and establishing the first necessary step towards supermassive black hole binary coalescence. Finding similar evidence between pc and kpc scales requires more resolution and necessarily leads to a smaller population to be observed. Despite these technical hurdles more recent observational efforts take on this challenge in a variety of different ways [4].

In many black hole (BH) systems, the mass accretion rate is at intermediate levels causing the accretion flow to be of a geometrically thin, optically thick type. In principle, by modeling the thermal radiation one can probe the location of the innermost stable circular orbit (ISCO) and hence the BH spin via the so-called continuum fitting method. However, this technique is far more powerful for stellar-mass systems and challenging to apply to active galactic nuclei (AGN) [5]. In addition, AGNs in the thin-disk regime can exhibit strong X-ray emission in the polar regions above and below the black hole that irradiates the colder disk material. This process can excite atomic transitions and gives rise to fluorescence in the process. In particular, the astrophysically-abundant iron coupled with its high fluorescent yield leads to a strong Fe Kα line.

Spectroscopic efforts sport an impressive spectral resolution and sensitivity [6], which allows detailed measurements of Fe Kα emission line profiles, which is affected by several relativistic effects. Many inferences on BH spin [5] via relativistic effects have been achieved with this method in what we believe to be single black hole systems. It is expected that Athena [7] (launch date 2031) will greatly boost the prospects of this research field. The same method can also be applied to binary black hole sources. The spectral profile of such a line is then very sensitive to the relativistic motion of the emitting gas producing asymmetric double-horned line profiles. Modeling such line profiles and comparing to observations offers a probe of the spacetime structure of the BHs in the strong-field regime. Secular trends may even be used to uncover orbital motion via an offset between the broad and narrow-line regions on the orbital time scale [8] or spectral imprints due to a radially migrating secondary [9].

Complementary efforts that also do not rely on spatially resolving the source seek to reveal periodicities in light curves and attempt to model the line profile through relativistic orbital motion [4]. Spectroscopic surveys are likely to identify numerous, additional candidate sources in the gravitational wave (GW)-driven regime [10].

An interesting candidate for a supermassive black hole (SMBH) binary is the well-known blazar source OJ-287 [11–14], whose optical lightcurve extends back 150 years showing several periodic features. However, alternative interpretations are being put forward, see e.g., the recent work in [15], that identifies the variability not with binary motion but with a precessing jet. Ultimately a combination of improvements in theoretical modeling and more observational data can be expected to inform us on the true nature of this fascinating source.

Other candidate SMBH binaries based on X-ray data (with XMM-Newton) [16] or optical data [17–20] have been proposed. Typical orbital time scales inferred in these studies are on the order of years, which is comparable to the duration of monitoring. This creates what is sometimes referred to as red noise. The main obstacle in these studies is therefore the limited observing period.

It remains key to extend our observational data sets both to inform theoretical work and to better our understanding of the physical processes at work throughout this long process. The most relevant physics strongly varies with binary separation: For galaxy merger events occurring on kpc scales and also for stellar distances on pc scales Newtonian gravity and perhaps hydrodynamics (without magnetic fields) is an adequate description. On scales comparable to the gravitational radius of the black holes of order $1AU$ gravity and spacetime itself becomes the dominant force in determining the fate of the system. Therefore, the most realistic models in the regime where GW emission with an electromagnetic counterpart can be expected, solve the equations of General Relativity (GR) and magnetohydrodynamics (MHD). We now discuss each epoch of the binary's evolution in turn.

1.2. pc: *The Final-Parsec Problem, Key Physics: Newtonian Self-Gravity + Hydrodynamics*

At separations of $\sim$ 1pc relativistic aspects of gravity are still unimportant, but the local disk mass (self-gravity) cannot be ignored. A brief discussion of this Newtonian physics is in order, because all studies of relativistic aspects rest on the assumption that binaries in the GW-driven regime can form in the first place. The current understanding of SMBH binary evolution on pc scale orbital separation and below is necessarily largely based on theory thus far.

The theoretical description up until pc scale separations seems rather unambigious and uninterrupted by any major theoretical roadblocks. This situation changes when the binary reaches separations of order 1pc, because the known drivers of inward binary migration at larger scales, i.e., interaction of the binary with the surrounding stellar dynamics, may grind to a halt under some circumstances. At the heart of the problem is the fact, that only stars with the right orbital properties may come in close enough to interact with the binary and take energy out of the system. Within the assumptions of these pioneering studies, see e.g., [21,22], in particular spherically symmetric distribution of stars, purely gravitational N-body interactions, the supply of such stars with the suitable orbital properties become unavailable at pc scale separations. At these orbital separations gravitational radiation is not yet efficient enough to cause sufficient energy loss. Lacking a similarly obvious contender or other possible remedies at the time to drain orbital energy from the binary the term "final-parsec problem" was coined [21,22]. One unfortunate side-effect of this development was, that it generated a strong paradigm. It was suspected or assumed that SMBH binaries simply stalled at pc orbital separations, despite a quickly growing literature that started systematically questioning or relaxing the assumptions made [23–29] involving self-gravity of the hydrodynamic component, non-spherical distributions of stars, and others all showing great potential to drive binary inward migration. A full discussion of the evolution through the parsec scale is far beyond the scope of this review, but it seems increasingly more difficult to ignore the wealth of possible ways for nature to overcome the (theoretical, but not actual) parsec scale barrier. It will be of greatest interest in the context of the Laser Interferometric Space Antenna (LISA) [30,31] mission to understand in detail when and at what rate supermassive black hole binaries merge as a function of cosmic time. Here only a rough sketch is given of various phases in the binary evolution on different length scales and the dominant physics that must be modeled. See [32] for a more detailed and comprehensive discussion.

1.3. sub-pc-r_g: Gravitational Wave-Driven Regime, Key Physics: GR + Magnetohydrodynamics (MHD)

At orbital separations a somewhat below the pc scale or more quantitatively

$$a_{GW} \sim 2 \times 10^{-3} f(e)^{1/4} \frac{q^{1/4}}{\sqrt{1+q}} \left(\frac{M_{BHBH}}{10^6 M_\odot} \right)^{3/4} pc \tag{1}$$

$$f(e) \equiv \left(1 + \frac{73}{24} e^2 + \frac{37}{96} e^4 \right)^{-7/2} \tag{2}$$

where e is the orbital eccentricity, q the mass ratio, and M_{BHBH} the total mass of the binary, the binary will coalesce in less than a Hubble time merely by energy and angular momentum losses due to GW emission losses which depend sensitively on the orbital eccentricity [33,34].

Gravitational radiation therefore becomes a critical element in determining the subsequent evolution of the system as a whole. It is this regime and everything that follows where the focus of this review lies.

At sub-pc length scales the impact of gravitational-wave-driven losses is still subdominant but in principle strong enough for a detection by Pulsar Timing Arrays (PTAs) [35], which are most sensitive at low ($f \sim nHz$) frequencies. However, only separations shorter than about milli pc separations ensure orbital periods of a few years or less and therefore avoid excessively long observational campaigns. This provides key information on population synthesis and the assembly of supermassive black holes through cosmic time [36–40]. Especially in light of recent surges in the number of proposed binary candidates, even current constraints from the unresolved GW background from PTAs are starting to become informative, favoring the hypothesis that false positives may be quite abundant in the current catalogues of proposed binary sources [41]. PTAs attempt to measure gravitational waves through detecting minute, but correlated changes in the time of arrival of pulses from a network of extremely well-timed pulsars. This galactic scale interferometer becomes more and more sensitive over time as more pulsars are discovered and the timing model of existing ones is improved further and further as more data is collected. On scales of several to hundreds of gravitational radii, the gravitational waves emitted will constitute a large signal for the LISA mission planned in the early 2030s, which e.g., would allow a pre-merger localization of the source [42].

1.4. Binary–Disk Decoupling

The rate of orbital decay of the binary due to the emission of gravitational waves [33,34] is

$$\frac{t_{GW}}{M} \sim 3000 \left(\frac{a}{10M}\right)^4 \frac{(1+q)^2}{4q}, \tag{3}$$

where $0 \leq q \leq 1$ is the binary mass ratio, and M the total mass[1]. Note, in particular, that t_{GW} scales differently with binary separation a than the response of the disk governed by the viscous time scale

$$\frac{t_{vis}}{M} = \frac{2R_{in}^2}{3\nu M} \sim 5000 \left(\frac{R_{in}}{16M}\right)^{3/2} \left(\frac{\alpha}{0.15}\right)^{-1} \left(\frac{H/R}{0.3}\right)^{-2}. \tag{4}$$

Here R_{in} is the location of the inner disk edge, α the Maxwell stress (describing the effective viscosity due to magneto-rotational instability (MRI) turbulence), and H/R the vertical scale height of the disk. As a result, there will be a separation, called the decoupling separation a_d, where the two time scales match:

$$\frac{a_d}{M} \approx 11.9 \left(\frac{R_{in}}{1.6a_d}\right)^{3/5} \left(\frac{\alpha}{0.15}\right)^{-2/5} \left(\frac{H/R}{0.3}\right)^{-4/5} \left(\frac{\tilde{\eta}}{1}\right)^{2/5} \tag{5}$$

$\tilde{\eta} \equiv \frac{4q}{(1+q)^2}$. For separations $d > a_d$ the disk can adjust to the shrinking orbit and follow the binary to smaller separations. For separations $d < a_d$ the binary orbital decay is becoming too rapid, thereby leaving the disk material behind. During these fast late inspiral phases the binary tidal torques also diminish quicker than the disk can respond. The disk material from this moment on drifts slowly radially inward.

When the black holes reach separations comparable to their size they coalesce forming one black hole. This regime marks the most significant departure from the Newtonian regime and only numerical relativity techniques can describe this regime. The merged black hole is larger than each progenitor, but the overall gravitational mass is lower than the combined mass of both progenitors due to significant energy that is radiated away in the form of gravitational waves. This reduction in total mass poses a sudden perturbation to the disk material and can cause transient effects [43,44]. If the binary configuration is not too symmetric the gravitational wave linear momentum will be radiated away in a non-spherically symmetric way imparting a GW recoil or "kick" onto the merger remnant that can either cause long term oscillatory motion between the black hole and the disk material or eject the central black hole altogether from the system. In most cases one expects the former [45–47] and the system will settle on several viscous time scales down to the classical accreting single black hole configuration.

1.5. Lessons Learned from Single Black Hole Accretion

In many respects, the study of the evolution of black hole *binaries* orbiting in a gaseous environment, can benefit greatly from insights gained from the study of *single* black hole accretion. The latter is a far more mature field of research with many independent groups world-wide actively contributing to solidify our knowledge of these systems. However, it is still fair to say that rather major theoretical uncertainties still prevail in some regimes, e.g., hot, puffy radiatively inefficient accretion flows [48] and the Eddington–Super Eddington regime [49], especially when it comes to their observational appearance.

It has long become the norm in single black hole accretion simulations to treat General Relativistic effects, e.g., to include the horizon and correct gravitational field properly. Studies of magnetized

---

[1]   We have normalized to a separation of $10M$ as was used in studies with full general relativistic magnetohydrodynamics (GRMHD). This value is only slightly smaller than the decoupling radius in Equation (5). After all, these time scales only give crude estimates and due to the steep scaling of t_{GW} with a the resulting inspiral time scales can seem prohibitively large.

accretion flows onto black hole binaries by extension also must be modeled with the effects of general relativity included or risk an uncontrolled assumption in the models. In particular, a proper description of the magnetically dominated outflows, the orbital decay (and eventually decoupling) of the binary, or to incorporate a region where no stable circular orbits are possible demands general relativistic effects to be included.

2. Results

A previous review article [50] focussed on Newtonian aspects of accreting black hole binaries. Since then several studies were conducted that have taken into account relativistic aspects of gravity. These findings have advanced our understanding of the role of accreting black hole binaries as electromagnetic counterparts to gravitational wave sources. Therefore an update to [50] is in order.

Initially, relativistic studies of black hole binaries in gaseous environments did not include magnetic fields due to numerical/technical issues arising from highly magnetized regions near the black hole event horizons, which is inherently more difficult to handle in the dynamic spacetime case compared to a time-independent spacetime like a Kerr black hole. The potential for electromagnetic (EM) counterparts to a strong GW source based on purely hydrodynamic studies [51–54] were underestimated due to the neglect of magnetic fields [55–58]. In the end, the role of binary tidal torques on driving accretion was overestimated. Instead, effective viscosity due to turbulent motion driven by the magneto-rotational instability [59,60] by far dominates angular momentum transport. This can most clearly be seen in the side-by-side comparison of a pure hydrodynamic and a magnetized model with the same code and grid setup, see Figure 6 in [57]. In fact, there were already strong indications for this in an earlier Newtonian study [61]. Now with magnetic field dynamics and relativistic gravity included, it is becoming ever more clear that accretion rates onto black hole binaries can be of the same magnitude as is known in single black hole systems, albeit with a different, certainly non-axisymmetric structure composed of two dense accretion streams. The parameter space describing an accreting binary black hole system is naturally higher dimensional and hence more computationally expensive to explore than the single black hole case.

First explorations in mass ratio [57,58], different separations [62] and disk height dependence [63] have led to an enhanced understanding of these systems.

2.1. New Structural Features in the Binary Case

The dominance of magnetically-induced, effective shear viscosity in driving accretion does not mean, however, that binary torques are unimportant. Quite to the contrary: Binary tidal torques play a dominant role in the structure of the inner disk and shape of the accretion flow, giving rise to a number of structures that are not seen in accreting systems composed of only a single black hole.

2.1.1. Cavity and Pile-Up

The very first theoretical discovery of a potential signature due to a binary, that is imparted on the structure of the accretion flow, is a cavity [64,65] inside which the density of the gas is much lower than elsewhere in the accretion flow. Such a cavity is formed where binary tidal torques dominate the angular momentum budget, thereby pushing gas elements away from the binary orbit leaving behind a region of substantially lowered density. A close analogy to planet formation in protoplanetary disks [66–68] can be drawn here: Depending on the mass ratio of the star-giant planet system binary tidal torques may clear out a gap in the disk thereby influencing the radial motion ("migration") of the binary. The border of the cavity/gap, i.e., the inner edge of the circumbinary disk features a pile-up, an axisymmetric overdensity caused by material being pushed from both radial directions towards this boundary. This cavity can have important observational consequences in the form of altering the main locations of emission and introducing coherent, i.e., ordered but not necessarily periodic, time variability.

2.1.2. Non-Axisymmetric Structures: Streams and Lumps

A feature also known from protoplanetary disks are overdense accretion streams across the secondary binary component. Two such dense accretion streams are readily identified in circumbinary accretion disks around SMBH binaries, too, see Figure 2. A persistent, non-axisymmetric overdensity, termed a "lump" and its implications were studied in detail in [55,61,69,70], see Figure 1. Such a feature is thought to arise due to non-accreting matter being repeatedly sloshed back towards the inner edge of the circumbinary disk. The gas that is driven back out, is ejected in a directionally preferred orientation, thereby gradually building up a somewhat local overdensity that then orbits near the inner edge of the disk. The lump exhibits overdensities of a factor of 4 over the average density in the rest of the disk and covers 1–2 radians in the azimuthal domain [69]. Crucially, in the relativistic treatment an $m = 1$ mode appeared to dominate over $m = 2$ modes seen in Newtonian studies [71]. Such a structure can similarly to the accretion streams produce coherent time variable signatures.

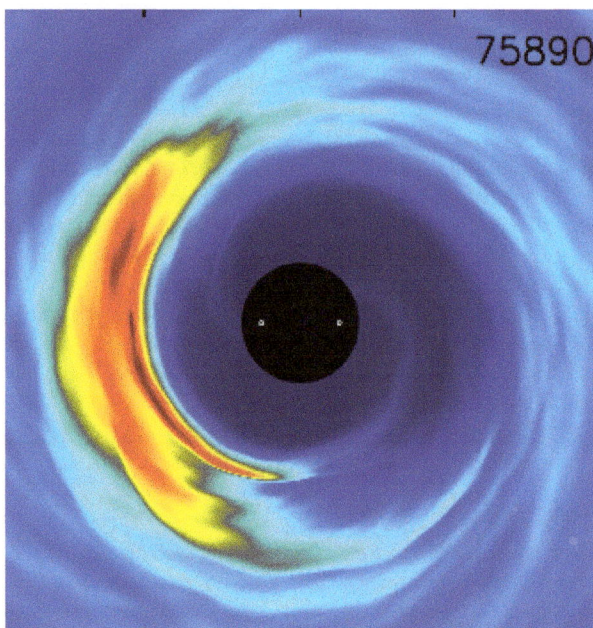

Figure 1. Rest-mass density (linear color scale) from [55] a GRMHD simulation in a spacetime framework similar to [72]. A persistent non-axisymmetric overdensity termed a "lump" develops in the bulk of the disk.

2.1.3. Mini-Disks and Mass Sloshing

At large enough binary separations a region analogous to the Newtonian Hill-sphere in which gravity is dominated by the near-by binary component allows for mini-disks to form. This phenomenon was investigated in [62,69]. In total the system can therefore in principle feature three disks: (i) A mini disk around the primary, (ii) another mini disk around the secondary and (iii) the circumbinary disk. In general, these three disks are interacting with each other and lead to episodic dynamics that affect the accretion onto the black holes. Some qualitative conclusions can immediately be drawn. For instance, if there is a BH driven jet or a dual jet structure emerging from the system, it will be fueled by the material that is closest to the black hole horizon, i.e., the mini disks if present, not the circumbinary disk. The mini-disk-driven accretion flows around each black hole move at relativistic velocities with the binary orbit, which has great potential to leave clear smoking-gun signatures in observations.

One important relativistic aspect, is that mini disks cannot persist towards merger and in fact cannot last as long as one would expect from Newtonian estimates. While in Newtonian gravity test particles can orbit arbitrarily close to a central gravitating object forever, test particles around black holes in General Relativity can only do so outside a finite radius, the radius of the innermost stable circular orbit. The argument is strict for single BHs, but can be expected to at least qualitatively carry over for the binary case as well. As a result, mini disks can only exist when the extent of the Hill sphere extends beyond the innermost stable circular orbits around each BH. As the binary then shrinks due to the emission of gravitational waves, so do the Hill spheres (linearly with binary separation) until the Hill spheres are too small for matter to find stable trajectories around the near horizon regions. When this occurs the mini disks should be accreted on a comparatively short time scale. As first pointed out by [58], it is this mechanism that explains why in the full relativistic studies published so far [56–58] no persistent mini-disks were seen, while mini disks seem to be a generic feature in studies at larger binary separation, e.g., [55,69].

One particular process identified in these mini-disk systems is a mass sloshing from one mini-disk to the other with cycles of depletion and refilling of the mini disks that are fed by the circumbinary disk [62,69,70], see Figure 2.

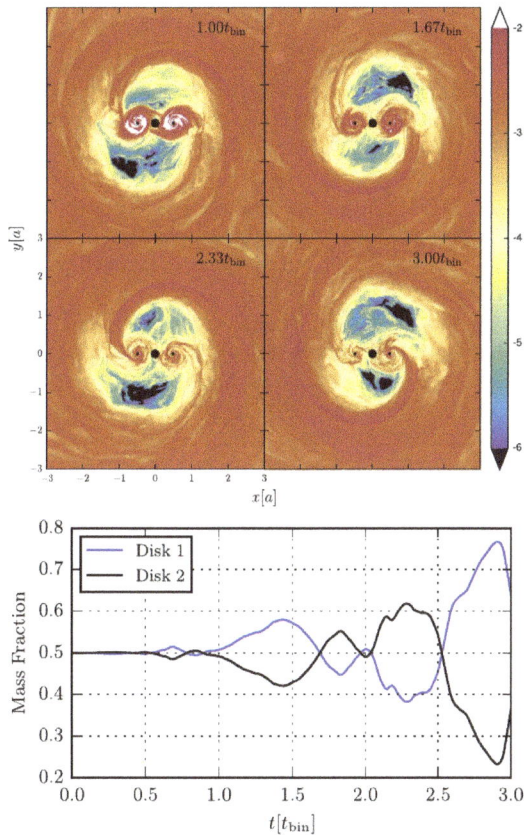

Figure 2. Rest-mass density (linear color scale) from a GRMHD simulation in a spacetime framework similar to [69,72] (**upper panel**) and individual disk masses as a function of orbital phase (**lower panel**). One can clearly see in the lower panel, that mass is sloshing back and forth between the two mini disks. This effect is shown to persist over longer time scales in the follow-up work [70].

2.1.4. Enhanced Variability

Robust features in all theoretical models of circumbinary accretion onto SMBH binaries like accretion streams are promising candidates to produce coherent time dependent signatures in the electromagnetic output of the system.

Similarly, the resulting dynamical impact of a lump-like structure leads to periodic or quasi-periodic variability as the orbital period of the lump and its close passages with the accretion streams can beat against each other [69].

There are unambiguous findings of preferred time scales and quasi-periodic signatures in the mass accretion rate at binary separations where the orbital decay is noticable but slow, see e.g., in [62,69]. Such relatively clean frequencies in the matter dynamics and mass accretion rate are not seen in fully relativitistic regimes [57,58], most likely because no persistent mini-disks are formed. It is not fully clear whether or not further differences in the models, such as different matter configurations, see e.g., [73], thermodynamics, or something else, are responsible. However, the combined theoretical efforts by the community thus far, indicate that accreting BH binaries, at least those surrounded by puffy hot accretion flows, do reveal quasi-periodic behavior, but only up until some finite binary separation shortly before merger. This transition from quasi-periodic to generically variable (non-quasi-periodic) behavior could be one of the most robust EM precursors to the merger event.

2.2. Connecting Theoretical Models and Observational Data

It is generally extremely important to distinguish between the behavior of intrinsic, dynamic quantities such as the fluid density and the observational appearance in a given EM waveband. A first and substantial advancement in this direction was given in [74]. This study was carried out at a larger binary separations of $d = 20M$, which is in the predecoupling regime, adopting a treatment for the spacetime metric based on [72], which is particularly relevant for potential Pulsar Timing Array sources This study incorporates, for the first time, direct information from radiative transfer effects. These studies constitute the best basis yet for judging electromagnetic emission from supermassive black hole binaries embedded in a magnetized, gaseous environment in the GW-driven regime. Figure 3 shows several ray-traced images of the system. One can nicely see how the strong lensing effects lead to major distortions in the images compared to a flat space treatment. In Figure 4 one can further see theoretical spectra computed from the GRMHD simulations used in [74]. One can clearly see the different contributing components to the broad band spectral energy distribution (SED). The thermal emission from the mini disks dominate over the streams and circumbinary disk emission at high-energy emission. The accretion streams and circumbinary disk on the other hand dominate at lower frequencies and appear to be difficult to distinguish at least based on spectral grounds and for the cases considered.

If system parameters and EM waveband lead to a situation where the flow is too optically thick, then information on the binary motion will at best be smeared out when viewed from outside the system. For optically thin cases, on the other hand, outgoing radiation can probe directly the binary orbital motion. It may still be possible to detect binary imprints like periodic features in light curves, non-axisymmetric structures, pile-ups etc. However the most desirable regime to gain information on the binary motion is a regime where the source is optically thin. One practical consideration is that optically thin emission is less efficient and typically produces dimmer sources which are then rarer due to an observational bias.

Also, specific features, such as mini-disks, although seen in GRMHD studies, need not necessarily shine strong enough to contribute to the EM output. At least in some accretion regimes with certain thermodynamical properties of the gas it is conceivable that jet emission may dominate over disk emission.

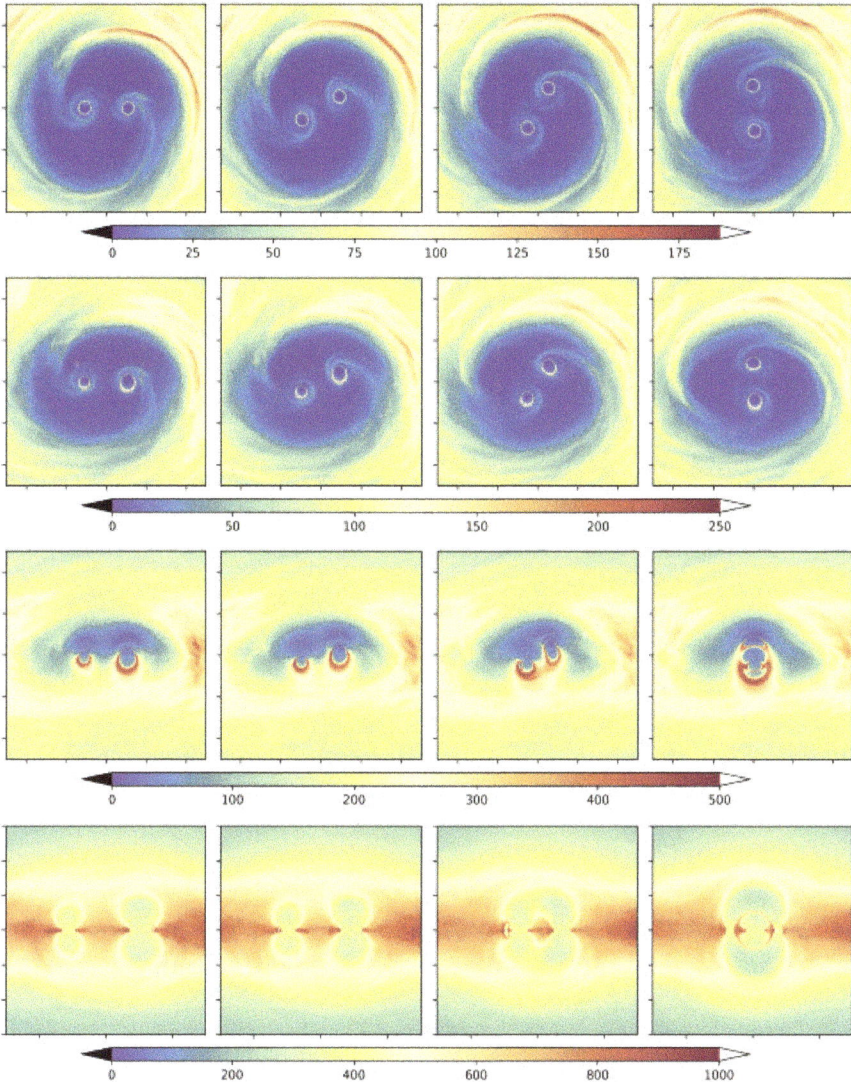

Figure 3. Images of optical depth obtained by properly tracing light rays through the spacetime for different times (columns) from circumbinary accretion around supermassive black holes from [74] for inclination angles 0°, 39°, 71°, 90° (from top to bottom row). Strong lensing effects introducing complicated image distortions are clearly visible.

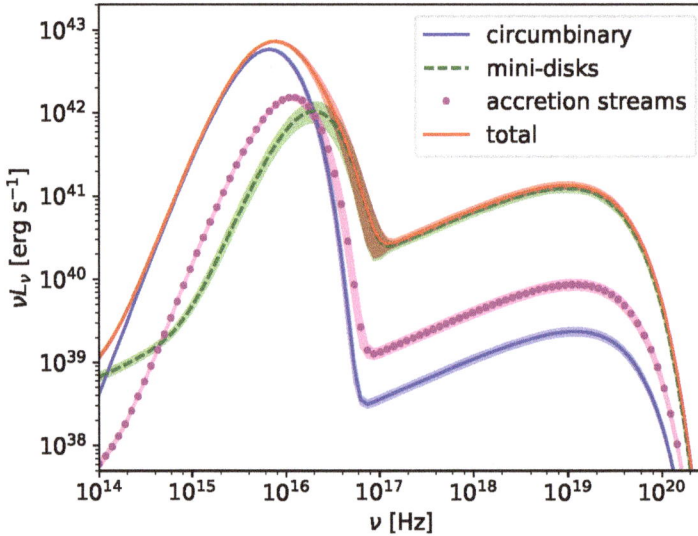

Figure 4. Different components from the various disks present in the system of [74] contributing to the spectrum. Mini disks reveal their emission at higher energies whereas emission from the circumbinary disk at larger radii contributes more at lower frequency. Emission from the accretion streams falls in the middle of these two regimes and based on this analysis may be difficult to distinguish from the other contributions.

2.3. GRMHD in Dynamic Spacetime

Full GR studies [56–58,73,75] incorporate dynamic spacetime evolution with spatially resolved black hole horizons. These ingredients are not only appropriate, but in fact essential to enable any prediction for the emergence of magnetized, incipient jets resulting from twisting the magnetic field lines that are anchored both in the accretion flow and the orbiting and (potentially also) spinning black hole horizons, see Figure 5. The formation of jets in these systems is one of many purely relativistic aspects of this problem, see also the force-free study in GR [75]. These studies focus on the accretion flow and jet dynamics in the dynamic spacetime. Solving the full set of Einstein's equations on non-uniform grids along with the equations of ideal MHD is a challenging task especially for long computational times as is required for accreting systems. Long term stability was achieved in [56] using an innovative approach to evolve the magnetic vector potential with a new gauge condition, the damped Lorenz gauge. This gauge condition damps EM gauge waves towards zero, thereby minimizing spurious amplifications of magnetic field strength at AMR boundaries. Only indirect diagnostics such as Poynting flux or approximative thermal luminosity based on an artificial emissivity prescriptions were given.

Just like for single BHs the system has to be evolved for long enough such that the system can respond on a viscous time scale to settle into an inflow outflow quasi-equilibrium. Typically such an equilibration takes place "inside-out", because the viscous time scale grows (steeply) with distance from the center. As a result, the longer a GRMHD simulation is run, the larger its inner region of validity where the flow has forgotten about its somewhat artificial initial state.

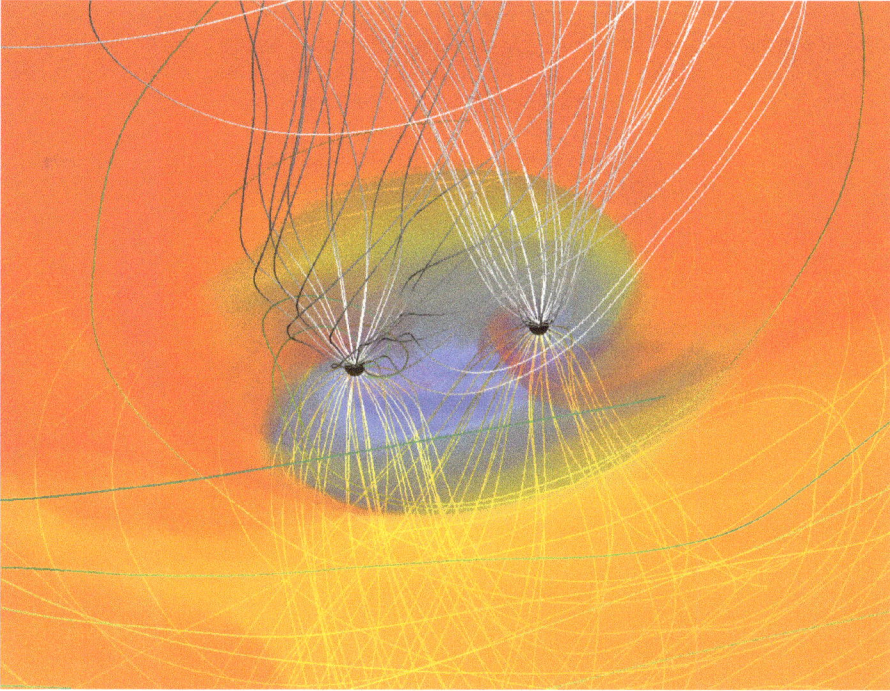

Figure 5. Rest-mass density (color scale) and magnetic field structure (white lines) of a fully relativistic GRMHD simulation in dynamical spacetime from [58].

2.4. During and Post Merger

In the last few orbits and merger the directed GW emission is increasing sharply with time and does not average out over the course of an orbit. The linear momentum that was emitted in the form of GWs is compensated by a recoil on the merger remnant to conserve overall momentum [76–78]. The resulting kick velocity depends on the binary parameters, especially the magnitude and relative orientation of the BHs spin w.r.t. their orbital angular momentum and varies from a modest (but still impressively non-pedestrian) 200 km/s to 1800 km/s and perhaps as large as 4000 km/s for maximally spinning black holes [79]. This range of "kick" velocities exceeds typical galaxy escape velocities. Therefore, depending on the kick velocity attained, two qualitatively distinct possibilities arise: (i) For kick velocities lower than the escape velocity, the merger remnant will oscillate relative to the bulk of the disk dragging with it only the matter in the immediate vicinity of the BH. (ii) For kick velocities exceeding the escape velocity of the galaxy hosting the system, the black hole will take off into space again dragging only the material in its nearest vicinity with it and leaving behind the bulk of the disk without a central gravitating object. Which of those scenarios is more common is an interesting and still somewhat open topic that finally awaits observational data. Indications are that most mergers will lead to bound BH remnants [45–47,80]. Observationally such BH remnants that were kicked out of their host galaxy or have been displaced from the gravitational center could be detectable through spectral signatures, see e.g., [81] and by using H_2O megamasers [82].

The scenario following a merger with high kick velocities has been modeled [43,44] by Lorentz boosting the disk material but not the black hole. This treatment does not take into account the various remnant structures imparted in the system by the past binary system, but allows one to study the system under these idealized assumptions in a systematic and more efficient way.

After a typical merger with low (or even hardly any kick velocity) the accretion flow is substantially different from that around a single BH. Consequently, there will be a period of adjustment until the source transitions into the classical single BH accretion state on a few viscous time scales of the accretion flow. This phase is expected to lead to a slow brightening of the source over time as more matter diffuses into the cavity that was carved out by the past binary tidal torques. This phase has not been simulated yet and is different from increases in jet luminosity found in [58]. As discussed before the expected level of density decrease in the cavity has seen some variation recently, especially in the context of studies that include the dynamics of magnetic fields and the resulting strong effective viscosities. So, it is not clear how much of a rebrightening one truly expects and how the phenomenology scales with decoupling radius (dictated by disk height).

2.5. Distinguishability of Single vs. Binary AGN

In many respects current studies only scratch the surface of many important aspects of accretion onto binary black holes. For instance, one key development in the field will be to improve our general understanding on how different binary black hole systems are from single black hole systems. This is a complex question not only because of the already large spread in phenomenology even in single BH systems. So the differences for any given binary system to a corresponding single BH system will depend on many things including but not limited to, the mass supply, the thermodynamic regime of the gas and the EM waveband one observes in. Theoretical implications already point to clear differences between accreting single and binary black holes in the form of different density distributions or mass accretion rates, see [57] for a single vs binary black hole comparison with the same code and setup. The circumbinary disks used in various GRMHD studies differ greatly in their spatial extent and also in the amount of magnetic flux on the horizon at late times. It is well known that the magnetic flux is one key parameter in accreting single black holes [83–85]. Therefore, by extension understanding the influence of different levels of magnetic flux will have to be an important aspect of future studies also in the circumbinary disk case.

On the other hand, several discoveries such as the lump, the presence of a cavity, accretion streams, and mini disks along with characteristic variability are promising indications that such a distinction will be possible at least for some physical regimes. Key steps towards making theoretical predictions more relevant to what is actually observed [74,86] have been started and should constitute a solid foundation for future model building efforts to extract SMBH binary science out of observational data in the near future.

3. Discussion/Conclusions

The last few years have seen major progress in our understanding of accreting black hole binaries that are in the gravitational wave driven regime. Most importantly there is unanimous agreement among theoretical models from various groups that magnetically-driven effective viscosity dominates the angular momentum flux, causing accretion rates onto the binary that are very much comparable to rates encountered in single black hole systems with the same total mass. For the same reason, a decoupling of the supermassive black hole binary from the circumbinary disk long before merger as originally discussed in [21,64] can in fact occur only very few orbits prior to merger at least for disks with a sufficiently large scale height $H/R \gtrsim 0.3$. In some respects, the sources share many similarities with and may in fact not always appear as different to single accreting BH sources as one would like. One key difference is that the mass inflow is not as axisymmetric near the black holes but instead very concentrated into the two dense accretion streams as well as the mini disks.

On the other hand there are equally clear and robust differences that are expected to yield distinct signatures that point to the binary nature of a source. The differences involve characteristic structural changes in the accretion flow (cavities, streams, lumps) which give rise to image, spectral (lack of emission due to cavity, enhanced emission due to overdensities) and time-dependent (periodicities, beating, sloshing) signatures. On top of these effects from the matter distributions and dynamics additional effects such as Doppler boosting will further enable us to extract scientific predictions by matching theory to observations. In [87] possible detection rates are estimated under the assumption of two variability models: Doppler boost related variations and periodic variability due to accretion flow dynamics. It is predicted that the Catalina Real-Time Survey (CRTS) [88] might see a few binary sources but 5 years of the Large Synoptic Survey Telescope (LSST) [89] (first engineering light 2019, science runs in 2021) can detect tens to hundreds of them.

The first key steps towards first theoretical model predictions closer to observational data have been taken [74] with radiative transfer calculation. It will be a major undertaking in the field in the future to advance such calculations and at the same time start to cover the vast parameter space of these sources, which is even larger than the combined parameter sets known from binary black hole coalescences (in vacuum). In addition the phenomenology of just the accretion flows onto supermassive black hole binaries themselves is truly immense, the accretion flow structure being even more feature rich than for single BHs and radiation possibly originating from these separate regions and via distinct physical mechanisms. Aside from the emission from shocks other potential emission mechanisms are expected to operate. One main engine for (polarized) synchrotron emission has for a long time been, magnetic energy as provided by MHD turbulent dynamics [48,60,85] accompanied by dissipation effects e.g., magnetic reconnection at small scales of its turbulent cascade. Recently, however, indications for a more ordered larger scale magnetic field structure more akin to the magnetically arrested state [90–92] of accretion seems favored at least in Sgr A* but [93,94] maybe also in other systems.

These are fascinating times for understanding accretion physics in strong field gravity. Theoretical efforts to provide us with testable predictions as well as observational data that probe the relativistic regime around black holes are around the corner. Gravitational waves which have famously been observed in the stellar mass black hole case will be detected for supermassive black hole binaries eventually, either by PTAs or a space-based laser interferometer like LISA. In the meantime electromagnetic emission across several wavebands can be used to refine our understanding of black holes and how they feed in their natural habitat. In particular, the Event Horizon Telescope [85,95–99] has reached resolutions capable of spatially resolving not only the two largest black hole event horizons as seen on the sky (in Sgr A* and M87) along with accretion flows governed by the physics of general relativistic magnetohydrodynamics but also sub-pc supermassive black hole binaries. Such resolved detections would not suffer from limitations of periodicity searches and be obtained even when the orbital period is large compared to the observational campaign. In fact, some binary candidate sources such as Ark 120, OJ-287 and others may already be spatially resolved. More generally, targeted very-long-baseline interferometry (VLBI) efforts will be able to resolve such binaries, see [100], see also a similar idea in [101]. At the current configuration of the Event Horizon Telescope *any* binary at a separation larger $\gtrsim 0.1$pc can be spatially resolved *throughout the universe*. This is possible thanks to recent technical improvements, the availability of very long baselines at 230 GHz (and 345 GHz in the near future), and the fact that angular separation projected on the sky is not a monotonic function of redshift in the universe we seem to live in. Obviously, the ability to spatially resolve a source is only one of the inherent challenges. The main technical challenge will therefore be to meet the sensitivity requirements for potentially very distant sources. Fortunately, this is an area where recent technical developments, such as bandwidth increases in the VLBI backends, have caused a surge in sensitivity improvements. From the theory side we can expect that radio emission from binary sources is comparable to single AGN sources which are successfully detected with radio VLBI routinely out to considerable distances.

Funding: This work is supported in part by the ERC synergy grant "BlackHoleCam: Imaging the Event Horizon of Black Holes" (Grant No. 610058).

Conflicts of Interest: The author declares no conflict of interest.

References

1. Begelman, M.C.; Blandford, R.D.; Rees, M.J. Massive black hole binaries in active galactic nuclei. *Nature* **1980**, *287*, 307. [CrossRef]
2. Roos, N. Galaxy mergers and active galactic nuclei. *Astron. Astrophys.* **1981**, *104*, 218–228.
3. Mohamed, Y.H.; Reshetnikov, V.P. Interacting galaxies in deep fields of the Hubble Space Telescope. *Astrophysics* **2011**, *54*, 155–161. [CrossRef]
4. Eracleous, M.; Boroson, T.A.; Halpern, J.P.; Liu, J. A Large Systematic Search for Close Supermassive Binary and Rapidly Recoiling Black Holes. *Astrophys. J. Suppl. Ser.* **2012**, *201*, 23. [CrossRef]
5. Reynolds, C.S. Observing black holes spin. *Nat. Astron.* **2019**, *3*, 41–47. [CrossRef]
6. Harrison, F.A.; Craig, W.W.; Christensen, F.E.; Hailey, C.J.; Zhang, W.W.; Boggs, S.E.; Stern, D.; Cook, W.R.; Forster, K.; Giommi, P.; et al. The Nuclear Spectroscopic Telescope Array (NuSTAR) High-energy X-ray Mission. *Astrophys. J.* **2013**, *770*, 103. [CrossRef]
7. Ayre, M.; Bavdaz, M.; Ferreira, I.; Wille, E.; Lumb, D.; Linder, M. ATHENA: System design and implementation for a next generation X-ray telescope. UV, X-Ray, and Gamma-Ray Space Instrumentation for Astronomy XIX. *Proc. SPIE* **2015**, *9601*, 96010L. [CrossRef]
8. Decarli, R.; Dotti, M.; Fumagalli, M.; Tsalmantza, P.; Montuori, C.; Lusso, E.; Hogg, D.W.; Prochaska, J.X. The nature of massive black hole binary candidates—I. Spectral properties and evolution. *Mon. Not. R. Astron. Soc.* **2013**, *433*, 1492–1504. [CrossRef]
9. McKernan, B.; Ford, K.; Kocsis, B.; Haiman, Z. Ripple effects and oscillations in the broad FeKa line as a probe of massive black hole mergers. *Mon. Not. R. Astrono. Soc.* **2013**, *32*, 1468–1482. [CrossRef]
10. Pflueger, B.J.; Nguyen, K.; Bogdanović, T.; Eracleous, M.; Runnoe, J.C.; Sigurdsson, S.; Boroson, T. Likelihood for Detection of Subparsec Supermassive Black Hole Binaries in Spectroscopic Surveys. *Astrophys. J.* **2018**, *861*, 59. [CrossRef]
11. Sillanpaa, A.; Haarala, S.; Valtonen, M.J.; Sundelius, B.; Byrd, G.G. OJ 287—Binary pair of supermassive black holes. *Astrophys. J.* **1988**, *325*, 628–634. [CrossRef]
12. Lehto, H.J.; Valtonen, M.J. OJ 287 Outburst Structure and a Binary Black Hole Model. *Astrophys. J.* **1996**, *460*, 207. [CrossRef]
13. Valtonen, M.J.; Lehto, H.J.; Nilsson, K.; Heidt, J.; Takalo, L.O.; Sillanpää, A.; Villforth, C.; Kidger, M.; Poyner, G.; Pursimo, T.; et al. A massive binary black-hole system in OJ287 and a test of general relativity. *Nature* **2008**, *452*, 851–853. [CrossRef] [PubMed]
14. Valtonen, M.; Ciprini, S. OJ 287 Binary Black Hole System. In Proceedings of the Multifrequency Behaviour of High Energy Cosmic Sources, Sicily, Italy, 23–28 May 2011.
15. Britzen, S.; Fendt, C.; Witzel, G.; Qian, S.J.; Pashchenko, I.N.; Kurtanidze, O.; Zajacek, M.; Martinez, G.; Karas, V.; Aller, M.; et al. OJ287: Deciphering the 'Rosetta stone of blazars'. *Mon. Not. R. Astron. Soc.* **2018**, *478*, 3199–3219. [CrossRef]
16. Liu, F.K.; Li, S.; Komossa, S. A Milliparsec Supermassive Black Hole Binary Candidate in the Galaxy SDSS J120136.02+300305.5. *Astrophys. J.* **2014**, *786*, 103. [CrossRef]
17. Graham, M.J.; Djorgovski, S.G.; Stern, D.; Glikman, E.; Drake, A.J.; Mahabal, A.A.; Donalek, C.; Larson, S.; Christensen, E. A possible close supermassive black-hole binary in a quasar with optical periodicity. *Nature* **2015**, *518*, 74–76. [CrossRef] [PubMed]
18. Liu, T.; Gezari, S.; Heinis, S.; Magnier, E.A.; Burgett, W.S.; Chambers, K.; Flewelling, H.; Huber, M.; Hodapp, K.W.; Kaiser, N.; et al. A Periodically Varying Luminous Quasar at z = 2 from the Pan-STARRS1 Medium Deep Survey: A Candidate Supermassive Black Hole Binary in the Gravitational Wave-driven Regime. *Astrophys. J. Lett.* **2015**, *803*, L16. [CrossRef]
19. Charisi, M.; Bartos, I.; Haiman, Z.; Price-Whelan, A.M.; Graham, M.J.; Bellm, E.C.; Laher, R.R.; Márka, S. A population of short-period variable quasars from PTF as supermassive black hole binary candidates. *Mon. Not. R. Astron. Soc.* **2016**, *463*, 2145–2171. [CrossRef]
20. Runnoe, J.C.; Eracleous, M.; Pennell, A.; Mathes, G.; Boroson, T.; Sigurdsson, S.; Bogdanović, T.; Halpern, J.P.; Liu, J.; Brown, S. A large systematic search for close supermassive binary and rapidly recoiling black holes—III. Radial velocity variations. *Mon. Not. R. Astron. Soc.* **2017**, *468*, 1683–1702. [CrossRef]
21. Milosavljević, M.; Merritt, D. Formation of Galactic Nuclei. *Astrophys. J.* **2001**, *563*, 34–62. [CrossRef]
22. Milosavljević, M.; Merritt, D. The Final Parsec Problem. *AIP Conf. Proc.* **2003**, *686*, 201–210. [CrossRef]

23. Milosavljević, M.; Merritt, D. Long-Term Evolution of Massive Black Hole Binaries. *Astrophys. J.* **2003**, *596*, 860–878. [CrossRef]

24. Lodato, G.; Nayakshin, S.; King, A.R.; Pringle, J.E. Black hole mergers: Can gas discs solve the 'final parsec' problem? *Mon. Not. R. Astron. Soc.* **2009**, *398*, 1392–1402. [CrossRef]

25. Cuadra, J.; Armitage, P.J.; Alexander, R.D.; Begelman, M.C. Massive black hole binary mergers within subparsec scale gas discs. *Mon. Not. R. Astron. Soc.* **2009**, *393*, 1423–1432. [CrossRef]

26. Roedig, C.; Dotti, M.; Sesana, A.; Cuadra, J.; Colpi, M. Limiting eccentricity of subparsec massive black hole binaries surrounded by self-gravitating gas discs. *Mon. Not. R. Astron. Soc.* **2011**, *415*, 3033–3041. [CrossRef]

27. Khan, F.M.; Holley-Bockelmann, K.; Berczik, P.; Just, A. Supermassive Black Hole Binary Evolution in Axisymmetric Galaxies: The Final Parsec Problem is Not a Problem. *Astrophys. J.* **2013**, *773*, 100. [CrossRef]

28. Khan, F.M.; Fiacconi, D.; Mayer, L.; Berczik, P.; Just, A. Swift Coalescence of Supermassive Black Holes in Cosmological Mergers of Massive Galaxies. *Astrophys. J.* **2016**, *828*, 73. [CrossRef]

29. Ryu, T.; Perna, R.; Haiman, Z.; Ostriker, J.P.; Stone, N.C. Interactions between multiple supermassive black holes in galactic nuclei: A solution to the final parsec problem. *Mon. Not. R. Astron. Soc.* **2018**, *473*, 3410–3433. [CrossRef]

30. Danzmann, K. LISA: Laser interferometer space antenna for gravitational wave measurements. *Classical Quant. Grav.* **1996**, *13*, A247, [CrossRef]

31. Amaro-Seoane, P.; Audley, H.; Babak, S.; Baker, J.; Barausse, E.; Bender, P.; Berti, E.; Binetruy, P.; Born, M.; Bortoluzzi, D.; et al. Laser Interferometer Space Antenna. *arXiv* **2017**, arXiv:1702.00786.

32. Dotti, M.; Sesana, A.; Decarli, R. Massive black hole binaries: Dynamical evolution and observational signatures. *Adv. Astron.* **2012**, *2012*, 940568. [CrossRef]

33. Peters, P.C.; Mathews, J. Gravitational radiation from point masses in a Keplerian orbit. *Phys. Rev.* **1963**, *131*, 435–440. [CrossRef]

34. Peters, P.C. Gravitational Radiation and the Motion of Two Point Masses. *Phys. Rev.* **1964**, *136*, B1224–B1232. [CrossRef]

35. Hobbs, G.; Archibald, A.; Arzoumanian, Z.; Backer, D.; Bailes, M.; Bhat, N.D.R.; Burgay, N.; Burke-Spolaor, S.; Champion, D.; Cognard, I.; et al. The international pulsar timing array project: Using pulsars as a gravitational wave detector. *Class. Quantum Grav.* **2010**, *27*, 084013. [CrossRef]

36. Sesana, A.; Volonteri, M.; Haardt, F. LISA detection of massive black hole binaries: Imprint of seed populations and extreme recoils. *Class. Quantum Grav.* **2009**, *26*, 094033. [CrossRef]

37. Haiman, Z.; Kocsis, B.; Menou, K. The Population of Viscosity- and Gravitational Wave-Driven Supermassive Black Hole Binaries Among Luminous AGN. *Astrophys. J.* **2009**, *700*, 1952–1969. [CrossRef]

38. Sesana, A.; Roedig, C.; Reynolds, M.T.; Dotti, M. Multimessenger astronomy with pulsar timing and X-ray observations of massive black hole binaries. *Mon. Not. R. Astron. Soc.* **2012**, *420*, 860–877. [CrossRef]

39. Tanaka, T.; Haiman, Z.; Menou, K. Electromagnetic counterparts of supermassive black hole binaries resolved by pulsar timing arrays. *Mon. Not. R. Astron. Soc.* **2012**, *420*, 705–719. [CrossRef]

40. Komossa, S.; Baker, J.G.; Liu, F.K. Growth of Supermassive Black Holes, Galaxy Mergers and Supermassive Binary Black Holes. *Proc. Int. Astron. Union* **2016**, *29*, 292–298. [CrossRef]

41. Sesana, A.; Haiman, Z.; Kocsis, B.; Kelley, L.Z. Testing the Binary Hypothesis: Pulsar Timing Constraints on Supermassive Black Hole Binary Candidates. *Astrophys. J.* **2018**, *856*, 42. [CrossRef]

42. Kocsis, B.; Haiman, Z.; Menou, K. Pre-Merger Localization of Gravitational-Wave Standard Sirens With LISA: Triggered Search for an Electromagnetic Counterpart. *Astrophys. J.* **2008**, *684*, 870–888. [CrossRef]

43. Corrales, L.R.; Haiman, Z.; MacFadyen, A. Hydrodynamical response of a circumbinary gas disc to black hole recoil and mass loss. *Mon. Not. R. Astron. Soc.* **2010**, *404*, 947–962. [CrossRef]

44. Zanotti, O.; Rezzolla, L.; Del Zanna, L.; Palenzuela, C. Electromagnetic counterparts of recoiling black holes: General relativistic simulations of non-Keplerian discs. *Astron. Astrophys.* **2010**, *523*, A8. [CrossRef]

45. Lousto, C.O.; Campanelli, M.; Zlochower, Y.; Nakano, H. Remnant masses, spins and recoils from the merger of generic black hole binaries. *Classical Quant. Grav.* **2010**, *27*, 114006. [CrossRef]

46. Schnittman, J.D.; Buonanno, A. The Distribution of Recoil Velocities from Merging Black Holes. *Astrophys. J. Lett.* **2007**, *662*, L63–L66. [CrossRef]

47. Kesden, M.; Sperhake, U.; Berti, E. Relativistic Suppression of Black Hole Recoils. *Astrophys. J.* **2010**, *715*, 1006–1011. [CrossRef]

48. Yuan, F.; Narayan, R. Hot Accretion Flows Around Black Holes. *Annu. Rev. Astron. Astrophys.* **2014**, *52*, 529–588. [CrossRef]

49. Abramowicz, M.A.; Fragile, P.C. Foundations of Black Hole Accretion Disk Theory. *Living Rev. Relat.* **2013**, *16*, 1. [CrossRef]

50. Schnittman, J.D. Astrophysics of super-massive black hole mergers. *Class. Quantum Grav.* **2013**, *30*, 244007. [CrossRef]

51. Bode, T.; Haas, R.; Bogdanovic, T.; Laguna, P.; Shoemaker, D. Relativistic Mergers of Supermassive Black Holes and their Electromagnetic Signatures. *Astrophys. J.* **2010**, *715*, 1117–1131. [CrossRef]

52. Farris, B.D.; Liu, Y.T.; Shapiro, S.L. Binary black hole mergers in gaseous disks: Simulations in general relativity. *Phys. Rev.* **2011**, *D84*, 024024. [CrossRef]

53. Bode, T.; Bogdanovic, T.; Haas, R.; Healy, J.; Laguna, P.; Shoemaker, D. Mergers of Supermassive Black Holes in Astrophysical Environments. *Astrophys. J.* **2011**, *744*, 45. [CrossRef]

54. Bogdanović, T.; Bode, T.; Haas, R.; Laguna, P.; Shoemaker, D. Properties of accretion flows around coalescing supermassive black holes. *Class. Quantum Grav.* **2011**, *28*, 094020. [CrossRef]

55. Noble, S.C.; Mundim, B.C.; Nakano, H.; Krolik, J.H.; Campanelli, M.; Zlochower, Y.; Yunes, N. Circumbinary MHD Accretion into Inspiraling Binary Black Holes. *Astrophys. J.* **2012**, *755*, 51. [CrossRef]

56. Farris, B.D.; Gold, R.; Paschalidis, V.; Etienne, Z.B.; Shapiro, S.L. Binary black hole mergers in magnetized disks: Simulations in full general relativity. *Phys. Rev. Lett.* **2012**, *109*, 221102. [CrossRef] [PubMed]

57. Gold, R.; Paschalidis, V.; Etienne, Z.B.; Shapiro, S.L.; Pfeiffer, H.P. Accretion disks around binary black holes of unequal mass: General relativistic magnetohydrodynamic simulations near decoupling. *Phys. Rev. D* **2014**, *89*, 064060. [CrossRef]

58. Gold, R.; Paschalidis, V.; Ruiz, M.; Shapiro, S.L.; Etienne, Z.B.; Pfeiffer, H.P. Accretion disks around binary black holes of unequal mass: General relativistic MHD simulations of postdecoupling and merger. *Phys. Rev. D* **2014**, *90*, 104030. [CrossRef]

59. Balbus, S.A.; Hawley, J.F. A powerful local shear instability in weakly magnetized disks. I—Linear analysis. II—Nonlinear evolution. *Astrophys. J.* **1991**, *376*, 214–233. [CrossRef]

60. Balbus, S.A.; Hawley, J.F. Instability, turbulence, and enhanced transport in accretion disks. *Rev. Mod. Phys.* **1998**, *70*, 1–53. [CrossRef]

61. Shi, J.M.; Krolik, J.H.; Lubow, S.H.; Hawley, J.F. Three-dimensional Magnetohydrodynamic Simulations of Circumbinary Accretion Disks: Disk Structures and Angular Momentum Transport. *Astrophys. J.* **2012**, *749*, 118. [CrossRef]

62. Bowen, D.B.; Campanelli, M.; Krolik, J.H.; Mewes, V.; Noble, S.C. Relativistic Dynamics and Mass Exchange in Binary Black Hole Mini-disks. *Astrophys. J.* **2017**, *838*, 42. [CrossRef]

63. Khan, A.; Paschalidis, V.; Ruiz, M.; Shapiro, S.L. Disks around merging binary black holes: From GW150914 to supermassive black holes. *Phys. Rev. D* **2018**, *97*, 044036. [CrossRef] [PubMed]

64. Macfadyen, A.I.; Milosavljevic, M. An Eccentric Circumbinary Accretion Disk and the Detection of Binary Massive Black Holes. *Astrophys. J.* **2008**, *672*, 83–93. [CrossRef]

65. Kocsis, B.; Haiman, Z.; Loeb, A. Gas pile-up, gap overflow, and Type 1.5 migration in circumbinary disks: Application to supermassive black hole binaries. *Mon. Not. R. Astron. Soc.* **2012**, *427*, 2680–2700. [CrossRef]

66. Armitage, P.J. Physical processes in protoplanetary disks. *arXiv* **2015**, arXiv:1509.06382.

67. Armitage, P.J.; Natarajan, P. Eccentricity of supermassive black hole binaries coalescing from gas rich mergers. *Astrophys. J.* **2005**, *634*, 921–928. [CrossRef]

68. Armitage, P.J.; Natarajan, P. Accretion during the merger of supermassive black holes. *Astrophys. J.* **2002**, *567*, L9–L12. [CrossRef]

69. Bowen, D.B.; Mewes, V.; Campanelli, M.; Noble, S.C.; Krolik, J.H.; Zilhão, M. Quasi-periodic Behavior of Mini-disks in Binary Black Holes Approaching Merger. *Astrophys. J. Lett.* **2018**, *853*, L17. [CrossRef]

70. Bowen, D.B.; Mewes, V.; Noble, S.C.; Avara, M.; Campanelli, M.; Krolik, J.H. Quasi-Periodicity of Supermassive Binary Black Hole Accretion Approaching Merger. *arXiv* **2019**, arXiv:1904.12048.

71. Ryan, G.; MacFadyen, A. Minidisks in Binary Black Hole Accretion. *Astrophys. J.* **2017**, *835*, 199. [CrossRef]

72. Ireland, B.; Mundim, B.C.; Nakano, H.; Campanelli, M. Inspiralling, nonprecessing, spinning black hole binary spacetime via asymptotic matching. *Phys. Rev. D* **2016**, *93*, 104057. [CrossRef]

73. Giacomazzo, B.; Baker, J.G.; Miller, M.C.; Reynolds, C.S.; van Meter, J.R. General Relativistic Simulations of Magnetized Plasmas around Merging Supermassive Black Holes. *Astrophys. J.* **2012**, *752*, L15. [CrossRef]

74. d'Ascoli, S.; Noble, S.C.; Bowen, D.B.; Campanelli, M.; Krolik, J.H.; Mewes, V. Electromagnetic Emission from Supermassive Binary Black Holes Approaching Merger. *Astrophys. J.* **2018**, *865*, 140. [CrossRef]

75. Palenzuela, C.; Lehner, L.; Liebling, S.L. Dual Jets from Binary Black Holes. *Science* **2010**, *329*, 927–930. [CrossRef] [PubMed]

76. Sperhake, U.; Berti, E.; Cardoso, V.; Pretorius, F.; Yunes, N. Superkicks in ultrarelativistic encounters of spinning black holes. *Phys. Rev. D* **2011**, *83*, 024037. [CrossRef]

77. Lousto, C.O.; Zlochower, Y. Hangup Kicks: Still Larger Recoils by Partial Spin-Orbit Alignment of Black-Hole Binaries. *Phys. Rev. Lett.* **2011**, *107*, 231102. [CrossRef] [PubMed]

78. Lousto, C.O.; Zlochower, Y. Nonlinear gravitational recoil from the mergers of precessing black-hole binaries. *Phys. Rev. D* **2013**, *87*, 084027. [CrossRef]

79. Campanelli, M.; Lousto, C.; Zlochower, Y.; Merritt, D. Large Merger Recoils and Spin Flips from Generic Black Hole Binaries. *Astrophys. J. Lett.* **2007**, *659*, L5–L8. [CrossRef]

80. Bogdanović, T.; Reynolds, C.S.; Miller, M.C. Alignment of the Spins of Supermassive Black Holes Prior to Coalescence. *Astrophys. J.* **2007**, *661*, L147–L150. [CrossRef]

81. Komossa, S. Recoiling Black Holes: Electromagnetic Signatures, Candidates, and Astrophysical Implications. *Adv. Astron.* **2012**, *2012*, 364973. [CrossRef]

82. Pesce, D.W.; Braatz, J.A.; Condon, J.J.; Greene, J.E. Measuring Supermassive Black Hole Peculiar Motion Using H_2O Megamasers. *Astrophys. J.* **2018**, *863*, 149. [CrossRef]

83. Blandford, R.D.; Znajek, R.L. Electromagnetic extraction of energy from Kerr black holes. *Mon. Not. R. Astron. Soc.* **1977**, *179*, 433–456. [CrossRef]

84. Tchekhovskoy, A.; Narayan, R.; McKinney, J.C. Efficient generation of jets from magnetically arrested accretion on a rapidly spinning black hole. *Mon. Not. R. Astron. Soc.* **2011**, *418*, L79–L83. [CrossRef]

85. The Event Horizon Telescope Collaboration. First M87 Event Horizon Telescope Results. V. Physical Origin of the Asymmetric Ring. *Astrophys. J.* **2019**, *875*, L5. [CrossRef]

86. Kelly, B.J.; Baker, J.G.; Etienne, Z.B.; Giacomazzo, B.; Schnittman, J. Prompt electromagnetic transients from binary black hole mergers. *Phys. Rev. D* **2017**, *96*, 123003. [CrossRef]

87. Charisi, M.; Haiman, Z.; Schiminovich, D.; D'Orazio, D.J. Testing the relativistic Doppler boost hypothesis for supermassive black hole binary candidates. *Mon. Not. R. Astron. Soc.* **2018**, *476*, 4617–4628. [CrossRef]

88. Drake, A.J.; Djorgovski, S.G.; Mahabal, A.; Prieto, J.L.; Beshore, E.; Graham, M.J.; Catalan, M.; Larson, S.; Christensen, E.; Donalek, C.; et al. The Catalina Real-time Transient Survey. *Proc. Int. Astron. Union* **2011**, *7*, 306–308. [CrossRef]

89. LSST Science and LSST Project Collaborations. LSST Science Book, Version 2.0. 2009. Available online: https://www.lsst.org/scientists/scibook (accessed on 29 May 2019).

90. Narayan, R.; Igumenshchev, I.V.; Abramowicz, M.A. Magnetically Arrested Disk: An Energetically Efficient Accretion Flow. *Publ. ASJ* **2003**, *55*, L69–L72. [CrossRef]

91. McKinney, J.C.; Dai, L.; Avara, M.J. Efficiency of super-Eddington magnetically-arrested accretion. *Mon. Not. R. Astron. Soc.* **2015**, *454*, L6–L10. [CrossRef]

92. Marshall, M.D.; Avara, M.J.; McKinney, J.C. Angular momentum transport in thin magnetically arrested discs. *Mon. Not. R. Astron. Soc.* **2018**, *478*, 1837–1843. [CrossRef]

93. Johnson, M.D.; Fish, V.L.; Doeleman, S.S.; Marrone, D.P.; Plambeck, R.L.; Wardle, J.F.C.; Akiyama, K.; Asada, K.; Beaudoin, C.; Blackburn, L.; et al. Resolved magnetic-field structure and variability near the event horizon of Sagittarius A*. *Science* **2015**, *350*, 1242–1245. [CrossRef]

94. Gold, R.; McKinney, J.C.; Johnson, M.D.; Doeleman, S.S. Probing the Magnetic Field Structure in Sgr A* on Black Hole Horizon Scales with Polarized Radiative Transfer Simulations. *Astrophys. J.* **2017**, *837*, 180. [CrossRef]

95. The Event Horizon Telescope Collaboration. First M87 Event Horizon Telescope Results. I. The Shadow of the Supermassive Black Hole. *Astrophys. J.* **2019**, *875*, L1. [CrossRef]

96. The Event Horizon Telescope Collaboration. First M87 Event Horizon Telescope Results. II. Array and Instrumentation. *Astrophys. J.* **2019**, *875*, L2. [CrossRef]

97. The Event Horizon Telescope Collaboration. First M87 Event Horizon Telescope Results. III. Data Processing and Calibration. *Astrophys. J.* **2019**, *875*, L3. [CrossRef]

98. The Event Horizon Telescope Collaboration. First M87 Event Horizon Telescope Results. IV. Imaging the Central Supermassive Black Hole. *Astrophys. J.* **2019**, *875*, L4. [CrossRef]

99. The Event Horizon Telescope Collaboration. First M87 Event Horizon Telescope Results. VI. The Shadow and Mass of the Central Black Hole. *Astrophys. J.* **2019**, *875*, L6. [CrossRef]

100. Burke-Spolaor, S.; Blecha, L.; Bogdanovic, T.; Comerford, J.M.; Lazio, T.J.W.; Liu, X.; Maccarone, T.J.; Pesce, D.; Shen, Y.; Taylor, G. The Next-Generation Very Large Array: Supermassive Black Hole Pairs and Binaries. *arXiv* **2018**, arXiv:1808.04368.

101. D'Orazio, D.J.; Loeb, A. Imaging Massive Black Hole Binaries with Millimeter Interferometry: Measuring black hole masses and the Hubble constant. *arXiv* **2017**, arXiv:1712.02362.

galaxies

MDPI

Article

Relativistic Jet Simulations of the Weibel Instability in the Slab Model to Cylindrical Jets with Helical Magnetic Fields

Ken-Ichi Nishikawa [1,*], Yosuke Mizuno [2], José L. Gómez [3], Ioana Duţan [4], Athina Meli [5,6], Jacek Niemiec [7], Oleh Kobzar [7], Martin Pohl [8,9], Helene Sol [10], Nicholas MacDonald [11] and Dieter H. Hartmann [12]

[1] Department of Physics, Chemistry and Mathematics, Alabama A&M University, Normal, AL 35762, USA
[2] Institute for Theoretical Physics, Goethe University, D-60438 Frankfurt am Main, Germany; mizuno@th.physik.uni-frankfurt.de
[3] Instituto de Astrofísica de Andalucía, CSIC, Apartado 3004, 18080 Granada, Spain; jlgomez@iaa.csic.es
[4] Institute of Space Science, Atomistilor 409, RO-077125 Bucharest-Magurele, Romania; idutan@spacescience.ro
[5] Department of Physics and Astronomy, University of Gent, Proeftuinstraat 86, B-9000 Gent, Belgium; ameli@ulg.ac.be
[6] Space Sciences & Technologies for Astrophysics Research (STAR) Institute Universite de Liege, Sart Tilman, 4000 Liège, Belgium
[7] Institute of Nuclear Physics PAN, ul. Radzikowskiego 152, 31-342 Kraków, Poland; Jacek.Niemiec@ifj.edu.pl (J.N.); oleh.kobzar@ifj.edu.pl (O.K.)
[8] Institut fur Physik und Astronomie, Universität Potsdam, 14476 Potsdam-Golm, Germany; pohlmadq@gmail.com
[9] DESY, Platanenallee 6, 15738 Zeuthen, Germany
[10] Laboratory for the Universe and THeory, Observatore de Paris-Meudon, 5 Place Jules Jansen, 92195 Meudon CEDEX, France; helene.sol@obspm.fr
[11] Max-Planck-Institut für Radioastronomie, Auf dem Hügel 69, D-53121 Bonn, Germany; nmacdona@mpifr-bonn.mpg.de
[12] Department of Physics and Astronomy, Clemson University, Clemson, SC 29634, USA; hdieter@g.clemson.edu
* Correspondence: kenichi.nishikawa@aamu.edu or knishika27@gmail.com; Tel.: +1-256-883-5446

Received: 14 November 2018; Accepted: 21 January 2019; Published: 30 January 2019

Abstract: The particle-in-cell (PIC) method was developed to investigate microscopic phenomena, and with the advances in computing power, newly developed codes have been used for several fields, such as astrophysical, magnetospheric, and solar plasmas. PIC applications have grown extensively, with large computing powers available on supercomputers such as Pleiades and Blue Waters in the US. For astrophysical plasma research, PIC methods have been utilized for several topics, such as reconnection, pulsar dynamics, non-relativistic shocks, relativistic shocks, and relativistic jets. PIC simulations of relativistic jets have been reviewed with emphasis placed on the physics involved in the simulations. This review summarizes PIC simulations, starting with the Weibel instability in slab models of jets, and then focuses on global jet evolution in helical magnetic field geometry. In particular, we address kinetic Kelvin-Helmholtz instabilities and mushroom instabilities.

Keywords: particle-in-cell simulations; relativistic jets; the Weibel instability; kink-like instability; mushroom instability; global jets; helical magnetic fields; recollimation shocks

1. Introduction

Relativistic jets are collimated outflows of ionized matter powered by black holes. Sites for such jets include the collapse of the core of a massive star forming a neutron star or a black hole, the merger of binary neutron stars, supermassive black holes associated with active galactic nuclei (AGN), gamma-ray bursts (GRBs), and pulsars (e.g., [1]). GRBs and blazars produce the brightest electromagnetic phenomena in the universe (e.g., [2]). Despite extensive observational, theoretical, and simulation studies, the understanding of their formation, their interaction with interstellar mediums, and consequently, their observable properties, such as spectra, variability, and polarization (e.g., [3]), remain quite limited.

Astrophysical jets are ubiquitous and exhibit a wide range of plasma phenomena, such as propagation in the interstellar medium, generation/decay of magnetic fields, magnetic reconnection, and turbulence. In these dynamic environments, particle acceleration may be able to achieve the highest level of energies observed in cosmic rays. Many of the processes that determine the evolution of global relativistic jets are very complex, and they occur on small spatial and short temporal scales associated with plasma kinetic effects. It is especially challenging to integrate microscopic physics into global, large-scale dynamics, which is crucial to understand the full dynamics of the jets. Kinetic plasma simulations are traditionally performed using particle-in-cell (PIC) codes, with the intent of addressing particle acceleration and kinetic magnetic reconnection, which cannot be investigated with fluid models (i.e., relativistic magnetohydrodynamic (RMHD) simulations). In particular, PIC simulations indicate that particle acceleration occurs due to kinetic instabilities, such as electron and ion Weibel instabilities (e.g., [4–26]).

In general, these simulations confirm that the Weibel instability is dominant among kinetic instabilities develped in weak or nonmagnetized plasma [27]. These instabilities, which develop in relativistic outflows, also lead to multiple shock structures. Dynamically changing current filaments and magnetic fields (e.g., [28]) accelerate electrons (e.g., [12]) and cosmic rays, which affect the pre-shock medium [29]. In order to model a shock, a relativistic plasma flow is injected from one end of the computational grid and reflected from a boundary at the opposite end. Such simulations are performed by the following: 1D simulations by [30,31], 2D simulations by (e.g., [14,15,19,25,26,32]), and 3D simulations by [33,34]. This method creates two identical counter-streaming beams which collide and interact. This approach also simplifies the numerical method, but leads to the drawbackwhere only one forward-moving shock (FS) is generated. In these settings, the backward (reverse) shock (RS) is indistinguishable from FS. There is another method where a jet is injected into an ambient plasma where FS and RS shock structures are fully modelled. Contact discontinuity (CD) is generated due to deceleration of the jet flow by the ambient plasma. The CD is the location where the electromagnetic field, the velocity of the jet, and the ambient plasmas are similar, but the density changes. FS and RS propagate away from the CD into the jet and ambient plasmas (in the CD frame) [18,21–23]. Ardaneh et al. [22] showed that FS, RS, and CD separate the jet and ambient plasma into four regions: (1) the unshocked ambient, (2) shocked ambient, (3) shocked jet, and (4) unshocked jet. In this way, the jet-to-ambient density ratio was selected as the appropriate plasma conditions of AGN and GRB jets. The shock formation processes can be investigated temporally and spatially. A leading and trailing shock system develops with strong electromagnetic fields accompanying the trailing shock. PIC simulations where jets are reflected at the simulation boundary were reviewed, including the generation of high-energy particles, by [35].

In this review, we briefly summarize our previous studies from the slab jet case to the global cylindrical jet case and present new three-dimensional simulation results for an electron-positron jet injected into an electron-positron plasma using a long simulation grid in the jet-propagation direction. We also present the results of a new study of global relativistic jets containing helical magnetic fields. The global simulation results, including velocity shears (this time) using a small simulation system, validate the use of the simulation code for the research project.

2. PIC Simulations in a Slab Model

It is natural to start to perform PIC simulations in a slab model where jets are injected into the whole simulation system. Since we use the periodic boundary conditions in the transverse direction to the jets, we are simulating a part of the jets without taking into account the boundary between the jets and the ambient plasmas. The instabilities generated between jets and ambient plasmas are described later.

2.1. Simulation of the Weibel Instability

The Weibel instability is a plasma instability which occurs in homogeneous or nearly homogeneous plasmas, where an anisotropy in the momentum (velocity) space exists [27]. The Weibel instability is often referred to as a filamentation instability [36].

The mechanisms of Weibel instability growth are explained as the following: Suppose a field $\mathbf{B} = B_z \cos k_y$ is spontaneously generated by thermal fluctuation. Here, k is a wave number, the x, y, and z are the coordinates, and electrons travel along the x-direction. The Lorentz force ($-e\mathbf{v} \times \mathbf{B}$) then bends the electron trajectories (travelling along the x-direction) along the y-direction, resulting in congregation of the electrons. The resultant current $\mathbf{j} = -en\mathbf{v}_e$ sheets (filaments) create a magnetic field, which enhances the original field and thus grows perturbation [28]. The Weibel instability is also common in astrophysical plasmas, such as collisionless shock formation in jets, supernova remnants, and GRBs.

2.1.1. Simulation Settings

The code used in this study is an MPI-based parallel version of the relativistic particle-in-cell (RPIC) code, TRISTAN [5,37,38]. The simulations have been performed using a grid with $(L_x, L_y, L_z) = (4005, 131, 131)$ cells and a total of $\sim$1 billion particles (12 particles/cell/species for the ambient plasma) in the active grid. The electron skin depth is $\lambda_s = c/\omega_{pe} = 10.0\Delta$, where $c = 1$ is the speed of light and $\omega_{pe} = (e^2 n_a/\epsilon_0 m_e)^{1/2}$ is the electron plasma frequency, and the electron Debye length λ_D is half of the cell size, Δ. This computational system length is six times longer than that used in the previous simulations [12,39]. The jet-electron number density in the simulation reference frame is $0.676n_a$, where n_a is the ambient electron density, and the jet Lorentz factor is $\gamma_{jt} = 15$. The jet-electron/positron thermal velocity is $v_{j,th} = 0.014c$ in the jet reference frame. The electron/positron thermal velocity in the ambient plasma is $v_{a,th} = 0.05c$. As in our previous work (e.g., [12]), the jet is injected in a plane across the computational grid located at $x = 25\Delta$ in order to eliminate artificial effects associated with the boundary at $x = x_{min}$. Radiating boundary conditions are used on the planes at $x = x_{min}$ and $x = x_{max}$ and periodic boundary conditions on all transverse boundaries [37]. The jet makes contact with the ambient plasma at a two-dimensional interface spanning the whole computational domain in the $y - z$ plane. In this way, only a small portion of whole jets is studied; that is, the simulation includes the spatial development of nonlinear saturation and dissipation from the injection point to the jet front composed of the fastest-moving jet particles. Therefore, the boundary between jets and ambient plasma is not taken into account, which will be described later.

2.1.2. Simulation Results

Figure 1a,b shows the average (in the $y - z$ plane) of (a) the jet (red), the ambient (blue), and the total (black) electron density, and (b) the electromagnetic field energy divided by the total jet kinetic energy ($E_t^j = \sum_{i=e,p} m_i c^2 (\gamma_{jt} - 1)$) at $t = 3250\omega_{pe}^{-1}$. Here, "e" and "p" denote the electron and positron. Positron density profiles are similar to the electron profiles, as both particles have the same mass. However, for the electron-ion jets, the densities of the electrons and the ions are slightly different, giving rise to double layers in the plasma [21–23]. As a result, ambient particles are dragged by the motion of the jet particles up to $x/\Delta \sim 500$. By $t = 3250\omega_{pe}^{-1}$, the ambient density has evolved into a two-step plateau behind the jet front, which is similar to the electron-ion jet cases [21–23].

The maximum density in this shocked region is about three times the initial ambient density. The jet-particle density remains nearly constant up to near the front of the jet. Careful comparisons reveal the differences between the pair jets and the electron-ion jets [21–23]. The differences arise due to the double layers generated in the trailing and leading edges in the electron-ion jets.

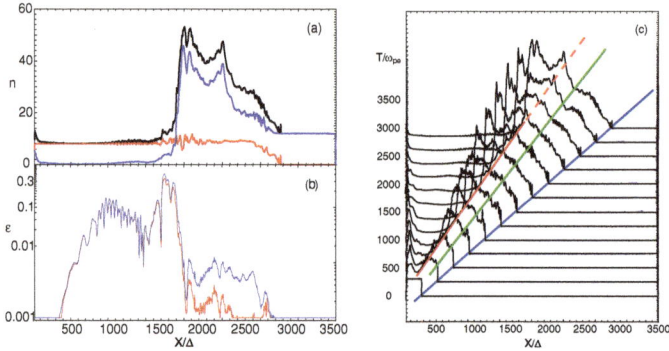

Figure 1. Averaged values of (**a**) the jet (red), the ambient (blue), and the total (black) electron density, and (**b**) electric (red) and magnetic (blue) field energy divided by the jet kinetic energy at $t = 3250\omega_{pe}^{-1}$. Panel (**c**) shows the evolution of the total electron density in time intervals of $\delta t = 250\omega_{pe}^{-1}$. Diagonal lines indicate the motion of the jet front (blue: $\leq c$), the predicted contact discontinuity (CD) speed (green: $\sim 0.76c$), and the trailing density jump (red: $\sim 0.56c$). Adapted from Figure 1 in [18].

The growth of the Weibel instability creates current filaments and strong electromagnetic fields in the trailing shock region. Since the nonlinear stage is formed in this simulation, the electromagnetic fields are about four times larger than those seen previously in simulations with a much shorter grid system ($L_x = 640\Delta$). At the simulation time $t = 3250\omega_{pe}^{-1}$, the electromagnetic fields have the highest intensity at $x/\Delta \sim 1700$, which then declines by about one order of magnitude beyond $x/\Delta = 2300$ in the shocked region [12,39].

Figure 1c shows the total electron density plotted at time intervals of $\delta t = 250\omega_{pe}^{-1}$. The jet front propagates with the initial jet speed ($\leq c$). Since anomalous resistivity exists in PIC simulations, sharp RMHD-simulation shock surfaces are not generated (e.g., [40]). A leading shock region (where the linear density increases) moves with a speed between that of the fastest moving jet particles $\leq c$ and a predicted CD value of $\sim 0.76c$. A CD region consisting of mixed ambient and jet particles moves at a speed which is between $\sim 0.76c$, and the trailing density jump speed $\sim 0.56c$. The modest density increase just behind the large trailing density jump should be taken note of. Similar shock structures and their velocities for the electron-ion jets are discussed in [21–23].

It is important to show the differences between the reflection and the injection models. The shock is set up by reflecting a cold "upstream" flow from a conducting wall located at $x = 0$ (Figure 1). The interaction between the incoming beam (that propagates along $-x$) and the reflected beam triggers the formation of a shock, which moves away from the wall along $+x$ [33]. This setup is equivalent to the head-on collision of two identical plasma shells, which would form a forward and reverse shock and contact discontinuity as an injection scheme. However, the forward and reverse shocks are not distinguished as in the injection scheme. Furthermore, the conducting wall corresponds to the contact discontinuity. The simulation is performed in the "wall" frame, where the "downstream" plasma behind the shock is stationary—and on the contrary, in the injection scheme, FS, RS, and CD are moving in the same direction.

In 3D, periodic boundary conditions are employed both in y- and in z-directions. Each computational cell is initialized with four particles (two per species) in 2D and two particles (one per species) in 3D. They have performed limited number of experiments with a larger number of particles per cell (up to eight per species in 2D), though essentially obtaining the same results.

Their 3D structure is shown in Figure 2, for a relativistic electron-positron shock with magnetization $\sigma = B^2/n_e m_e \gamma_{jt} c^2 = 0$ (top panel) and $\sigma = 10^{-3}$ (bottom panel). The background magnetic field B_0 is initially set along the z-direction, in the same way as for our 2D simulations. The yz slice of the magnetic energy fraction in Figure 2c shows that for $\sigma = 10^{-3}$, the magnetic field ahead of the shock is primarily organized in pancakes stretched in the direction orthogonal to the background magnetic field (i.e., along y). This can be easily understood, considering that the Weibel instability is seeded by focusing the counter-streaming particles into two channels of charge and current. In the absence of a background magnetic field, the currents tend to be organized into cylindrical filaments, as demonstrated by Spitkovsky [41] and shown in the yz slice of the top panel in Figure 2. In the presence of an ordered magnetic field along z, the particles will preferentially move along the magnetic field (rather than orthogonal), so that their currents will more likely be focused at certain locations of constant z, into sheets elongated along the xy plane. This explains the structure of the magnetic turbulence ahead of the shock in the bottom panel of Figure 2, common to all the cases of weakly magnetized shocks they have investigated (i.e., $0 < \sigma \le 10^{-1}$).

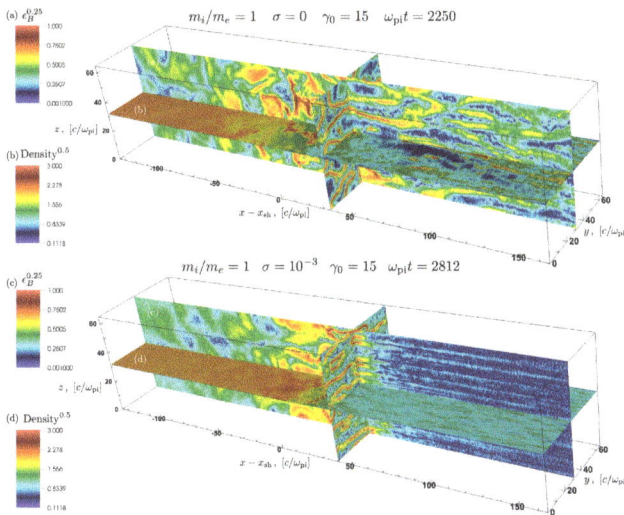

Figure 2. Structure of the flow, from the 3D simulation of an electron-positron shock with magnetization $\sigma = 0$ (**top**) or $\sigma = 10^{-3}$ (**bottom**). Panels (**b**) and (**d**) show the particle density in the xy slice (with color scale stretched for clarity), whereas Panels (**a**) and (**c**) show the the magnetic energy fraction ϵ_B in the xz and yz slices (with color scale stretched for clarity). Adapted from Figure 5 in [33].

Figure 3 shows the phase-space distribution of the jet (red) and the ambient (blue) electrons at $t = 3250\omega_{pe}^{-1}$ and confirms the shock-structure interpretation. The electrons injected with $\gamma_{jt}v_x \sim 15$ become thermalized due to the Weibel instability, which is induced by interactions. The swept-up ambient electrons (blue) are heated by interaction with the jet electrons. Some ambient electrons are strongly accelerated.

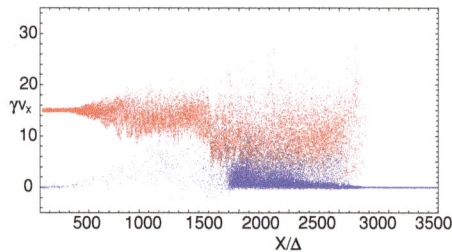

Figure 3. Phase-space distribution of the jet (red) and the ambient (blue) electrons at $t = 3250\omega_{pe}^{-1}$. About 18,600 electrons of both species are selected randomly. Adapted from Figure 2 in [18].

This simulation shows that the shocks are excited through the injection of a relativistic jet into ambient plasma, leading to two distinct shocks (referred to as the trailing shock and the leading shock) and contact discontinuity. It should be noted that the simulations where jets are reflected on the simulation boundary do not show the structure of a leading shock, contact discontinuity, and a trailing (reverse) shock.

For the electron-ion jet case, the mass ratio is $m_i/m_e = 16$ and, therefore, the evolution of density (shock) structures are different to those in the electron-positron jet ($m_i/m_e = 1$) [22,23]. Furthermore, the double layers generated in the trailing and leading edges further accelerate the electrons up to the ion kinetic energy [23].

2.2. Simulation of Jets with Velocity-Shears

The generation of shocks in slab jet models have been studied extensively; however, the velocity shears between the jet and the ambient medium still need to be taken into account, where the outflow interaction with an ambient medium induces velocity shearing.

In particular, the Kelvin–Helmholtz instability (KHI) has been investigated on the macroscopic level as a means to generate magnetic fields in the presence of strong relativistic velocity shears in AGN and in GRB jets (e.g., [42–46]). Recently, PIC simulations have been employed to study magnetic field generation and particle acceleration in velocity shears at the microscopic level using counter-streaming setups. Here, the shear interactions are associated with kinetic Kelvin–Helmholtz instability (kKHI), also referred to as electron-scale Kelvin–Helmholtz instability (ESKHI; e.g., [47–53]).

Alves et al. [54] presented the shear surface instability that occurs in the plane perpendicular to that of the ESKHI. These new unstable modes explain the transverse dynamics and the plasma parameter structures similar to those observed in the PIC simulations performed by [47,49,52,55]. They named this effect "mushroom instability" (MI), due to the mushroom-like structures that emerge in the electron density, and the 2D simulation in particular. In 3D simulations, the shape of mushrooms cannot seen clearly; nevertheless, they grow to be good and strong [56].

Multi-dimensional PIC simulations confirm the analytic results and further show the appearance of mushroom-like electron density structures in the nonlinear stage of the instability, similar to those observed in the Rayleigh Taylor instability, despite the great disparity in scales and different underlying physics [54,56]. This transverse electron-scale instability may play an important role in relativistic and supersonic sheared flow scenarios, which are stable at the (magneto)hydrodynamic level. This aspect will be discussed later in the case of a cylindrical relativistic jet. Macroscopic (dimensional scale $\gg c/\omega_{pe}$) fields are shown to be generated by this microscopic shear instability, which are relevant for the generation of a DC electric field and toroidal magnetic field (B_ϕ), acceleration of particles, and emission, as well as seeding magnetohydrodynamic processes at long time-scales [54,56].

Spine-Sheath (Two-Components) Jet Setup

Next, we consider the simulation of a jet with a spine-sheath (two-component) plasma jet structure, which was studied using the counter-streaming plasma setup implemented in simulations by [47,49–53]. In the setup, a jet spine (core) with velocity γ_{core} propagates in the positive x-direction in the middle of the computational box. The upper and lower quarters of the numerical grid contain a sheath plasma that can be stationary or moving with velocity v_{sheath} in the positive x-direction [48,55]. This model is similar to that used in the RMHD simulations [44] containing a cylindrical jet spine (core).

Nishikawa et al. [55] performed 3D PIC simulations of the kKHI and the MI for both $e^{\pm}$ and $e^{-} - p^{+}$ plasmas. The processes studied here are inspired from the jets from AGN and GRBs that are expected to have velocity shears between a faster spine (core) and a slower sheath wind (stationary ambient plasmas). In these simulations, large velocity shears were studied with relative Lorentz factors of 1.5, 5, and 15.

Figure 4a shows the structure of the B_y component of the magnetic field in the $y - z$ plane (jet flows out of the page) at the midpoint of the simulation box, where $x = 500\Delta$. Figure 4b depicts 1D cuts along the z axis showing the magnitude and direction of the magnetic field components at the midpoint of the simulation box, where $x = 500\Delta$ and $y = 100\Delta$ for the $e^{-} - p^{+}$ case at the simulation time $t = 300\omega_{pe}^{-1}$, with $\gamma_{jt} = 15$ [55]. In the $e^{-} - p^{+}$ case, magnetic fields appear relatively uniform at the velocity shear surfaces along the transverse y-direction, just as it had been at the velocity shear surfaces along the parallel x-direction, with almost no transverse fluctuations visible in the magnetic field structure. Small fluctuations in the y-direction over distances on the order of $\sim10\Delta$ are visible in the currents, whereas small longitudinal mode fluctuations in the x-direction occur over distances $\sim100\Delta$. This behavior indicates that the MI generates DC fields in the transverse direction, a fact that has also been seen in the results of global jet simulations without helical magnetic fields [56].

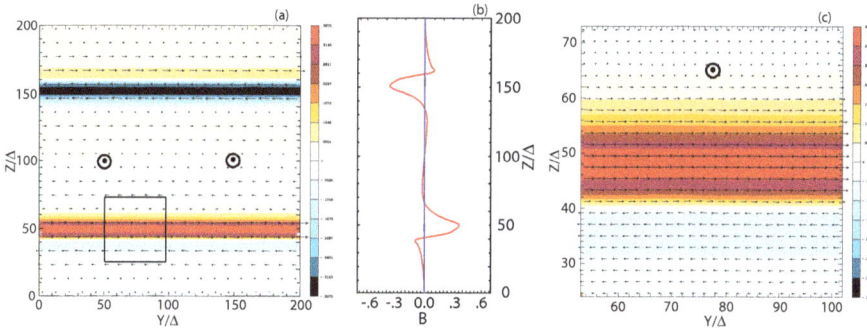

Figure 4. Magnetic field structure transverse to the flow direction for $\gamma_{jt} = 15$ is shown in the $y - z$ plane (jet flows out of the page) at the center of the simulation box, $x = 500\Delta$ for the $e^{-} - p^{+}$ case. The small arrows show the magnetic field direction in the transverse plane (the arrow length is not scaled to the magnetic field strength). 1D cuts along the z axis of magnetic field components B_x (black), B_y (red), and B_z (blue) are plotted at $x = 500\Delta$ and $y = 100\Delta$ for (**b**) the $e^{-} - p^{+}$ case. Note that the magnetic field strength scales in panels (**a**) (±0.367) are different. An enlargement of the shear surface structure in the $y - z$ plane contained within the squares in the left panels is shown in the panels (**c**) to the right. Adapted from Figure 6 in Nishikawa et al. [55].

For the $e^{\pm}$ case, the magnetic field alternates in both the y- and z-directions, and these transverse fluctuations occur over distances of the order of $\sim100\Delta$, whereas longitudinal mode fluctuations in the x-direction occur over distances $\sim100\Delta$ [55]. The 1D cuts show that (i) the B_y field component dominates in the $e^{-} - p^{+}$ case, (ii) the B_y field component is about an order of magnitude smaller for the $e^{\pm}$ case, and (iii) the B_z component is significant for the $e^{\pm}$ case. The 1D cuts also show that there is a sign reversal of the magnetic field on either side of the maximum, which is relatively small for the $e^{\pm}$ case but much more significant for the $e^{-} - p^{+}$ case. More details are revealed by the enlargement of the region contained in the squares, as it is shown in Figure 4c. For the $e^- - p^+$ case, the generated relatively uniform DC magnetic field is symmetric about the velocity shear surface—e.g., note that $B_y > 0$ immediately around the shear surface, and $B_y < 0$ in the jet and the ambient plasmas at somewhat larger distances from the shear surface. It should be noted that this DC magnetic field is generated by the MI and saturated at this time. The MI is also generated in the global $e^- - p^+$ jet, where this instability generates toroidal magnetic fields that pinch the jet plasma [56]. On the other hand, for the $e^\pm$ case, the generated AC magnetic field resides largely on the jet side of the velocity shear surface. This phenomenon is also found in the global jet simulation [56] and the outflow simulation [57].

The strong electric and magnetic fields in the velocity shear zone can also provide the right conditions for particle acceleration. Nevertheless, the simulations are too short for definitive statements on the efficacy of the process and the resulting spectra. Also, the organization of the field in compact regions will complicate the interpretation of emission spectra, and a spatially resolved treatment of particle acceleration and transport would be mandatory for a realistic assessment, which is beyond the scope of this review paper. Relativistic electrons, for example, can suffer little synchrotron energy loss outside the thin layer of the strong magnetic field. Thus, synchrotron emissivity can be dominated by the shear layer, and in general, this emissivity can depend on how efficiently electrons can flow in and out of the shear layer and be accelerated in the regions with strong magnetic fields. An immediate consequence for radiation modeling is that the energy-loss time of electrons cannot be calculated with the same mean magnetic field that is used to compute emission spectra, because the former includes the volume-filling factor of the strong-field regions.

3. PIC Simulations of Cylindrical Jets

Cylindrical geometry is the simplest form that can be used to model the relativistic jets. Therefore, cylindrical jets have been used to study the shear instabilities that occur at the interface between a jet and its ambient plasma, where the plasma is unmagnetized and composed of either $e^\pm$ or $e^- - p^+$. Moreover the jet was implemented in the ambient plasma along the x-direction (periodic along the x-direction). Figure 5 shows isocontour images of the x-component of the current, along with the magnetic field lines that are generated by the kinetic instabilities for both $e^\pm$ and $e^- - p^+$ jets. The isocontour images show that in the $e^- - p^+$ jet case, currents are generated in sheet-like layers and the magnetic fields are wrapped around the jet generated by the dominant MI. On the other hand, in the $e^\pm$ jet case, many distinct current filaments are generated near the velocity shear, and the individual current filaments are wrapped by the magnetic field. Since the growth rates of kinetic instabilities depend on the species of jets, dominant growing modes are different. The clear difference in the magnetic field structure between these two cases may make it possible to distinguish different jet compositions via differences in circular and linear polarization, which are seen clearly in the global jets injected into ambient plasmas [56].

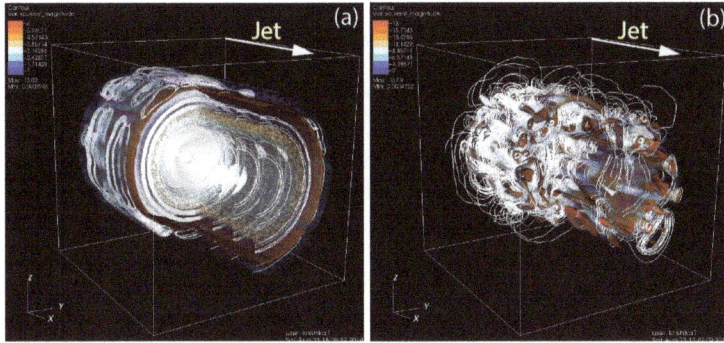

Figure 5. Isocontour plots of the J_x magnitude with magnetic filed lines (one fifth of the jet size) for (**a**) an $e^- - p^+$ and (**b**) an ee$^\pm$ jet at simulation time $t = 300\omega_{pe}^{-1}$. The 3D displays are clipped both along and perpendicular to the jet in order to view the interior. Adapted from Figure 4 in Nishikawa et al. [58].

Alves et al. [59] considered magnetic field profiles of the form $\mathbf{B}(r) = B_0(r/R_c)e^{1-r/R_c}\mathbf{e}_\phi + B_z\mathbf{e}_z$, where R_c is the cross-sectional radius of the jet spine. They also demonstrated that the toroidal magnetic field profiles decay as $r^{-\alpha}$ (with $\alpha \geq 1$) and determined that their overall findings are not sensitive to the structure of the magnetic field far from R_c. Near the black hole, the poloidal and the toroidal magnetic field components (B_z and B_α, respectively) are comparable to one another [60]. However, the ratio B_z/B_α decreases with the increase of the distance from the source, and it can be very small at a distance—relevant to astrophysical jets—of $\sim$100 pc. The characteristic magnetic field amplitude (henceforth denoted as B_0) at such distances, $B_0 \sim$mG, is quite strong in the sense that the ratio σ of the magnetic energy density to plasma rest-mass energy density may exceed unity. In this review, we would like to emphasise the importance of the macroscopic-like instabilities (as, for example, the kink instability), since strong helical magnetic fields can suppress the kinetic instabilities (such as the Weibel instability, kKHI, and MI) and a kink-like instability is more likely to occur, as it is shown in [61,62].

Recently, global relativistic PIC simulations have been performed where a cylindrical unmagnetized jet is injected into an ambient plasma in order to investigate shock (Weibel instability) and velocity shear instabilities (the kKHI and the MI) simultaneously [56]. Previously, these two processes have been investigated separately. For example, kKHI and MI have been investigated for sharp velocity shear slabs and cylindrical geometries extending across the computational grid (e.g., [55,58,59]).

4. Simulation Setups of Global Jet Simulations

Recently, global simulations have been performed while involving the injection of a cylindrical unmagnetized jet into an ambient plasma in order to simultaneously investigate shock (Weibel instability) and velocity shear instabilities (kKHI and MI) [56]. Previously, these two processes have been investigated separately. For example, kKHI and MI have been investigated for sharp velocity shear slab and cylindrical geometries extending across the computational grid (e.g., [55,58,59]). In this section, we present the results of this new study of global relativistic jets containing helical magnetic fields.

Jets generated from black holes and merging neutron stars, which are then injected into the ambient interstellar medium, are thought (in many cases) to carry helical magnetic fields (e.g., [1]). Since many GRMHD simulations of jet formations show that the generated jets carry helical magnetic fields (e.g., [63]), jets in PIC global simulations are injected into an ambient medium implementing helical magnetic fields near the jet orifice, (e.g., [61,64]). One of the key issues is how the helical magnetic fields affect the growth of the kKHI, the MI, and the Weibel instability. The RMHD simulations demonstrated that jets containing helical magnetic fields develop kink instability (e.g., [65–67]). Since the PIC simulations are large enough to include kink instability, a kink-like instability was found in the pair and $e^- - p^+$ jet cases (e.g., [61,64]).

4.1. Helical Magnetic Field Structure

In the simulations of [61,62], cylindrical jets containing a helical magnetic field were injected into an ambient plasma (see Figure 6a). The structure of the helical magnetic field was implemented like that in the RMHD simulations performed by Mizuno et al. [68], where a force-free expression of the field at the jet orifice was used; that is, the magnetic field was not generated self-consistently, e.g., from simulations of jet formation by a rotating black hole. For the initial conditions, the force-free helical magnetic field was used as described in Equations (1) and (2) of Mizuno et al. (2014) [65].

The following form was used for the poloidal (B_x) and the toroidal (B_ϕ) components of the magnetic field determined in the laboratory frame:

$$B_x = \frac{B_0}{[1 + (r/a)^2]^\alpha}, \quad B_\phi = \frac{B_0}{(r/a)[1 + (r/a)^2]^\alpha} \sqrt{\frac{[1 + (r/a)^2]^{2\alpha} - 1 - 2\alpha(r/a)^2}{2\alpha - 1}}, \tag{1}$$

where r is the radial coordinate in cylindrical geometry, B_0 parameterizes the magnetic field, a is the characteristic radius of the magnetic field (the toroidal field component has a maximum value at a, for a constant magnetic pitch), and α is the pitch profile parameter.

The expressions for describing the helical magnetic field used by [61,62] are written in Cartesian coordinates. Since $\alpha = 1$ Equation (1) was reduced to Equation (2), the magnetic field takes the form:

$$B_x = \frac{B_0}{[1 + (r/a)^2]}, \quad B_\phi = \frac{(r/a)B_0}{[1 + (r/a)^2]}. \tag{2}$$

The toroidal component of the magnetic field was created by a current $+J_x(y, z)$ in the positive x-direction, and it is defined in Cartesian coordinates as:

$$B_y(y, z) = \frac{((z - z_{jc})/a)B_0}{[1 + (r/a)^2]}, \quad B_z(y, z) = -\frac{((y - y_{jc})/a)B_0}{[1 + (r/a)^2]}. \tag{3}$$

Here, the center of the jet is located at (y_{jc}, z_{jc}) and $r = \sqrt{(y - y_{jc})^2 + (z - z_{jc})^2}$. The chosen helicity is defined through Equation (3), which has a left-handed polarity with positive B_0. At the jet orifice, the helical magnetic field is implemented without the motional electric fields. This corresponds to a toroidal magnetic field generated by jet particles moving along the $+x$-direction.

The poloidal (B_x: black) and the toroidal (B_ϕ: red) components of the helical magnetic field with a constant pitch ($\alpha = 1$) are shown in Figure 6b. The toroidal magnetic fields become zero at the center of the jet, as shown by red lines in Figure 6b. To date, simulations with a constant pitch ($\alpha = 1$) and with $b = 200$ have been performed using $r_{jet} = 20, 40, 80, 120\Delta$ [61,62]. Here, b is the dumping factor of the magnetic fields outside the jet.

It should be noted that the structure of the jet formation region is more complicated than what is implemented in the PIC simulations at the present time (e.g., [69,70]). Furthermore, so far these global jet simulations have been performed with the simplest kind of jet structure with a top-hat shape (flat-density profile). A more realistic jet structure needs to be implemented in a future simulation study.

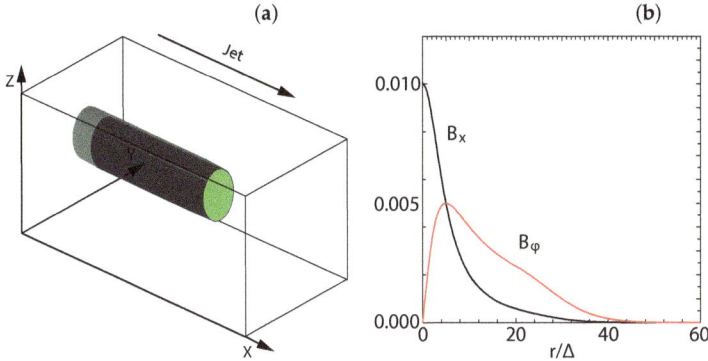

Figure 6. Panel (**a**) shows a schematic simulation setup; a global jet setup. The jet is injected at $x = 100\Delta$ with the jet radius r_{jet} at the center of the $y - z$ plane (not scaled). Panel (**b**) shows the helical magnetic fields, B_x(black), B_φ(red) with $B_0 = 0.01$ for the pitch profile $\alpha = 1.0$ with damping functions outside the jet with $b = 200.0$. The jet boundary is located at $r_{jet} = 20\Delta$ [61]. So far, simulations were performed with $r_{jet} = 20, 40, 80, 120\Delta$ [62].

4.2. Helically Magnetized Global Jet Simulations with Larger Jet Radii

In this section, we explore how the jet evolution is affected by the helical magnetic field using a short system before performing more large-scale simulations. A schematic of the simulation injection setup is shown in Figure 6b [61,62]. The initial jet and ambient (electron and ion) plasma number density measured in the simulation frame is $n_{jt} = 8$ and $n_{am} = 12$, respectively. This set of plasma parameters is used for obtaining the simulation results presented in [56,61,64].

In their simulations, the electron skin depth $\lambda_s = c/\omega_{pe} = 10.0\Delta$, where c is the speed of light ($c = 1$), $\omega_{pe} = (e^2 n_{am}/\epsilon_0 m_e)^{1/2}$ is the electron plasma frequency, and the electron Debye length for the ambient electrons is $\lambda_D = 0.5\Delta$. The jet–electron thermal velocity is $v_{jt,th,e} = 0.014c$ in the jet reference frame. The electron thermal velocity in the ambient plasma is $v_{am,th,e} = 0.03c$, and ion thermal velocities are smaller by $(m_i/m_e)^{1/2}$. Simulations were performed using an $e^\pm$ plasma or an $e^- - p^+$ (with $m_p/m_e = 1836$) plasma for the jet Lorentz factor of 15 and with the ambient plasma at rest ($v_{am} = 0$).

In these short system simulations, a numerical grid with $(L_x, L_y, L_z) = (645\Delta, 131\Delta, 131\Delta)$ (simulation cell size: $\Delta = 1$) is used, imposing periodic boundary conditions in transverse directions with a jet radius of $r_{jet} = 20\Delta$ [61]. In this review, all simulation parameters are maintained as described above except for the jet radius and the size of the simulation grid (which is adjusted based on the jet radius) [62]. Therefore, the jet radius is increased from the value $r_{jet} = 20\Delta$ up to several values: $r_{jet} = 40\Delta, 80\Delta$, and 120Δ, which corresponds to a numerical grid with $(L_x, L_y, L_z) = (645\Delta, 257\Delta, 257\Delta), (645\Delta, 509\Delta, 509\Delta)$, and $(645\Delta, 761\Delta, 761\Delta)$, respectively. The cylindrical jet with jet radius $r_{jet} = 40\Delta, 80\Delta$, and 120Δ is injected into the middle of the $y - z$ plane $((y_{jc}, z_{jc}) = (129\Delta, 129\Delta), (252\Delta, 252\Delta), (381\Delta, 381\Delta))$ at $x = 100\Delta$. The largest jet radius ($r_{jet} = 120\Delta$) is larger than that ($r_{jet} = 100\Delta$) in [56], but the simulation length is much shorter ($L_x = 2005\Delta$).

Other parameters used in their simulations include the initial magnetic field amplitude parameter $B_0 = 0.1c$, where $\sigma = B^2/n_e m_e \gamma_{jt} c^2 = 2.8 \times 10^{-3}$ is used, and $a = 0.25 * r_{jet}$. The helical field structure inside the jet is defined by Equations (1) and (2). For the magnetic fields outside the jet, a damping function $\exp[-(r - r_{jet})^2/b]$ ($r \geq r_{jet}$) is imposed on Equations (1) and (2) with the tapering parameter $b = 200\Delta$. The final profiles of the helical magnetic field components are similar to those obtained in the case where the jet radius is $r_{jet} = 20\Delta$, with the only difference being that $a = 0.25 \cdot r_{jet}$, as it is shown in Figure 6b.

Figure 7 shows the *y*-component of the magnetic field (B_y) for two values of the jet radius with $r_{jet} = 20\Delta$ and 80Δ, respectively. In both cases, the initial helical magnetic field (left-handed; clockwise, viewed from the jet front) is enhanced and disrupted due to the plasma instabilities.

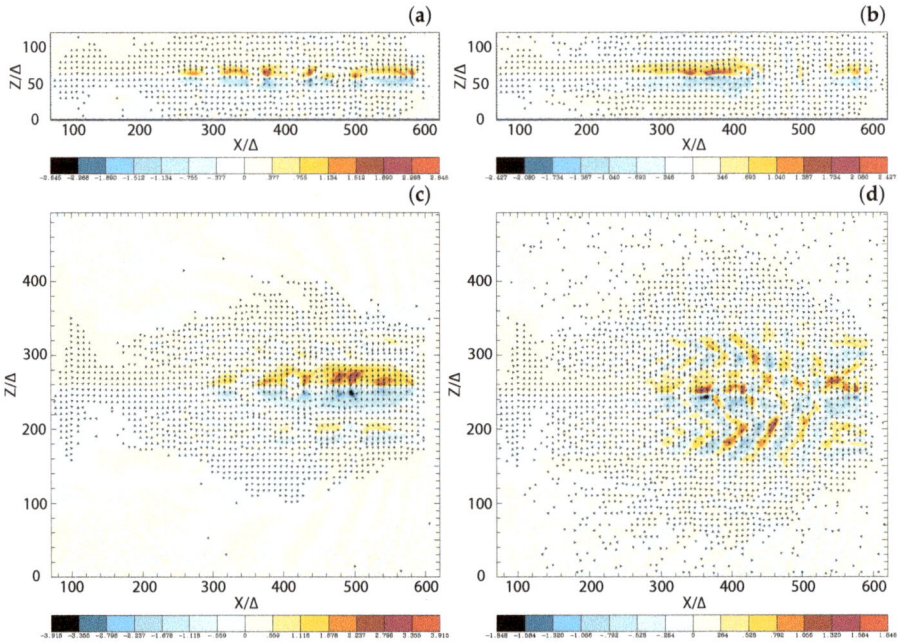

Figure 7. Isocontour plots of the azimuthal component of magnetic field B_y intensity at the center of the jets for $e^- - p^+$ (**a,c**) $e^\pm$ (**b,d**) jets; with $r_{jet} = 20\Delta$ (**a,b**) $r_{jet} = 80\Delta$ (**c,d**) at time $t = 500\omega_{pe}^{-1}$. The disruption of helical magnetic fields are caused by instabilities and/or reconnection. The max/min numbers of panels are (**a**) ±2.645, (**b**) ±2.427, (**c**) ±3.915, (**d**) ±1.848. Adapted from Figure 1 in Nishikawa et al. [62].

Thus, even when shorter simulation systems are used, growing instabilities are affected by the helical magnetic fields. The simple recollimation shock generated in the small jet radius is shown in Figure 7a,b. The currents generated by instabilities in the jets determine these complicated patterns of B_y, as it is shown in Figure 7. Using a larger jet radius adds more modes of growing instabilities in the jets, which make the jet structure more complicated. In order to investigate the full development of instabilities in jets with helical magnetic fields, longer simulations are required.

To illustrate the production of acceleration of the particles in the jet, the Lorentz factor of the jet electrons was plotted for the two cases of plasma type used ($e^- - p^+$ and $e^\pm$, respectively) when the jet radius is $r_{jet} = 120\Delta$, as it is shown in Figure 8. These observed patterns of the Lorentz factor coincide with the changing directions of the local magnetic fields in the *y*-direction, which are generated by kinetic instabilities like the Weibel instability, the kKHI, and the MI. The directions of the magnetic fields are indicated by the arrows (black spots) in the $x - z$ plane. (The arrows are better seen when the figure is magnified.) The directions of magnetic fields are determined by the generated instabilities. The structures at the edge of the jets are generated by the kKHI. Moreover, the plots of the Lorentz factor in the $y - z$ plane, which are not presented here, show the production of the MI at the circular edge of the jets.

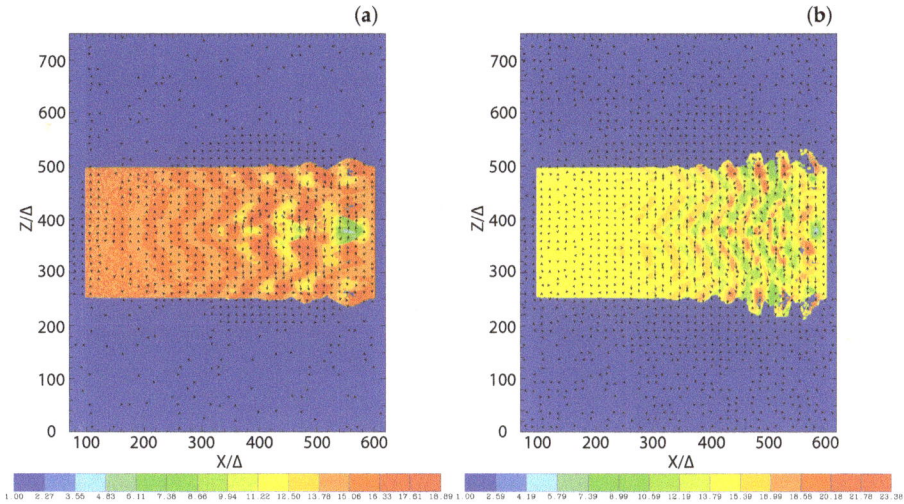

Figure 8. Panels (**a**,**b**) show the 2D plot of the Lorentz factor of jet electrons for $e^- - p^+$ (**a**) and $e^\pm$ (**b**) jet with $r_{jet} = 120\Delta$ at time $t = 500\omega_{pe}^{-1}$. The arrows (black spots) show the magnetic fields in the $x - z$ plane. Adapted from Figure 3 in Nishikawa et al. [62].

Figure 9 shows the isosurface of the Lorentz factor of the jet electrons for a plasma that is composed of (a) $e^- - p^+$ and (b) $e^\pm$. The 3D isosurface of the averaged jet electron Lorentz factor in a quadrant of the jet front ($320 \le x/\Delta \le 620$, $381 \le y, z/\Delta \le 531$) shows where jet electrons are accelerated (in reddish color) locally. The cross-sections and the surfaces of the jets show complicated patterns that are generated by mixed instabilities, where the fine lines represent the magnetic field lines.

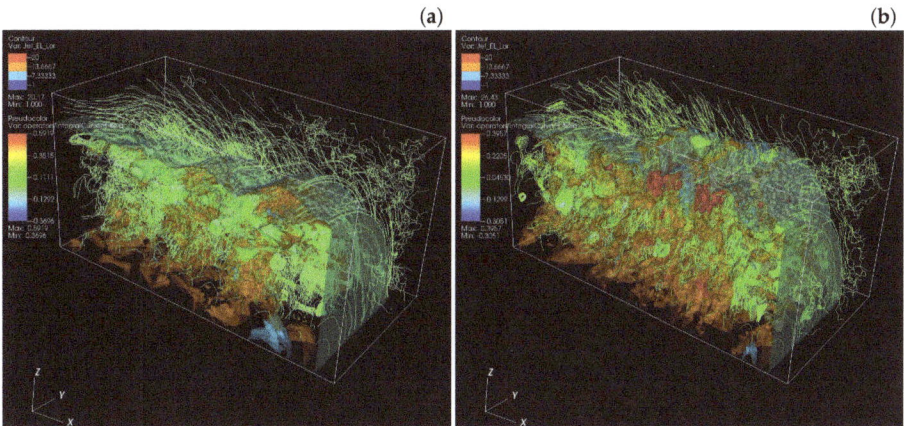

Figure 9. Panels show 3D isosurface plots of the Lorentz factor of the jet electrons for $e^- - p^+$ (**a**) and $e^\pm$ (**b**) jet with $r_{jet} = 120\Delta$ at time $t = 500\omega_{pe}^{-1}$. The lines show the magnetic field stream lines in the quadrant of the front part of the jets. The color scales for contour (upper left): red 20.0; orange 13.67; right blue 7.33. blue 1. The color scales of streaming lines (**a**) (5.92, 3.52, 0.174, -1.29, -3.70) $\times 10^{-1}$; (**b**) (3.96, 2.21, 0.453, -1.30, -3.05) $\times 10^{-1}$. Adapted from Figure 4 in Nishikawa et al. [62].

For the jet radii larger than $r_{jet} = 80\Delta$, the kKHI and the MI are generated at the jet surface, whereas inside the jet, the Weibel instability is generated together with a kink-like instability, particularly in the case of the $e^- - p^+$ plasma. Answering the question on how the growth of kink-like instabilities depends on the helical magnetic fields requires further investigations using different parameters, including *a*, which determines the structure of the helical magnetic field in Equations (2) and (3). Furthermore, an imprint on the plasma behavior of different values of the pitch parameter α is also necessary for investigation using Equation (1).

Recently, Dieckmann et al. [57] investigated the expansion of a cloud of electrons and positrons with the temperature 400 keV that propagates at the mean speed 0.9c (c: speed of light) through an initially unmagnetized $e^- - p^+$ plasma with PIC simulation. They found a mechanism that could collimate the pair cloud into a jet. The electrons and positrons of the cloud expanded rapidly due to their high temperature, which decreased the density of the cloud. A filamentation instability developed between the protons at rest and the moving positrons in the interval, where the latter were still dense. It is noted that it is difficult to distinguish the filament instability from the kKHI, which is shown in the simulation where the electron-positron jet was injected into an electron-positron ambient [56]. The instability expelled the protons from large areas, which were then filled with positrons. Magnetic fields grew only in those locations where protons and rapidly streaming positrons were present, which confined the magnetic field to a small spatial span. The effect of the filamentation instability and the resulting magnetic field were to push the protons away from the regions with no protons. The instability and the magnetic field followed the pushed protons and, hence, the filament grew in size. The largest filament grew along the reflecting boundary of their simulation, and the magnetic field that pushed the protons out became a stable magnetic piston. This filament is the largest one because the density of the cloud is largest where it is close to the boundary, and because it was aligned with the flow direction of the pair cloud. The large pool of directed flow energy was converted to magnetic field energy by the filament instability. Similar expansion of electron-positron jet plasmas was observed in the global jets without helical magnetic fields [56].

The filament was generated in a pair jet due to the separation by the generated magnetic field from the expelled and shocked ambient plasma. The front of the jet propagated with the speed 0.15c along the boundary and expanded laterally at a speed that amounted up to about 0.03c. The growth of the filament was limited by their simulation box size and by the limited cloud size; a decrease of the ram pressure would inevitably lead to a weakening of the filamentation instability and to a collapse of the jet. But it appears that, as long as the pair cloud has enough ram pressure, the filaments can grow to arbitrarily large sizes if the filamentation instability develops between a pair cloud and an electron-proton plasma, at least for plasma parameters similar to those used here. It should be noted that this simulation study shows the importance of kinetic processes of injected cylindrical plasma clouds using PIC simulation.

4.3. Reconnection in Jets with Helical Magnetic Fields

Reconnection is ubiquitous in solar and magnetosphere plasmas, and it is an important additional particle acceleration mechanism for AGN and GRB jets (e.g., [71]). Despite the extensive research on reconnection, most of all reconnection simulations have been performed with the Harris sheet [72]. where the unperturbed magnetic fields **B** are anti-parallel ($\mathbf{B} = -\tanh(x)\mathbf{e}_y$). The release of energy stored in helical magnetic fields and particle acceleration during reconnection have been proposed as a mechanism for producing high-energy emissions and cosmic rays (e.g., [71,73]). It should be noted that the stored magnetic field energy in anti-parallel magnetic fields in the slab model is not consistent with the helical magnetic fields in the relativistic jets; therefore, a realistic argument on particle acceleration due to reconnection requires consideration of the helical magnetic field in the jets.

The importance of reconnection in jets has been proposed previously, but no kinetic simulation of global jets with helical magnetic fields has been performed before, with the exception of our own simulations [61,62].

Figure 10 shows the vectors of magnetic fields in the quadrant of the front part of the jets. Unfortunately, these vectors do not show the changes in direction which may reveal reconnection sites. In order to find the reconnection region, it is necessary to analyze the critical points (CPs). These CPs or magnetic nulls are the points where the magnitude of the magnetic field vector vanishes [74]. These points may be characterized by the behaviour of nearby magnetic field curves or surfaces. The set of curves or surfaces that end on CPs is of special interest because it defines the behaviour of the magnetic field in the neighborhood of CP.

The usual magnetic field configuration satisfies the hyperbolic conditions in which the vector field system has a nonzero real part of eigenvalues. The bifurcation (the topological change) represents the magnetic reconnection in the magnetic field. Thus, the particular sets of CPs, curves, and surfaces can be used to define a skeleton that uniquely characterizes the magnetic field [74]. In order to investigate the location of reconnection and its evolution, the method described by Cai, Nishikawa, and Lembege (2007) [74] needs to be employed in future work.

(a) (b)

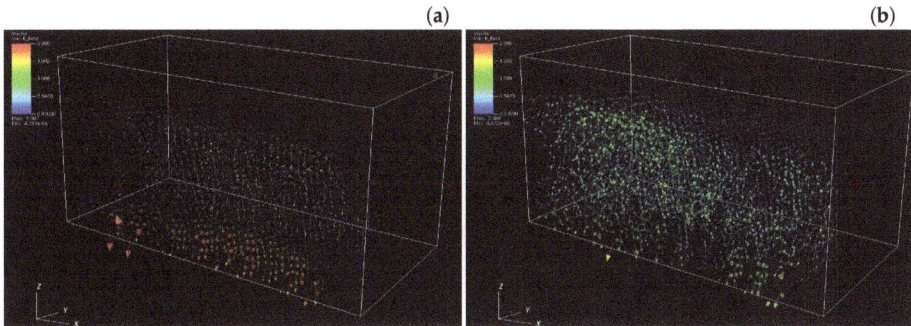

Figure 10. Panels show 3D vector plots of the magnetic fields for the (**a**) $e^- - p^+$ and (**b**) $e^\pm$ jets with $r_{jet} = 120\Delta$ at time $t = 500\omega_{pe}^{-1}$. The colors show the strength of the magnetic fields in the quadrant of the front part of the jets.

5. Discussion

In this paper, simulations of relativistic jets have been investigated extensively, starting from the study of the Weibel instability in slab mode, and continuing with simulations of instabilities in velocity-shears. Recently, a cylindrical geometry of the jets has been taken into account to be able to model the jet plasma more realistically.

The global jet simulations performed with large jet radii show the importance of a larger jet radius in PIC simulations for examining the macroscopic processes found in RMHD simulations. Due to the mixing of generated instabilities, the phase space plots of the jet electrons show little or no bunching in comparison to that when the jet radius is smaller, $r_{jet} = 20\Delta$. Consequently, recollimation-like shocks occur, rather in the center of the jets. Moreover, the recollimation-like shock structure is dependent on the value of the parameter of the helical magnetic field geometry a. To better understand the production of such recollimation-like shocks, further investigations of PIC simulations performed with even larger radii of the jets are needed.

The Weibel instability is ubiquitous in plasma flows, particularly when the plasma is unmagnetized. However, as shown in one of the simulations with global $e^- - p^+$ jets without helical magnetic fields, the Weibel instability is suppressed and the MI grows dominantly at the linear stage (see Figure 3a in Nishikawa et al. [56]). On the contrary, for $e^\pm$ jets, the Weibel instability grows with the kKHI and the MI.

So far, the global jet simulations have been performed only for two values of the ion-to-electron -mass ratio, $m_i/m_e = 1$ and 1836. The simulation results obtained even when $m_p/m_e = 1836$ indicate that a small grid system is not appropriate for studying the kinetic plasma instabilities altogether in a realistic way. At this time, these two cases will provide us clearer differences between two different cases with the maximum mass ratio. In the simulations performed by Nishikawa et al. [61,62], only a weak magnetization factor was used. Simulations with stronger helical magnetic fields were performed by us, and preliminary results show that MI grows stronger with stronger magnetic fields. However, further investigation is necessary with larger systems.

These simulations show that the excitation of kinetic instabilities like the kKHI, the MI, and the Weibel instability with kink-like instability release the energy stored in the helical magnetic field. Consequently, jet and ambient electrons are accelerated and magnetic fields become turbulent. Furthermore, the accelerating electrons emit radiation, and the turbulent magnetic field induces the polarization of the emitted radiation.

MacDonald & Marscher [3] have developed a radiative transfer scheme that allows the Turbulent Extreme Multi-Zone (TEMZ) code to produce simulated images of the time-dependent linearly and circularly polarized intensity at different radio frequencies. Using the PIC simulation output data as input parameters in the TEMZ code, synthetic polarized emission maps were obtained. These maps highlight the linear and circular polarization expected within the above PIC models. This algorithm is currently being refined to account for slow-light interpolation through the global PIC simulations reviewed here.

Simulations of global jets with helical magnetic fields are promising in regard to providing new insights into jet evolution and associated phenomena. However, at the present time, the length of the system is too small, and a much longer system is required in order to investigate a nonlinear stage. Possibly even when using larger systems, such as a numerical grid with $(L_x, L_y, L_z) = (2005\Delta, 1005\Delta, 1005\Delta)$, the jet radius 100Δ is not large enough to accommodate the microscopic processes, such as the gyro-motion of electrons and ions.

Therefore, these simulation results only provide some qualitative information which supplements those investigated by RMHD simulations. In the present simulations, jets were injected with a top-hat model. However, jets generated from black holes (either in AGN or in merging systems) have an opening angle and structured shapes. The helical magnetic fields used in the PIC simulations are not formed self-consistently as generated from rotating black holes like those performed in GRMHD simulations, and the initial setup with magnetic fields and the associated jet injection scheme need to be refined in future investigations. Furthermore, simulations of relativistic jets with large Lorentz factor particularly require the inclusion of radiation loss (e.g., [75]).

Since the power of supercomputers is growing rapidly, very large simulations of global jets could be performed, which will provide new insights on jet evolution, including reconnection and associated phenomena such as flares and high-energy particle generation.

Author Contributions: K.-I.N.: Perform simulations, analyze the data and prepare a manuscript; Y.M.: Understanding PIC simulation results based on RMHD simulations; J.L.G.: Contribute for comparing simulation results to observations; I.D.: Perform PIC simulations; A.M.: Perform simulations, critical reading and discussion on this research; J.N.: Contribute modifying the code for this research; O.K.: Modify the code for this simulation; M.P.: Overlook the simulation results; H.S.: Essential suggestions for this research; N.M.: Perform ray-tracing calculation for polarimetric images; D.H.H.: Useful discussions for this research.

Funding: This work is supported by NSF AST-0908010, AST-0908040, NASA-NNX09AD16G, NNX12AH06G, NNX13AP-21G, and NNX13AP14G grants. The work of J.N. and O.K. has been supported by Narodowe Centrum Nauki through research project DEC-2013/10/E/ST9/00662. Y.M. is supported by the ERC Synergy Grant "BlackHoleCam—Imaging the Event Horizon of Black Holes" (Grant No. 610058). M.P. acknowledges support through grant PO 1508/1-2 of the Deutsche Forschungsgemeinschaft. Simulations were performed using Pleiades and Endeavor facilities at NASA Advanced Supercomputing (NAS), and using Gordon and Comet at The San Diego Supercomputer Center (SDSC), and Stampede at The Texas Advanced Computing Center, which are supported by the NSF. This research was started during the program "Chirps, Mergers and Explosions: The Final Moments of Coalescing Compact Binaries" at the Kavli Institute for Theoretical Physics, which is supported by the National Science Foundation under grant No. PHY05-51164. The first velocity shear results using an electron−positron plasma were obtained during the Summer Aspen workshop "Astrophysical Mechanisms of Particle Acceleration and Escape from the Accelerators" held at the Aspen Center for Physics (1–15 September 2013).

Conflicts of Interest: The authors declare no conflict of interest.

References

1. Hawley, J.; Fendt, C.; Hardcastle, M.; Nokhrima, E.; Tchekhovskoy, A. Disks and Jets. *Galaxies* **2015**, *191*. [CrossRef]
2. Pe'er, A. Energetic and Broad Band Spectral Distribution of Emission from Astronomical Jets. *Space Sci. Rev.* **2014**, *183*, 371–403. [CrossRef]
3. MacDonald, N.R.; Marscher, A.P. Faraday Conversion in Turbulent Blazar Jets. *Astrophys. J.* **2018**, *862*, 58. [CrossRef]
4. Silva, L.O.; Fonseca, R.A.; Tonge, J.W.; Dawson, J.M.; Mori, W.B.; Medvedev, M.V. Interpenetrating Plasma Shells: Near-Equipartition Magnetic Field Generation and Nonthermal Particle Acceleration. *Astrophys. J. Lett.* **2003**, *596*, L121. [CrossRef]
5. Nishikawa, K.I.; Hardee, P.; Richardson, G.; Preece, R.; Sol, H.; Fishman, G.J. Particle Acceleration in Relativistic Jets Due to Weibel Instability. *Astrophys. J.* **2003**, *595*, 555–563. [CrossRef]
6. Frederiksen, J.T.; Hededal, C.B.; Haugbølle, T.; Nordlund, Å. Magnetic Field Generation in Collisionless Shocks: Pattern Growth and Transport. *Astrophys. J. Lett.* **2004**, *608*, L13. [CrossRef]
7. Hededal, C.B.; Haugbølle, T.; Frederiksen, J.T.; Nordlund, Å. Non-Fermi Power-Law Acceleration in Astrophysical Plasma Shocks. *Astrophys. J. Lett.* **2004**, *617*, L107. [CrossRef]
8. Hededal, C.B.; Nishikawa, K.I. The Influence of an Ambient Magnetic Field on Relativistic collisionless Plasma Shocks. *Astrophys. J. Lett.* **2005**, *623*, L89. [CrossRef]
9. Nishikawa, K.I.; Hardee, P.; Richardson, G.; Preece, R.; Sol, H.; Fishman, G.J. Particle Acceleration and Magnetic Field Generation in Electron-Positron Relativistic Shocks. *Astrophys. J.* **2005**, *622*, 927. [CrossRef]
10. Jaroschek, C.H.; Lesch, H.; Treumann, R.A. Ultrarelativistic Plasma Shell Collisions in γ-Ray Burst Sources: Dimensional Effects on the Final Steady State Magnetic Field. *Astrophys. J.* **2005**, *618*, 822. [CrossRef]
11. Nishikawa, K.I.; Hardee, P.E.; Hededal, C.B.; Fishman, G.J. Acceleration Mechanics in Relativistic Shocks by the Weibel Instability. *Astrophys. J.* **2006**, *642*, 1267. [CrossRef]
12. Nishikawa, K.I.; Hardee, P.; Hededal, C.; Richardson, G.; Preece, R.; Sol, H.; Fishman, G. Particle acceleration, magnetic field generation, and emission in relativistic shocks. *Adv. Space Res.* **2006**, *38*, 1316–1319. [CrossRef]
13. Nishikawa, K.I.; Mizuno, Y.; Fishman, G.J.; Hardee, P. Particle Acceleration, Magnetic Field Generation, and Associated Emission in Collisionless Relativistic Jets. *Int. J. Mod. Phys. D* **2008**, *17*, 1761–1767. [CrossRef]
14. Spitkovsky, A. On the Structure of Relativistic Collisionless Shocks in Electron-Ion Plasmas. *Astrophys. J. Lett.* **2008**, *673*, L39. [CrossRef]
15. Spitkovsky, A. Particle Acceleration in Relativistic Collisionless Shocks: Fermi Process at Last? *Astrophys. J. Lett.* **2008**, *682*, L5. [CrossRef]
16. Chang, P.; Spitkovsky, A.; Arons, J. Long-Term Evolution of Magnetic Turbulence in Relativistic Collisionless Shocks: Electron-Positron Plasmas. *Astrophys. J.* **2008**, *674*, 378. [CrossRef]
17. Dieckmann, M.E.; Shukla, P.K.; Drury, L.O.C. The Formation of a Relativistic Partially Electromagnetic Planar Plasma Shock. *Astrophys. J.* **2008**, *675*, 586–595. [CrossRef]
18. Nishikawa, K.I.; Niemiec, J.; Hardee, P.E.; Medvedev, M.; Sol, H.; Mizuno, Y.; Zhang, B.; Pohl, M.; Oka, M.; Hartmann, D.H. Weibel Instability and Associated Strong Fields in a Fully Three-Dimensional Simulation of a Relativistic Shock. *Astrophys. J. Lett.* **2009**, *698*, L10. [CrossRef]
19. Martins, S.F.; Fonseca, R.A.; Silva, L.O.; Mori, W.B. Ion Dynamics and Acceleration in Relativistic Shocks. *Astrophys. J. Lett.* **2009**, *695*, L189. [CrossRef]
20. Nishikawa, K.I.; Niemiec, J.; Medvedev, M.; Zhang, B.; Hardee, P.; Nordlund, A.; Frederiksen, J.; Mizuno, Y.; Sol, H.; Pohl, M.; et al. Radiation from relativistic shocks in turbulent magnetic fields. *Adv. Space Res.* **2011**, *47*, 1434–1440. [CrossRef]
21. Choi, E.J.; Min, K.; Nishikawa, K.I.; Choi, C.R. A study of the early-stage evolution of relativistic electron-ion shock using three-dimensional particle-in-cell simulations. *Phys. Plasmas* **2014**, *21*, 072905. [CrossRef]
22. Ardaneh, K.; Cai, D.; Nishikawa, K.I.; Lembége, B. Collisionless Weibel Shocks and Electron Acceleration in Gamma-Ray Bursts. *Astrophys. J.* **2015**, *811*, 57. [CrossRef]
23. Ardaneh, K.; Cai, D.; Nishikawa, K.I. Collisionless Electron-ion Shocks in Relativistic Unmagnetized Jet-ambient Interactions: Non-thermal Electron Injection by Double Layer. *Astrophys. J.* **2016**, *827*, 124. [CrossRef]

24. Grassi, A.; Grech, M.; Amiranoff, F.; Macchi, A.; Riconda, C. Radiation-pressure-driven ion Weibel instability and collisionless shocks. *Phys. Rev. E* **2017**, *96*, 033204. [CrossRef] [PubMed]

25. Iwamoto, M.; Amano, T.; Hoshino, M.; Matsumoto, Y. Precursor Wave Emission Enhanced by Weibel Instability in Relativistic Shocks. *Astrophys. J.* **2018**, *858*, 93. [CrossRef]

26. Takamoto, M.; Matsumoto, Y.; Kato, T.N. Magnetic Field Saturation of the Ion Weibel Instability in Interpenetrating Relativistic Plasmas. *Astrophys. J. Lett.* **2018**, *860*, L1. [CrossRef]

27. Weibel, E. Spontaneously Growing Transverse Waves in a Plasma Due to an Anisotropic Velocity Distr. *Phys. Rev. Lett.* **1959**, *2*, 83–84. [CrossRef]

28. Medvedev, M.V.; Loeb, A. Generation of Magnetic Fields in the Relativistic Shock of Gamma-Ray Burst Sources. *Astrophys. J.* **1999**, *526*, 697. [CrossRef]

29. Medvedev, M.V.; Zakutnyaya, O.V. Magnetic Fields and Cosmic Rays in GRBs: A Self-Similar Collisionless Foreshock. *Astrophys. J.* **2009**, *696*, 2269. [CrossRef]

30. Hoshino, M.; Shimada, N. Nonthermal Electrons at High Mach Number Shocks: Electron Shock Surfing Acceleration. *Astrophys. J.* **2002**, *572*, 880. [CrossRef]

31. Amano, T.; Hoshino, M. Electron Injection at High Mach Number Quasi-perpendicular Shocks: Surfing and Drift Acceleration. *Astrophys. J.* **2007**, *661*, 190. [CrossRef]

32. Amano, T.; Hoshino, M. Electron Shock Surfing Acceleration in Multidimensions: Two-Dimensional Particle-in-Cell Simulation of Collisionless Perpendicular Shock. *Astrophys. J.* **2009**, *690*, 244. [CrossRef]

33. Sironi, L.; Giannios, D. Sironi, L,; Spitkovsky, A.; Arons, J. The Maximum Energy of Accelerated Particles in Relativistic Collisionless Shocks. *Astrophys. J.* **2013**, *771*, 54.

34. Guo, X.; Sironi, L.; Narayan, R. Non-thermal Electron Acceleration in Low Mach Number Collisionless Shocks. I. Particle Energy Spectra and Acceleration Mechanism. *Astrophys. J.* **2014**, *794*, 153. [CrossRef]

35. Sironi, L.; Petropoulou, M.; Giannios, D. Relativistic jets shine through shocks or magnetic reconnection? *Mon. Not. R. Astron. Soc.* **2015**, *450*, 183–191. [CrossRef]

36. Bret, A.; Alvaro, E.P. Robustness of the filamentation instability as shock mediator in arbitrarily oriented magnetic field. *Phys. Plasmas* **2011**, *18*, 080706. [CrossRef]

37. Buneman, O. Tristan. In *Computer Space Plasma Physics: Simulation Techniques and Software*; TERRAPUB: Tokyo, Japan, 1993; Volume 1, pp. 67–100.

38. Niemiec, J.; Pohl, M.; Stroman, T.; Nishikawa, K.I. Production of Magnetic Turbulence by Cosmic Rays Drifting Upstream of Supernova Remnant Shocks. *Astrophys. J.* **2008**, *684*, 1174. [CrossRef]

39. Ramirez-Ruiz, E.; Nishikawa, K.I.; Hededal, C.B. $e^{\pm}$ Pair Loading and the Origin of the Upstream Magnetic Field in GRB Shocks. *Astrophys. J.* **2007**, *671*, 1877. [CrossRef]

40. Mizuno, Y.; Lyubarsky, Y.; Nishikawa, K.I.; Hardee, P.E. Three-Dimensional Relativistic Magnetohydrodynamic Simulations of Current-Driven Instability. I. Instability of a Static Column. *Astrophys. J.* **2009**, *700*, 684. [CrossRef]

41. Spitkovsky, A. Simulations of relativistic collisionless shocks: Shock structure and particle acceleration. In *Astrophysical Sources of High Energy Particles and Radiation*; Bulik, T., Rudak, B., Madejski, G., Eds.; American Institute of Physics Conference Series; Springer: Berlin, Germany, 2005; Volume 801, pp. 345–350, doi:10.1063/1.2141897.

42. D'Angelo, N. Kelvin-Helmholtz Instability in a Fully Ionized Plasma in a Magnetic Field. *Phys. Fluids* **1965**, *8*, 1748–1750. [CrossRef]

43. Gruzinov, A. GRB: Magnetic fields, cosmic rays, and emission from first principles? *arXiv* **2008**, arXiv:0803.1182.

44. Mizuno, Y.; Hardee, P.; Nishikawa, K.I. Three-dimensional Relativistic Magnetohydrodynamic Simulations of Magnetized Spine-Sheath Relativistic Jets. *Astrophys. J.* **2007**, *662*, 835–850. [CrossRef]

45. Perucho, M.; Lobanov, A.P. Kelvin-Helmholtz Modes Revealed by the Transversal Structure of the Jet in 0836+710. In *Extragalactic Jets: Theory and Observation from Radio to Gamma Ray*; Rector, T.A., De Young, D.S., Eds.; Astronomical Society of the Pacific Conference Series; Cornell University: San Francisco, CA, USA, 2008; Volume 386, pp. 381–387.

46. Zhang, W.; MacFadyen, A.; Wang, P. Three-Dimensional Relativistic Magnetohydrodynamic Simulations of the Kelvin-Helmholtz Instability: Magnetic Field Amplification by a Turbulent Dynamo. *Astrophys. J. Lett.* **2009**, *692*, L40. [CrossRef]

47. Alves, E.P.; Grismayer, T.; Martins, S.F.; Fiúza, F.; Fonseca, R.A.; Silva, L.O. Large-scale Magnetic Field Generation via the Kinetic Kelvin-Helmholtz Instability in Unmagnetized Scenarios. *Astrophys. J. Lett.* **2012**, *746*, L14. [CrossRef]

48. Nishikawa, K.I.; Hardee, P.; Zhang, B.; Duᴀ̊an, I.; Medvedev, M.; Choi, E.J.; Min, K.W.; Niemiec, J.; Mizuno, Y.; Nordlund, A.; et al. Magnetic field generation in a jet-sheath plasma via the kinetic Kelvin-Helmholtz instability. *Ann. Geophys.* **2013**, *31*, 1535–1541. [CrossRef]

49. Liang, E.; Boettcher, M.; Smith, I. Magnetic Field Generation and Particle Energization at Relativistic Shear Boundaries in Collisionless Electron-Positron Plasmas. *Astrophys. J. Lett.* **2013**, *766*, L19. [CrossRef]

50. Grismayer, T.; Alves, E.P.; Fonseca, R.A.; Silva, L.O. dc-Magnetic-Field Generation in Unmagnetized Shear Flows. *Phys. Rev. Lett.* **2013**, *111*, 015005. [CrossRef]

51. Grismayer, T.; Alves, E.P.; Fonseca, R.A.; Silva, L.O. Theory of multidimensional electron-scale instabilities in unmagnetized shear flows. *Plasma Phys. Control. Fusion* **2013**, *55*, 124031. [CrossRef]

52. Liang, E.; Fu, W.; Boettcher, M.; Smith, I.; Roustazadeh, P. Relativistic Positron-Electron-Ion Shear Flows and Application to Gamma-Ray Bursts. *Astrophys. J. Lett.* **2013**, *779*, L27. [CrossRef]

53. Alves, E.P.; Grismayer, T.; Fonseca, R.A.; Silva, L.O. Electron-scale shear instabilities: Magnetic field generation and particle acceleration in astrophysical jets. *New J. Phys.* **2014**, *16*, 035007. [CrossRef]

54. Alves, E.P.; Grismayer, T.; Fonseca, R.A.; Silva, L.O. Transverse electron-scale instability in relativistic shear flows. *Phys. Rev. E* **2015**, *92*, 021101. [CrossRef] [PubMed]

55. Nishikawa, K.I.; Hardee, P.E.; Duᴀ̊an, I.; Niemiec, J.; Medvedev, M.; Mizuno, Y.; Meli, A.; Sol, H.; Zhang, B.; Pohl, M.; et al. Magnetic Field Generation in Core-sheath Jets via the Kinetic Kelvin-Helmholtz Instability. *Astrophys. J.* **2014**, *793*, 60. [CrossRef]

56. Nishikawa, K.I.; Frederiksen, J.T.; Nordlund, Å.; Mizuno, Y.; Hardee, P.E.; Niemiec, J.; Gómez, J.L.; Pe'er, Å.; Dutan, I.; Meli, A.; et al. Evolution of Global Relativistic Jets: Collimations and Expansion with kKHI and the Weibel Instability. *Astrophys. J.* **2016**, *820*, 94. [CrossRef]

57. Dieckmann, M.; Sarri, G.; Folini, D.; Walder, R.; Borghesi, M. Cocoon formation by a mildly relativistic pair jet in unmagnetized collisionless electron-proton plasma. *Phys. Plasmas* **2018**, *25*, 112903. [CrossRef]

58. Nishikawa, K.I.; Hardee, P.; Dutan, I.; Zhang, B.; Meli, A.; Choi, E.J.; Min, K.; Niemiec, J.; Mizuno, Y.; Medvedev, M.; et al. Radiation from Particles Accelerated in Relativistic Jet Shocks and Shear-flows. *arXiv* **2014**, arXiv:1412.7064.

59. Alves, E.P.; Zrake, J.; Fiuza, F. Efficient Nonthermal Particle Acceleration by the Kink Instability in Relativistic Jets. *arXiv* **2018**, arXiv:1810.05154.

60. Blandford, R.D.; Znajek, R.L. Electromagnetic extraction of energy from Kerr black holes. *Mon. Not. R. Astron. Soc.* **1977**, *179*, 433–456. [CrossRef]

61. Nishikawa, K.I.; Mizuno, Y.; Niemiec, J.; Kobzar, O.; Pohl, M.; Gómez, J.L.; Dutan, I.; Pe'er, A.; Frederiksen, J.T.; Nordlund, Å.; et al. Microscopic Processes in Global Relativistic Jets Containing Helical Magnetic Fields. *Galaxies* **2016**, *4*, 38. [CrossRef]

62. Nishikawa, K.I.; Mizuno, Y.; Gómez, J.L.; Dutan, I.; Meli, A.; White, C.; Niemiec, J.; Kobzar, O.; Pohl, M.; Pe'er, A.; et al. Microscopic Processes in Global Relativistic Jets Containing Helical Magnetic Fields: Dependence on Jet Radius. *Galaxies* **2017**, *5*, 58. [CrossRef]

63. Tchekhovskoy, A. Launching of Active Galactic Nuclei Jets. In *The Formation and Disruption of Black Hole Jets*; Springer: Berlin, Germany, 2015; Volume 414, pp. 45–82.

64. Dutan, I.; Nishikawa, K.I.; Mizuno, Y.; Niemiec, J.; Kobzar, O.; Pohl, M.; Gómez, J.L.; Pe'er, A.; Frederiksen, J.T.; Nordlund, Å.; et al. Particle-in-cell Simulations of Global Relativistic Jets with Helical Magnetic Fields. *Proc. Int. Astron. Union* **2016**, *12*, 199–202. [CrossRef]

65. Mizuno, Y.; Hardee, P.E.; Nishikawa, K.I. Spatial Growth of the Current-driven Instability in Relativistic Jets. *Astrophys. J.* **2014**, *784*, 167. [CrossRef]

66. Singh, C.B.; Mizuno, Y.; de Gouveia Dal Pino, E.M. Spatial Growth of Current-driven Instability in Relativistic Rotating Jets and the Search for Magnetic Reconnection. *Astrophys. J.* **2016**, *824*, 48. [CrossRef]

67. Barniol Duran, R.; Tchekhovskoy, A.; Giannios, D. Simulations of AGN jets: Magnetic kink instability versus conical shocks. *Mon. Not. R. Astron. Soc.* **2017**, *469*, 4957–4978. [CrossRef]

68. Mizuno, Y.; Gomez, J.L.; Nishikawa, K.I.; Meli, A.; Hardee, P.E.; Rezzolla, L. Recollimation Shocks in Magnetized Relativistic Jets. *Astrophys. J.* **2015**, *809*, 38. [CrossRef]

69. Broderick, A.E.; Loeb, A. Imaging the Black Hole Silhouette of M87: Implications for Jet Formation and Black Hole Spin. *Astrophys. J.* **2009**, *697*, 1164. [CrossRef]
70. Mościbrodzka, M.; Dexter, J.; Davelaar, J.; Falcke, H. Faraday rotation in GRMHD simulations of the jet launching zone of M87. *Mon. Not. R. Astron. Soc.* **2017**, *468*, 2214–2221. [CrossRef]
71. Giannios, D.; Uzdensky, D.A.; Begelman, M.C. Fast TeV variability in blazars: Jets in a jet. *Mon. Not. R. Astron. Soc. Lett.* **2009**, *395*, L29–L33. [CrossRef]
72. Harris, E.G. On a plasma sheath separating regions of oppositely directed magnetic field. *Il Nuovo Cimento* **1962**, *23*, 115–121. [CrossRef]
73. Barniol Duran, R.; Leng, M.; Giannios, D. An anisotropic minijets model for the GRB prompt emission. *Mon. Not. R. Astron. Soc. Lett.* **2016**, *455*, L6–L10. [CrossRef]
74. Cai, D.; Nishikawa, K.I.; Lembege, B. Visualization of Tangled Vector Field Topology and Global Bifurcation of Magnetospheric Dynamics. In *Advanced Methods for Space Simulations*; TERRAPUB: Tokyo, Japan, 2007; pp. 145–166.
75. Cerutti, B.; Uzdensky, D.A.; Begelman, M.C. Extreme Particle Acceleration in Magnetic Reconnection Layers: Application to the Gamma-Ray Flares in the Crab Nebula. *Astrophys. J.* **2012**, *746*, 148. [CrossRef]

MDPI

Article

Progress in Multi-Wavelength and Multi-Messenger Observations of Blazars and Theoretical Challenges

Markus Böttcher

Centre for Space Research, North-West University, Potchefstroom 2520, South Africa;
Markus.Bottcher@nwu.ac.za; Tel.: +27-18-299-2418

Received: 6 November 2018; Accepted: 9 January 2019; Published: 18 January 2019

Abstract: This review provides an overview of recent advances in multi-wavelength and multi-messenger observations of blazars, the current status of theoretical models for blazar emission, and prospects for future facilities. The discussion of observational results will focus on advances made possible through the *Fermi Gamma-Ray Space Telescope* and ground-based gamma-ray observatories (H.E.S.S., MAGIC, VERITAS), as well as the recent first evidence for a blazar being a source of IceCube neutrinos. The main focus of this review will be the discussion of our current theoretical understanding of blazar multi-wavelength and multi-messenger emission, in the spectral, time, and polarization domains. Future progress will be expected in particular through the development of the first X-ray polarimeter, IXPE, and the installation of the Cherenkov Telescope Array (CTA), both expected to become operational in the early to mid 2020s.

Keywords: active galaxies; blazars; multi-wavelength astronomy; muti-messenger astronomy; neutrino astrophysics; polarization

1. Introduction

Blazars are the class of jet-dominated, radio-loud active galactic nuclei (AGN) whose relativistic jets point close to our line of sight. Due to this viewing geometry, all emissions from a region moving with Lorentz factor $\Gamma \equiv (1 - \beta_\Gamma^2)^{-1/2}$, where $\beta_\Gamma c$ is the jet speed, along the jet, at an angle θ_{obs} with respect to our line of sight, will be Doppler boosted in frequency by a factor $\delta = (\Gamma[1 - \beta_\Gamma \cos \theta_{obs}])^{-1}$ and in bolometric luminosity by a factor δ^4 with respect to quantities measured in the co-moving frame of the emission region. Any time scale of variability in the co-moving frame will be observed shortened by a factor δ^{-1}. These effects make blazars the brightest γ-ray sources in the extragalactic sky (e.g., [1]), exhibiting variability, in extreme cases, on time scales down to just minutes (e.g., [2–5]).

The broad-band continuum (radio through γ-ray) spectral energy distributions (SEDs) of blazars are typically dominated by two broad, non-thermal radiation components. The low-frequency component, from radio through optical/UV (in some cases, X-rays) is generally agreed to be synchrotron emission from relativistic electrons in the jet, as evidenced by the measurement of significant, variable linear polarization in the radio (e.g., [6]) and optical (e.g., [7]) wavebands. For the high-energy (X-ray through γ-ray) SED component, both leptonic (high-energy emission dominated by electrons and/or electron-positron pairs) and hadronic (high-energy emission dominated by ultrarelativistic protons) models are being considered (for a comparative discussion of both types of models with application to a sample of γ-ray blazas see, e.g., [8], and for a general review of emission models of relativistic jet sources, see, e.g., [9]).

The population of blazars consists of flat-spectrum radio quasars (FSRQs) and BL Lac objects, the latter distinguishing themselves by (nearly) featureless optical spectra with emission-line equivalent widths $EW < 5$ Å. An alternative classification based on the broad-emission-line luminosity (rather than EW) was proposed by [10]. The featureless optical continuum spectra of BL Lac objects often makes it difficult or impossible to determine their redshift. A more physical distinction between different blazar classes might be on the basis of the location of their SED peak frequencies. Low-synchrotron peaked blazars (LSP) are defined by having a synchrotron peak frequency $\nu_{sy} < 10^{14}$ Hz, intermediate-synchrotron peaked blazars (ISP) have 10^{14} Hz $\leq \nu_{sy} < 10^{15}$ Hz, while high-synchrotron peaked blazars (HSPs) have $\nu_{sy} > 10^{15}$ Hz [11]. Most HSP and ISP blazars have been classified as BL Lac objects based on their optical spectra, while the LSP class contains both FSRQs and low-frequency-peaked BL Lac objects.

In leptonic models, the high-energy emission is produced by Compton scattering of soft (IR–optical–UV) target photon fields by relativistic electrons. Target photon fields can be the co-spatially produced synchrotron radiation (synchrotron self-Compton or SSC; see, e.g., [12]) or external radiation fields, such as those from the accretion disk (e.g., [13]), the broad-line region (e.g., [14]) a dusty, infrared-emitting torus (e.g., [15]), or synchrotron emission from other, slower or faster moving regions of the jet, such as a slow sheath surrounding the highly relativistic spine in a radially stratified jet (e.g., [16]) or another (slower/faster) jet component in a decelerating jet flow (e.g., [17]). Even though in leptonic models the radiation output is dominated by electrons and/or pairs, it is generally believed that the jets also contain non- or mildly-relativistic protons. Due to their much larger mass, they will not contribute significantly to the radiative output, but they may still carry a significant (if not dominant) fraction of the momentum and kinetic power of the jet (see, e.g., [18]).

In hadronic models for blazar emission, it is assumed that protons are accelerated to ultra-relativistic energies so that they can dominate the high-energy emission through proton-synchrotron radiation (e.g., [19,20]) or through photo-pion production (e.g., [21,22]), with subsequent pion decay leading to the production of ultra-high-energy photons and pairs (and neutrinos!). These ultra-relativistic secondary electrons/positrons lose their energy quickly due to synchrotron radiation. Both these synchrotron photons and the initial π^0 decay photons have too high energy to escape $\gamma\gamma$ absorption in the source, thus initiating synchrotron-supported pair cascade. This typically leads to very broad emerging γ-ray spectra, typically extending into the X-ray regime (e.g., [23]).

This review will provide, in Section 2, an overview of recent observational highlights on blazars across the electromagnetic spectrum, including aspects of flux and polarization variability, as well as multi-messenger aspect, especially the recent likely identification of the blazar TXS 0506+056 as a source of very-high-energy neutrinos detected by IceCube. Necessarily, this review will need to focus on recent highlights for a few selected topics, also biased by the author's scientific interests. However, a balanced and fair review of relevant works on the selected topics is attempted.

Section 3 summarizes recent developments on the theory side, with a view towards inferences from these recent observational highlights. Section 4 prospects for future multi-wavelength and multi-messenger observations, especially towards addressing the following questions, where the author believes that major breakthroughs are possible through dedicated blazar observations within the next decade:

- What is the matter composition of blazar jets, and what is the dominant particle population responsible for the high-energy emission? Answering this question will allow major progress concerning physics of jet launching and loading, and the mode of acceleration of relativistic particles in jets.
- What is the structure of magnetic fields in the high-energy emission region and their role in the acceleration of relativistic particles? The answer to this question will aid in understanding the physics of jet collimation and stability and provide further clues to the physics of ultra-relativistic particle acceleration (magnetic reconnection vs. shocks vs. shear layers, ...)
- Where along the jet is high-energy and very-high-energy γ-ray emission predominantly produced? Different observational results currently point towards different answers (sub-pc vs. 10s of pc from the central black hole). The confident localization of the blazar γ-ray emission region will further constrain plausible radiation mechanisms and could possibly hint at beyond-the-standard-model

physics (if evidence suggests that $\gamma\gamma$ absorption in the radiation field of the broad line region is suppressed below standard expectations).

Throughout the text, physical quantities are parameterized with the notation $Q = 10^x Q_x$ in c.g.s. units.

2. Recent Observational Highlights

This section summarizes highlights of recent multi-wavelength and multi-messenger observations of blazars, focusing on results directly probing the high-energy emission region, which tends to be optically thick to radio wavelengths. Radio observations typically probe the larger-scale (pc-scale and larger) structure of jets and will not be discussed in this review. For recent reviews of radio observations of blazar jets, the reader is referred to, e.g., [24–27].

2.1. Flux Variability

Blazars are characterized by their significant variability across the entire electromagnetic spectrum, on all time scales ranging from years down to minutes. In addition to the very short variability time scales, a major challenge for our understanding of the optical through γ-ray emission is the fact that the variability patterns in different wavelength bands do not show a consistent behaviour of correlation (or non-correlation).

2.1.1. Minute-Scale Variability

The shortest variability time scales, down to minutes, have been found in very-high-energy γ-ray observations using ground-based Imaging Atmospheric Cherenkov Telescope (IACT) facilities (H.E.S.S., MAGIC, VERITAS). They were first identified in HSP blazars, such as PKS 2155-304 [2] and Mrk 501 [3], but later also in LSP blazars, namely the prototypical BL Lac object BL Lacertae [5] and the FSRQ PKS 1222+21 [4]. Remarkably, sub-hour very-high-energy (VHE) γ-ray variability was also seen by MAGIC in IC 310 [28], which, based on its large-scale radio structure had been classified as a radio galaxy, in which case relativistic beaming effects would be inefficient. However, its broadband SED and variability behaviour suggest a blazar-like orientation of the inner jet, as is probably also the case for the H.E.S.S.-detected radio galaxy PKS 0625-35 [29].

Rapid variability has also been seen at GeV energies with *Fermi*-LAT, although the limited photon statistics (because of the $\sim 10^5$ times smaller collection area of *Fermi*-LAT compared to IACTs) typically limits the time scales to $\gtrsim$ hours (e.g., [30,31]). Note, however, that minute-scale GeV variability has been detected with *Fermi*-LAT during the giang γ-ray outburst of 3C 279 in June 2015 [32] and, more recently, at 4.7 σ significance during the exceptional long-term outburst of the FSRQ CTA 102 in 2016–2017 (see Figure 1 [33]).

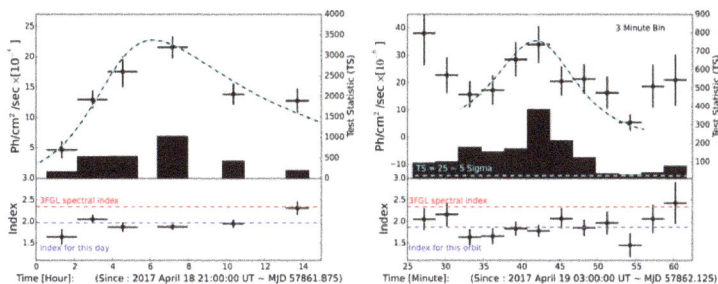

Figure 1. *Fermi*-LAT GeV γ-ray light curve of Cherenkov Telescope Array (CTA) 102 on 19 April 2017. **Left**: Orbit-binned light curve; **Right**: 3-min binned light curve (from] [33]). Reproduced with permission from the American Astronomical Society (AAS).

Due to causality arguments, the observed variability time scale t_{var}^{obs} imposes a limit on the size R of the emission region,

$$R \leq c\, t_{var}^{obs} \frac{\delta}{1+z} = 1.8 \times 10^{14} \left(\frac{t_{var}^{obs}}{5\,\text{min}} \right) \delta_1. \tag{1}$$

The observed γ-ray flux, modulo Doppler boosting, implies a luminosity, which can be translated into a lower limit on the photon energy density in the emission region, using the size constraint from Equation (1). Since there is, so far, no evidence for internal $\gamma\gamma$ absorption on the co-spatially produced low-energy (IR–X-ray) radiation field in the MeV–GeV γ-ray emission of blazars, the emission region must be optically thin to this process. This then implies a lower limit on the Doppler factor. γ-ray photons in the GeV regime interact primarily with target photons that are observed in the X-ray regime. Assuming an observed X-ray spectrum with an energy spectral index α_X and an integrated flux F_{0-1} in the range E_0 to E_1 (corresponding to normalized energies $\epsilon_{0,1} \equiv E_{0,1}/[m_e c^2]$), the limit on the Doppler factor can be derived as [34,35]

$$\delta \geq \left(10^{3\alpha_X} \frac{\sigma_T d_L^2}{3\, m_e c^4\, t_{var}^{obs}} F_{0-1} \frac{1 - \alpha_X}{\epsilon_1^{1-\alpha_X} - \epsilon_0^{1-\alpha_X}} [1+z]^{2\alpha_X} E_{GeV}^{\alpha_X} \right)^{\frac{1}{4+2\alpha_X}}. \tag{2}$$

where E_{GeV} is the maximum photon energy (in units of GeV) out to which there is no evidence for a spectral break due to $\gamma\gamma$ absorption (typically of the order of $\sim$10–100 GeV in the case of most Fermi blazars), d_L is the luminosity distance, and σ_T is the Thomson cross section. In the case of the 5-min variability observed in a few blazars, this, in fact, implies minimum Doppler factors of $\delta \gtrsim 50$ [36]. Such values are much higher than the Doppler factors of $\delta \sim 10$ typically inferred from superluminal motion speeds observed in radio Very Long Baseline Interferometry (VLBI) monitoring observations of blazars (e.g., [37,38])—a problem sometimes referred to as the *Doppler-factor crisis* [39]. Various suggested model solutions to this problem will be discussed in Section 3.1.

2.1.2. Multiwavelength Correlations

Another major challenge to our current understanding of the physical processes in blazar γ-ray emission regions is the fact that multi-wavelength variability patterns are sometimes correlated, sometimes not, among different wavelength bands. Even within the same object, the correlated/uncorrelated variability behaviour changes between different observation periods. Most blazar emission scenarios ascribe the entire IR–γ-ray emission to one single dominant emission region (single-zone models). In this case, one would naturally expect all radiating particles to be subject to the same acceleration and cooling mechanisms. As the radiative cooling time scales of particles are energy-dependent (scaling as $t_{cool} \propto \gamma^{-1}$ for synchrotron radiation and Thomson scattering), variability patterns at different energies are expected to show time delays, but still be correlated. In fact, if an observed time delay between the variability patterns at two frequencies $E_1 = E_{1,keV}$ keV and $E_2 = E_{2,keV}$ keV is related to different radiative cooling time scales of synchrotron-emitting electrons radiating at those energies, this can be used to place a lower limit on the magnetic field [40]. If electron cooling is dominated by synchrotron and Compton cooling in the Thomson regime, a synchrotron time delay $\tau_{1,2} \equiv \tau_h$ hr translates into a lower limit on the magnetic field of

$$B \gtrsim 0.9\, \tau_h^{-2/3} \delta^{-1/3} (1+k)^{-2/3} \left(E_{1,keV}^{-1/2} - E_{2,keV}^{-1/2} \right)^{2/3} \text{G} \tag{3}$$

where $k = L_C/L_{sy}$ is the Compton dominance parameter of the SED. Such arguments have, in several cases, constrained the magnetic fields in the emission regions of blazars to be of the order of $B \gtrsim 1$ G (e.g., [41]).

In the case of LSP blazars, leptonic single-zone emission models predict the optical synchrotron emission and GeV (Compton) γ-ray emission to be produced by electrons of approximately the same energy. The optical and γ-ray light curves are therefore expected to be closely correlated with very small time lags. In the case of HSP blazars, the same type of correlation is expected to exist between X-rays and VHE γ-rays. Such a close correlation is often observed, but there are also many examples in which the correlation is absent. The most striking cases are the so-called "orphan flares", in which either γ-ray flares occur with no visible counterpart in the synchrotron component ((e.g., [42,43]), or synchrotron (IR–optical–X-ray) flares without γ-ray counterpart.

A remarkable and unusual case of an orphan flare was detected by H.E.S.S. from the FSRQ 3C 279 on 27–28 January 2018 [44]. H.E.S.S. observations had been triggered by a *Fermi*-LAT detected GeV γ-ray flare of 3C 279 around 16 January 2018. While H.E.S.S. did not detect the source around the time of the *Fermi*-LAT flare, a significant detecton ($\sim$11σ) resulted with a delay of $\sim$11 days with respect to the GeV flare (see Figure 2). Also the optical (e.g., from the Steward Observatory Blazar Monitoring Program[1]) and X-ray (e.g., *Swift*-XRT[2]) light curves showed no renewed activity at the time of the H.E.S.S. VHE flare detection. Such orphan flares appear to strongly argue against simple single-zone emission models, and possible alternatives will be discussed in Section 3.1.

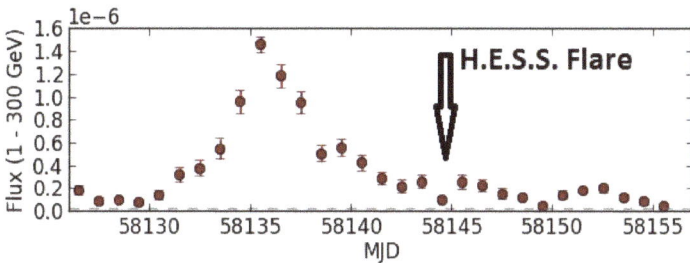

Figure 2. *Fermi*-LAT GeV γ-ray light curve of 3C 279 around the flare of January 2018. The heavy arrow marks the night of the H.E.S.S. VHE flare detection [44].

2.1.3. Periodicities?

AGN activity is often associated with recent galaxy mergers (e.g., [45–48]). If this is true, then one might expect binary supermassive black hole (SMBH) systems, instead of a single SMBH, to be present in the centers of at least some AGN. The orbital modulation as well as Lense-Thirring precession of the dominant accretion disk (and likely also the jet), might then lead to periodic or quasi-periodic modulations of the multi-wavelength emissions of blazars. However, the only case in which quasi-periodicity, likely related to the presence of a binary SMBH system, is clearly established, is the BL Lac object OJ 287, where a dominant SMBH of $\sim$ a few $\times 10^9 \, M_\odot$ appears to be in a $\sim$12 year orbit with a smaller SMBH, which intercepts the primary accretion disk twice per orbit [49–51].

Given the long time lines of many (especially optical and radio) blazar monitoring campaigns and the now $\sim$10 years of operations of *Fermi*, continuous γ-ray and well-sampled multi-wavelength light curves now exist for a large number of blazars, allowing for efficient searches for periodicities on yearly time scales. The most promising candidate for such periodicities appears to be the BL Lac object PG 1553+113, where a period of $\sim$2.2 years has been identified in *Fermi*-LAT and multi-wavelength data [52], possibly including secondary "twin peaks" symmetrically spaced around the main peaks [53]. A systematic search for periodicities in *Fermi*-LAT blazar light curves by [54], however, finds no significant evidence for periodicities beyond 95% confidence in any of the 10 sources studied, including

---

[1] http://james.as.arizona.edu/ psmith/Fermi/.
[2] https://www.swift.psu.edu/monitoring/.

PG 1553+113. Thus, one may conclude that the question concerning periodicities and the existence of binary SMBHs in blazars (beyond the case of OJ 287) is still controversial.

2.2. Polarization—Variability

The radio through optical emission from blazars has long been known to be polarized with significant variability both in the degree of polarization (Π) and the polarization angle (PA). In this review, we will focus on optical polarization measurements, as those might be the best probes of the magnetic field structure in the high-energy emission region. The optical polarization in blazars varies from virtually unpolarized sources/states to highly polarized states with polarization degrees of $\Pi \lesssim 50\%$ (e.g., [55]). In addition to photopolarimetric and spectropolarimetric monitoring of blazars by, e.g., the Steward Observatory blazar monitoring program, a systematic study of the polarization variability of a large sample of blazars was performed with the RoboPol polarimeter[3] on the 1.3 m telescope of the Skinakas Observatory [56].

The average degree of polarization has been found to be systematically larger for γ-ray loud blazars compared to γ-ray quiet ones [57], which is likely due to the fact that γ-ray loud blazars appear more strongly Doppler boosted and thus more strongly synchrotron dominated in the optical spectrum. The average degree of optical polarization systematically decreases as a function of increasing synchrotron peak frequency within the RoboPol sample [57]. This may be attributed to the fact that in LSP blazars, the optical range is at or near the peak of the synchrotron emission, thus reflecting freshly accelerated electrons in a presumably very confined region in the jet, while in HSP blazars, optical-synchrotron-emitting electrons have already cooled substantially and are most likely distributed over a larger portion of the jet (see Figure 3).

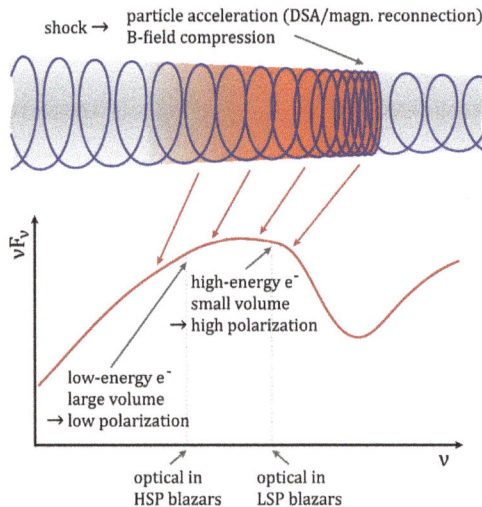

Figure 3. Sketch to explain the dependence of the optical polarization angle (PA) on the synchrotron peak frequency: In low-synchrotron peaked blazars (LSP) blazars, the optical spectrum is near the peak of the synchrotron emission, thus reflecting freshly accelerated electrons. In high-synchrotron peaked blazars (HSP) blazars, the optical range is far below the synchrotron peak frequency, thus reflecting electrons that have already cooled substantially after the initial acceleration. From [57]. Reproduced with kind permission from Oxford University Press and the Royal Astronomical Society.

---

[3] http://robopol.org/.

While the PAs in blazars typically exhibit erratic small-angle variations, they occasionally undergo systematic PA swings exceeding 180°, typically over the course of a few days [58–60]. These PA swings are often associated with γ-ray and multi-wavelength flares. A central goal of the RoboPol project was a systematic study of such PA swings in a large sample of blazars [7,56,61–64]. Key results of this study were that (a) PA rotations do not occur in all blazars, but whether a blazar shows PA rotations or not does not appear to depend on its sub-class (FSRQ/BL Lac object/LSP/HSP) or its average fractional polarization [7], and (b) PA rotations are statistically correlated with γ-ray flares detected by *Fermi*-LAT, while the reverse is not true, i.e., not all *Fermi*-LAT flares in a blazar showing PA rotations are actually associated with such a rotation ([63] see Figure 4). Notably, also, there does not appear to be a preferred direction of PA rotations, i.e., PA swings can occur in either direction in any given object.

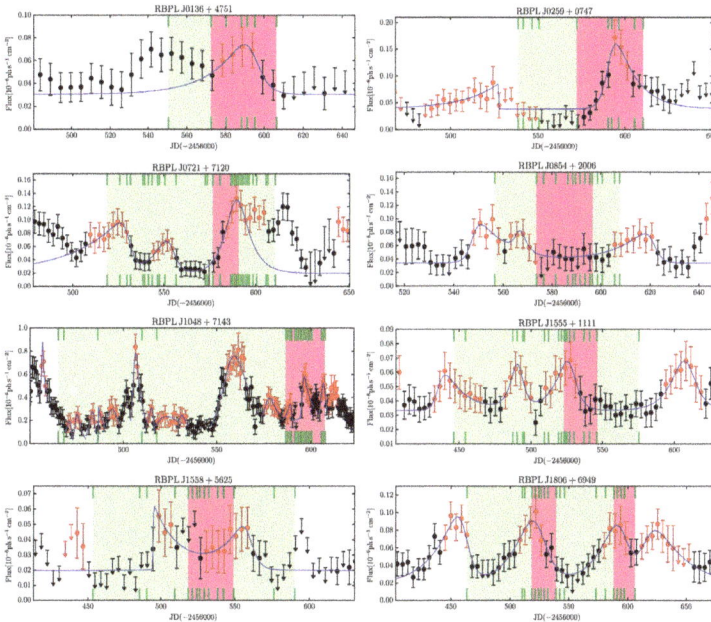

Figure 4. *Fermi*-LAT γ-ray light curves of a representative sample of 8 blazars monitored by RoboPol. The green areas show the period of RoboPol monitoring; red areas indicate periods of PA rotations. From [63]. Reproduced with kind permission from Oxford University Press and the Royal Astronomical Society.

The observations of such PA variability, and especially the PA swings, have spurred a large number of theoretical works attempting to explain them. They will be discussed in Section 3.2.

2.3. Multi-Messenger Observations—Neutrinos

The past few years (2015–2018) marked the birth of multi-messenger astronomy, with the first direct detection of gravitational waves from a binary-black-hole merger [65], the first confirmed multi-messenger detection of gravitational waves from a binary-neutron-star merger and the associated short gamma-ray burst [66], and the first strong hint for a blazar as the source of astrophysical high-energy neutrinos [67,68].

The jets of AGN have long been considered a prime candidate for the sites of acceleration of high-energy cosmic rays and the production of high-energy neutrinos, as detected by IceCube [69,70]. Such neutrino emission is expected in hadronic models for the γ-ray emission from blazars (e.g., [21,23,71–75]). However, until 2017, all searches for electromagnetic counterparts of the IceCube astrophysical neutrinos remained inconclusive (e.g., [76,77]), except for the identification of the blazar PKS B1424-418 as the possible source of the PeV IceCube neutrino event HESE-35 (aka "Big Bird"), which was detected during an extended multi-wavelength outburst of the blazar in 2012–2013 [78].

This picture changed with the detection of the ~290 TeV neutrino IceCube-170922A from a direction consistent with the blazar TXS 0506+056 [67] on 22 September 2017. The blazar was in an extended GeV γ-ray flaring state in September–October 2017 (see Figure 5), as detected by *Fermi*-LAT, and was subsequently also detected in VHE γ-rays by MAGIC. This single neutrino, however, only had a ~50% likelihood of actually being of astrophysical origin (due to its moderate energy), thus, by itself, providing only marginal evidence for the association. Furthermore, being only one single event, it allowed for the calculation of only flux upper limits (see Figure 6). In an archival search for additional neutrino events from the direction of TXS 0506+056, however, the IceCube collaboration found evidence for an excess of 13 ± 5 astrophysical high-energy muon-neutrinos from that location during an extended ~5 month long period in 2014–2015 [68], henceforth termed the "neutrino flare" (see Figure 7). This provided the first strong hint for TXS 0506+056 being a source of high-energy neutrinos, and allowed for the first calculation of a measured high-energy neutrino flux from an astrophysical source, corresponding to an all-flavour fluence (after correcting for neutrino oscillations) of $4.2^{+2.0}_{-1.4} \times 10^{-3}$ erg cm^{-2} with a spectrum between 32 TeV and 3.6 PeV fitted by a power-law with spectral index $\gamma = 2.1 \pm 0.3$.

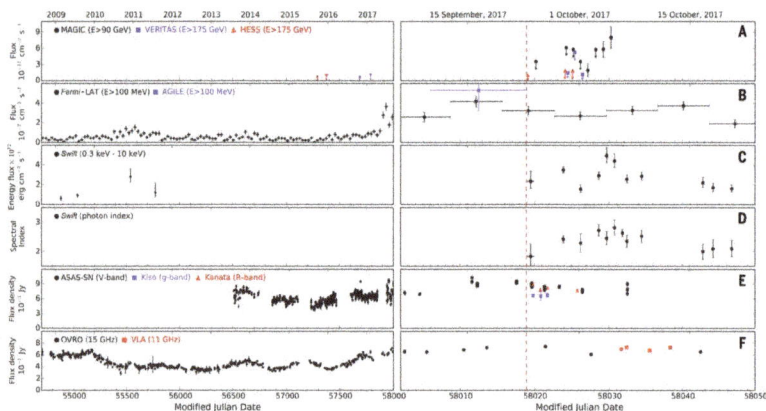

Figure 5. Multi-wavelength lightcurves of TXS 0506+056. The vertical red dashed line indicates the time of the IceCube-170922A event. Note also the absence of γ-ray activity during the period 2014–2015 of the neutrino flare. From [67]. Reproduced with permission by the American Association for the Advancement of Science (AAAS).

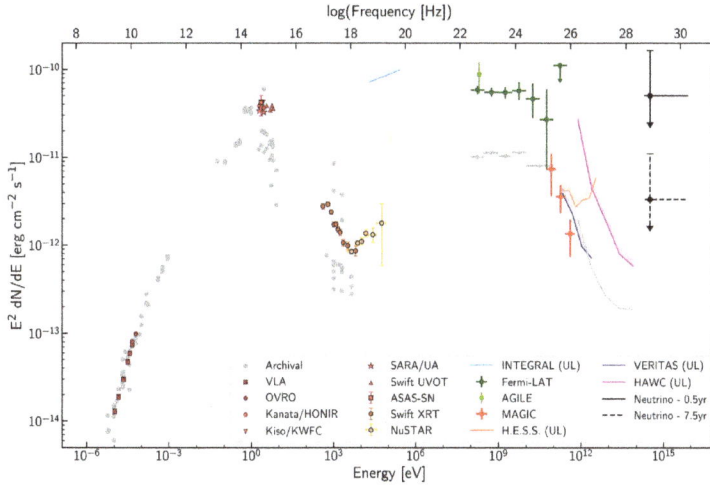

Figure 6. Multi-wavelength spectral energy distributions (SED) of TXS 0506+056, including neutrino flux upper limits corresponding to the IceCube-170922A event, assuming one event in 0.5 years (solid black) and one event in 7 years (dashed black). From [67]. Reproduced with permission by the American Association for the Advancement of Science (AAAS).

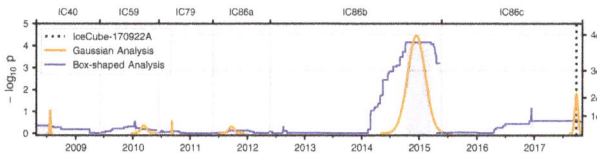

Figure 7. Results of a time-dependent search for an IceCube neutrino excess from the direction of TXS 0506+056. The orange and blue curves show the results of analyses using a Gaussian and box-shaped time profile for the neutrino emission, respectively. The vertical dashed blue line in IC86c indicates the time of the IceCube-170922A event. From [68]. Reproduced with permission by the American Association for the Advancement of Science (AAAS).

Most notably, the γ-ray and multi-wavelength flux of TXS 0506+056 during the time of this neutrino flare showed no evidence of enhanced activity. Note, however, that [79] identified a nearby blazar, PKS 0502+049, which was in a γ-ray flaring state for several weeks before and after the 2014–2015 neutrino flare (see Figure 8), but disfavour this source as the potential counterpart of IceCube-170922A as it was not flaring during the neutrino flare, but TXS 0506+056 was in a historical hard-spectrum (but low-flux) state at that time. The substantial body of theoretical developments spurred by this association will be discussed in Section 3.3.

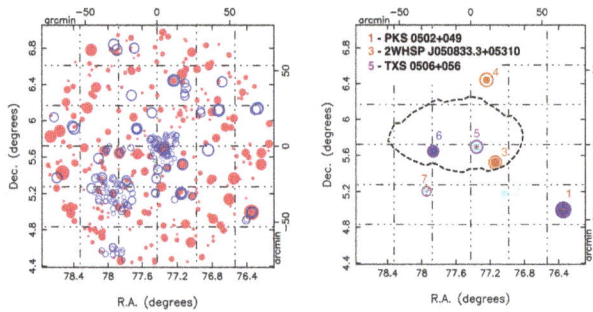

Figure 8. Left: Radio (red) and X-ray (blue) sources within 80 arcmin of the position of the IceCube-170922A event. **Right**: Known and candidate blazars in the same field. Dark blue circles represent low-frequency peaked BL Lac candidates, cyan symbols represent intermediate BL Lac objects, and orange symbols indicate high-frequency peaked BL Lacs. The dashed line shows the 90% error contour of the IceCube-170922 event. From [79]. Reproduced with kind permission from Oxford University Press and the Royal Astronomical Society.

3. Theoretical Developments

This section summarizes some of the most recent developments in the modeling and interpretation of the multi-wavelength and multi-messenger emission from blazars, specifically addressing the observational highlights described in the previous section. For a general review of leptonic and hadronic blazar emission models, see, e.g., [8,34,80].

3.1. Models of Flux Variability

As elaborated in Section 2.1, observed blazar variability patterns pose at least two major challenges to currently existing blazar radiation models: (a) the rapid, minute-scale (γ-ray) variability, and (b) the inconsistent cross-correlation patterns, with emission in different wavelength bands being sometimes correlated, sometimes not, including occurrences of orphan flares. This section will provide a brief overview over various blazar variability models currently "on the market", discussing how they may have the potential to address the issues mentioned in the preivous section.

3.1.1. Causes of Variability

Variability of blazar emission can, in principle, be caused in a variety of ways, by which different models may be classified:

1. **Shock-in-jet models**: In these models (also termed "internal shock models") inhomogeneities in the jet flow produce mildly relativistic shocks travelling through the (relativistically moving) jet plasma, leading to the acceleration of particles, most plausibly through Diffusive Shock Acceleration (DSA; see, e.g., [81–85]). The shock-in-jet model for blazars was first suggested in the seminal work by Marscher & Gear [86], and subsequently refined in a large number of works, mostly in the framework of leptonic emission scenarios (e.g., [87–98]).

In shock-in-jet models, particle acceleration occurs in a single region around the shock, equally affecting all radiating particles (leptonic or hadronic). Thus, these models generally predict correlated multi-wavelength variability with inter-band time lags reflecting energy-dependent electron (or proton) cooling (and/or acceleration) time scales (e.g., [87]). In the case of a leptonic emission scenario, for FSRQs, the optical and GeV γ-ray emissions are produced by electrons of similar energies and therefore are expected to be correlated with close to zero time lag. Radio and X-rays are produced by lower-energy electrons, expected to exhibit a delayed response compared to the optical and γ-ray emissions. Figure 9 shows an example resulting from a shock-in-jet simulation representing a multiwavelength flare of the FSRQ 3C 279 (flare C from [30]), where the radio and X-ray emissions are expected to lag behind the optical and VHE γ-ray emissions by $\sim$10 h. In the case of HSP blazars, the X-ray and γ-ray emissions are expected to be closely correlated, with GeV γ-ray and optical emissions lagging behind the X-ray and VHE variations.

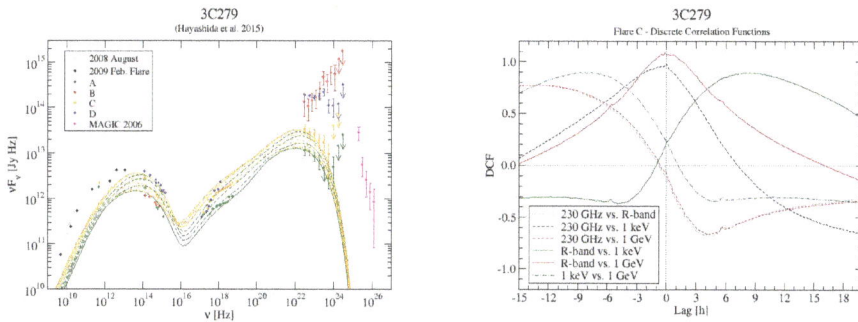

Figure 9. Shock-in-jet simulation of a flare of the FSRQ 3C279. Data are from [30]. **Left**: Snap-shot SEDs resulting from the time-dependent simulaton; **Right**: Cross-correlations between various frequency bands. Radio and X-ray variations are expected to lag behind optical and GeV γ-ray variations by $\sim$10 h. From Böttcher & Baring (2019, in preparation).

Uncorrelated variability amongst the two SED components could be achieved, in the framework of shock-in-jet models, with the assumption of the emission from the dominant shock being strongly synchrotron- or high-energy (Compton in leptonic models) dominated, enhancing the quiescent emission from the larger-scale jet only in a narrow frequency band (e.g., [99,100]). Alternatively, in a hadronic shock-in-jet scenario, vastly different acceleration time scales of electrons and protons might give the impression of uncorrelated variability due to the long time delay between the radiation components produced by electrons and protons (e.g., see [101] Figure 10).

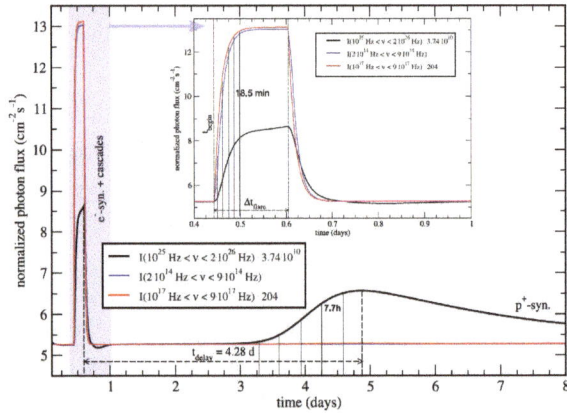

Figure 10. Multi-band (optical = blue, X-ray = red, γ-ray = black) light curves from a lepto-hadronic model with gradual, stochastic acceleration of particles. The acceleration time scale for protons is substantially longer than for electrons, leading to a delayed γ-ray orphan flare due to proton-synchrotron emission. From [101]. Reproduced with permission from ESO.

Shock-in-jet models naturally assume that the shock affects the entire cross section of the jet, with a radius of typically $R_\perp \sim 10^{15}$–10^{16} cm. As discussed in Section 2.1, this constrains the variability time scale to $t_{var} \gtrsim R_\perp (1+z)/(\delta c) \sim 9 R_{\perp,16} (1+z)/\delta_1$ hours, which is very difficult to reconcile with minute-scale variability, unless a very small jet cross section and/or a very large Doppler factor are assumed.

It is well known that the formation of strong shocks and efficient particle acceleration at shocks is suppressed in the presence of a dominant magnetic field (e.g., [102]). The low magnetizations ($u_B/u_e \sim 0.1$–10^{-3}) typically inferred from broadband SED modeling of blazars (see, e.g., [8,103]) therefore seem to support the hypothesis of shocks being the dominant particle acceleration sites in blazars.

2. **Turbulence / Magnetic Reconnection:** The relativistic flows of AGN jets are likely to develop turbulence, which may trigger magnetic reconnection. This has been studied in a large number of works in recent years (e.g., [104–108]). It has been shown that magnetic reconnection produces hard power-law spectra of relativistic electrons, $n_e(\gamma) \propto \gamma^{-p}$, including spectral indices approaching a value of 1 (e.g., [109]), which, however, is also achievable with oblique relativistic, but still subluminal shocks [85].

In view of the minute-scale variability problem discussed in Section 2.1, magnetic-reconnection models are particularly appealing as they provide the possibility to produce small-scale ultrarelativistic flows within the reconnection region, the so-called "jets-in-a-jet" scenario [110–112]. The resulting ultrarelativistic bulk motion of small plasmoids provides additional Doppler boosting, resulting in very short, bright flares. With modest magnetization ($\sigma \sim$ a few), this model is capable of producing the observed fast (minute-scale) variability (see [110,112–115] Figure 11) without requiring ultrarelativistic ($\Gamma \gg 10$) bulk motions of the entire jet material.

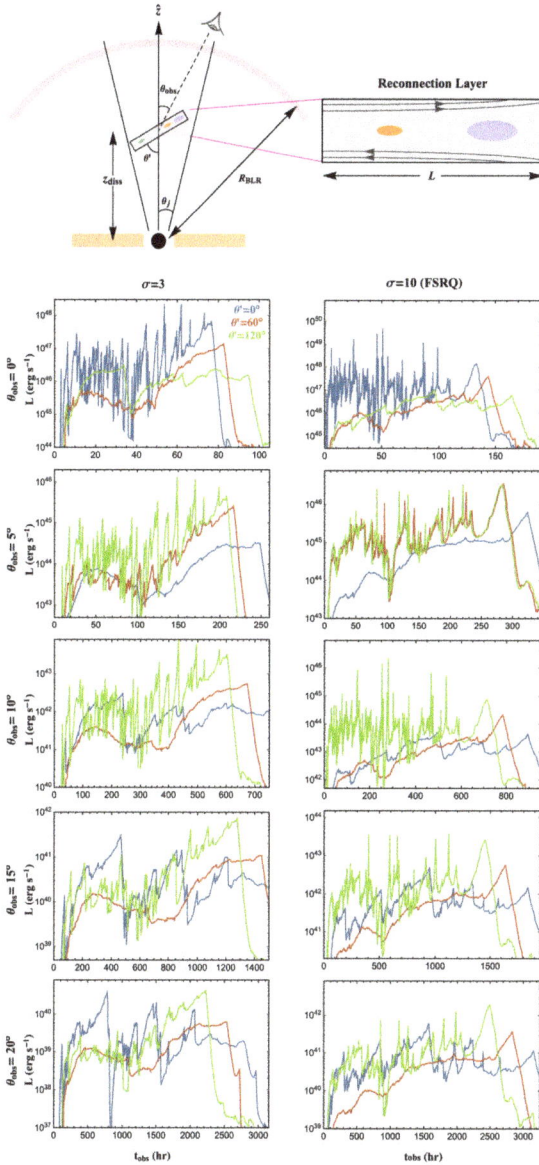

Figure 11. **Top**: Sketch of a jet-in-a-jet scenario where particle acceleration results from pasmoid-dominated magnetic reconnection. **Bottom**: Simulated 0.1–300 GeV γ-ray light curves at different viewing angles (top to bottom: $\theta_{obs} = 0°$, ..., 20°) for two different plasma magnetization values of $\sigma = 3$ (**left**) and $\sigma = 10$ (**right**). Different colours indicate different degrees of alignment of the plasmoids with the jet axis. The light curves illustrate the rapid, large-amplitude variability that can be produced in such a scenario. From [113]. Reproduced with kind permission from Oxford University Press and the Royal Astronomical Society.

3. **External Sources of Variability:** Another class of models attributes variability to interactions of the jet (or the high-energy emission region within the jet) with matter external to the jet, either by direct collisions or by means of radiative interactions.

    Examples of the former models include jet–star/cloud collision models (e.g., [116–120]), where the jet interacts with and (at least partially) disrupts a star or a gas cloud, leading to the formation of a strong shock with subsequent particle acceleration. The natural duration of flares in such a scenario is expected to be of the order of the crossing time of the cloud or star through the jet, typically of the order of days to weeks or months. For example, a model of cloud ablation by a blazar jet has recently been proposed by [121] to model the ~4 months long giant multi-wavelength outburst of the FSRQ CTA 102 in 2016–2017, see Figure 12. Note that [119] argue that a jet-star interaction may also produce rapid, minute-scale variability due to the acceleration of fragments of the stellar envelope to ultra-relativistic bulk speeds.

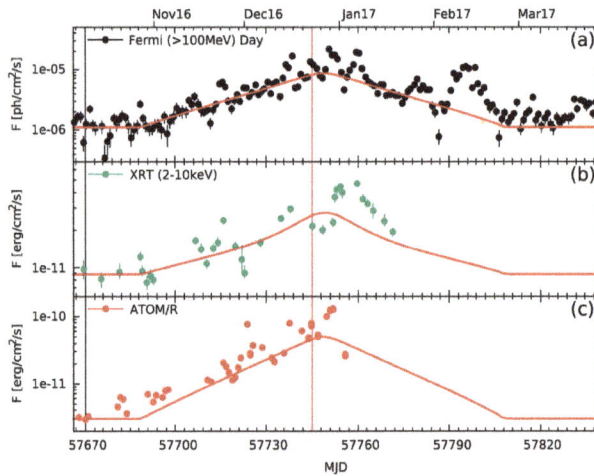

Figure 12. Fits to the ~4 months-long multi-wavelength outburst of CTA 102 in 2016–2017 in γ-rays (**a**), X-rays (**b**), and optical (**c**) using a model of cloud ablation by the blazar jet. From [121]. Reproduced with permission from the American Astronomical Society (AAS).

Variability models based on radiative interactions of the jet with external medium include, in particular, synchrotron mirror models, in which the synchrotron emission produced within the jet is reflected off an external obstacle (the "mirror", which could, e.g., be a cloud of the Broad Line Region or a stationary feature within the jet) (e.g., [122,123] see Figure 13). It thereby appears as an intense target photon field either for Compton scattering (leptonic models—see Figure 14) or pγ pion producton (hadronic models) for a short time around the passage of the high-energy emission region by the mirror. A hadronic synchrotron mirror model has been proposed by [122] to explain the orphan TeV flare of 1ES 1959+650. The reflected synchrotron X-ray emission in this case would be an inefficient target for Compton scattering due to Klein-Nishina suppression, leaving pγ interactions off relativistic protons as the dominant signature of the mirror process.

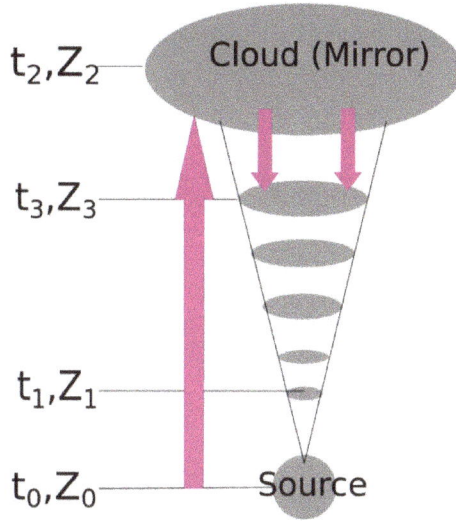

Figure 13. Sketch of a synchrotron mirror model. From Oberholzer & Böttcher (2019, in preparation).

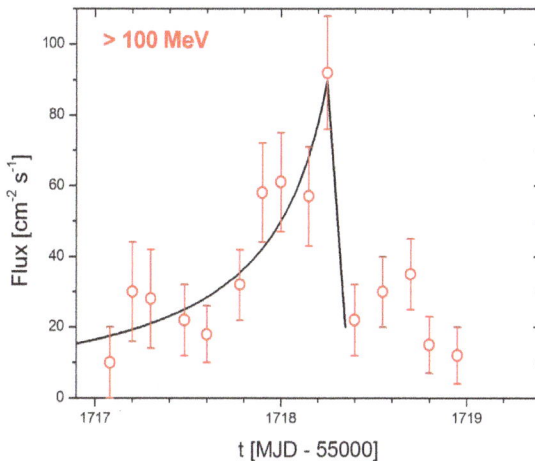

Figure 14. Simulated γ-ray light curve in a leptonic synchrotron mirror model. From [123]. Reproduced with permission from the American Astronomical Society (AAS).

Into the same category falls the "Ring of Fire" model by [124,125], in which the high-energy emission region passes a stationary external source of seed photons for Compton scattering and which has also been proposed as a model for "orphan flares" [125].

4. **Geometric Models:** Variability models based on bending or helical jets invoke a change of the viewing geometry as the dominant source of variability, due to a change in the Doppler factor (e.g., [126–130]). Bending or helical jet structures are often observed in radio VLBI monitoring (e.g., [131]). Models based on Doppler-factor variability typically predict correlated, almost acromatic variability, except for a small shift in frequency by the changing Doppler factor, unless

one assumes that different portions of the electromagnetic spectrum are not produced co-spatially and may be affected by different Doppler-factor variations (e.g., [129]). Such models have also had some success in reproducing optical polarization variability, as discussed further in the next sub-section.

Another class of models that may be categorized as geometric invokes particle acceleration triggered by the kink instability in jets (e.g., [132,133]). In these models, variability is caused by particle acceleration due to magnetic energy dissipation in the development of the instability, which is expected to be accompanied by significant changes in the polarization signatures, as discussed further below.

3.1.2. Numerical Approaches

Modeling of blazar variability requires the time-dependent treatment of both the distributions of relativistic particles and the radiation fields in the emission region. In leptonic models, the relevant particle populations are only the electrons (and positrons, which are usually not distinguished from electrons, as they cool and radiate identically, and pair annihilation is irrelevant for highly relativistic electrons); in hadronic models, in principle, electrons/positrons need to be evolved simultaneously with ultrarelativistic protons, pions, and muons. In almost all models currently available in the literature, the particle momentum distributions are assumed to be isotropic in the rest frame of the high-energy emission region. This is a critical simplifying assumption which makes the models tractable, as one needs to keep track only of the particles' energy distributions (in one dimension), and current models based on isotropic particle distributions have met with significant success in representing SEDs and variability patterns of blazars. However, realistically, neither relativistic shock acceleration (e.g., [85]) nor particle acceleration at relativistic shear layers [134] appear to produce isotropic particle distributions in any frame. In particular, [134] have shown that relativistic shear layer acceleration produces highly beamed particle distributions in the direction of the shear flow, possibly leading to much more strongly beamed radiation patterns than the standard $1/\Gamma$ scaling resulting from relativistic aberration of emission produced isotropically in an emission region moving at Lorentz factor Γ (see Figure 15).

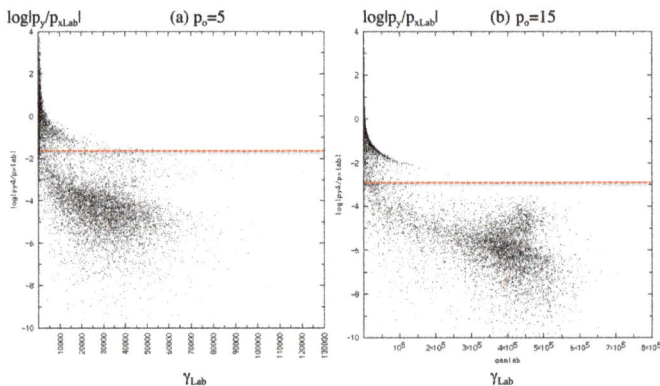

Figure 15. Anisotropic particle acceleration in relativistic shear layers from particle-in-cell simulations. Plotted is the log of the angle of the particles' motion with respect to the jet axis, vs. particle energy, indicating that high-energy particles are beamed forward much more strongly than the standard $1/\Gamma$ characteristic from relativistic aberration of an isotropic distribution (red dashed lines). $p_0 = \Gamma_{cm} \beta_{\Gamma,cm}$ is the dimensionless momentum in the center-of-momentum frame, in which spine and layer move with equal velocity $\beta_{\Gamma,cm} c$ in opposite directions. (**a**) For $p_0 = 5$; (**b**) for $p_0 = 15$. From [134]. Reproduced with permission from the American Astronomical Society (AAS).

An interesting contribution to the discussion about anisotropy of particle distributions was published by [135]. These authors argue that gygoresonant pitch-angle scattering, which might isotropize accelerated particles in the emission region, is effective only out to some isotropization energy γ_{iso}, typically much smaller than the break energy γ_b due to inefficient acceleration and/or escape. Particles with $\gamma_e > \gamma_b$ are expected to move primarily along the magnetic field and will therefore not contribute to synchrotron emission. However, particles in the range $\gamma_{iso} < \gamma_e < \gamma_b$ may still efficiently contribute to Compton scattering. This removes the need for strongly sub-equipartition magnetic fields, which contradict the paradigm of dynamically important B-fields in the jets of AGN.

Accepting the isotropic particle distribution approximation, as done in most of the current literature, the evolution of particle energy distributions is usually done by means of an isotropic Fokker-Planck equation of the form (see, e.g., [136–143] for leptonic models, and [71,101,144,145] for hadronic models):

$$\frac{\partial n_i(\gamma, t)}{\partial t} = Q_i(\gamma, t) - \frac{\partial}{\partial \gamma}\left(\dot{\gamma}\, n_i[\gamma, t]\right) - \frac{n_i(\gamma, t)}{t_{esc}} + \frac{\partial}{\partial \gamma}\left(D[\gamma]\frac{\partial n_i[\gamma, t]}{\partial \gamma}\right) - \frac{n_i(\gamma, t)}{\gamma\, t_{decay}}. \tag{4}$$

Here, i indicates the particle species (electrons/positrons, protons, pions, muons). $Q_i(\gamma, t)$ is an injection term, which is often used to describe rapid particle acceleration, such as first-order Fermi acceleration. This is because the first-order Fermi acceleration time scale increases with energy as $t_{acc,1} \propto \gamma^{\alpha}$, typically with $\alpha \geq 1$, while radiative cooling time scales (at least for synchrotron and Compton scattering) scale as $t_{cool} \propto \gamma^{-1}$. Thus, for energies below the maximum energy where $t_{acc,1}(\gamma_{max}) = t_{cool}(\gamma_{max})$, typically $t_{acc} << t_{cool}$. Thus, first-order fermi acceleration is well approximated by an instantaneous injection term. Depending on particle species, $Q_i(\gamma, t)$ may also include pair production by $\gamma\gamma$ absorption and particle production through the decay of pions or muons, thus directly coupling to the evolution of those parent particles. The term $\dot{\gamma}$ describes systematic energy losses or gains (if not already included in Q_i), in particular radiative losses. t_{esc} is the (possibly energy-dependent) escape time scale. The fourth term on the right-hand side describes diffusion in momentum space, leading to second-order Fermi acceleration where $D(\gamma)$ is the momentum diffusion coefficient. The last term describes the decay of unstable particles (pions, muons).

For a self-consistent solution, in the case of hadronic models, the Fokker-Planck Equation (4) for all species have to be solved simultaneously, along with the radiation transfer problem (see below). The most efficient way to achieve stable numerical solutions to Equation (4) is through implicit Crank-Nichelson schemes (e.g., [88,146]).

There are two main approaches to solving the radiation transfer problem in most blazar emission codes. Most commonly employed are direct solutions to a photon continuity equation of the form

$$\frac{\partial n_{ph}(\epsilon)}{\partial t} = \frac{4\pi\, j_{\epsilon}}{\epsilon\, m_e c^2} - \alpha_{\epsilon}\, c\, n_{ph}(\epsilon) - \frac{n_{ph}(\epsilon)}{t_{esc}}. \tag{5}$$

Here, $\epsilon = h\nu/(m_e c^2)$ is the dimensionless photon energy, j_{ϵ} is the emissivity due to the various radiation mechanisms, α_{ϵ} is the absorption coefficient, primarily due to synchrotron self-absorption and $\gamma\gamma$ absorption (wich then feeds back into the electron/positron Fokker-Planck equation through pair production), and t_{esc} is the photon escape time scale. Such schemes are typically numerically inexpensive, but they are appropriate only for very simple (typically homogeneous, one-zone) geometries. There are, however, a few attempts to apply such schemes also to inhomogeneous multi-zone models, in particular shock-in-jet [91,92] and extended-jet [147–149] models.

An alternative method to solve the radiation transfer problem is through Monte-Carlo simulations (e.g., [88,89,97,98,150]). Such schemes are much more flexible in terms of geometries, they allow straightforward time-tagging of photons and polarization-dependent ray tracing [97,98]. However, time-dependent multi-zone simulations quickly become extremely time consuming due to the large number of photons that need to be tracked in order to achieve meaningful photon statistics.

An innovative new approach to solving time-dependent particle and photon evolution in blazar models has been developed by [151–153]. These authors solve the Fokker-Planck and radiation transfer equations in Fourier space and re-convert them into observable light curves and cross correlations. With this approach, it was, for the first time, feasible to model Fourier-frequency-dependent time lags between hard and soft X-rays, as observed from Mrk 421 (see Figure 16).

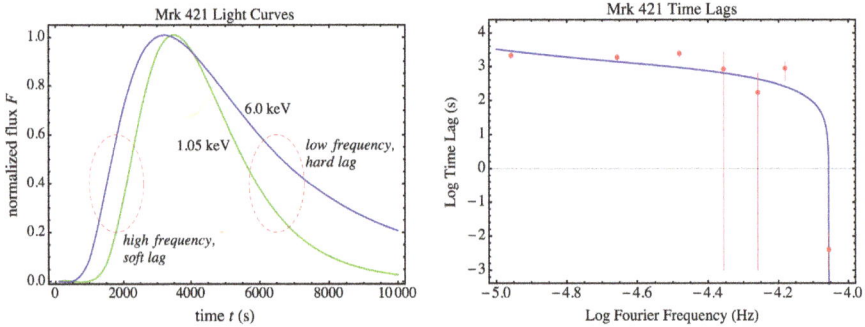

Figure 16. Left: Re-constructed X-ray light curves at 1.05 and 6 keV from a Fourier-space solution to time-dependent electron acceleration and synchrotron emission in Mrk 421. **Right**: The resulting Fourier-frequency-dependent time lags, compared to *Beppo*SAX data from [154]. From [153]. Reproduced with permission from the American Astronomical Society (AAS).

3.2. Multi-Wavelength Polarization Modeling

In this section, various models advanced to explain the large-angle optical PA swings in flaring blazars, as well as predictions for future high-energy polarimeters will be discussed.

3.2.1. Optical polarization Angle Swings

The large-angle optical PA rotations associated with multi-wavelength flares discussed in Section 2.2 have spurred a large number of theoretical works to interpret these events. The large degree of optical polarization is a clear indication that the non-thermal emission is synchrotron radiation in partially ordered magnetic fields. A non-thermal synchrotron spectrum with energy index $\alpha = (p-1)/2$, where p is the underlying non-thermal electron spectral index, can be maximally polarized by a degree of

$$\Pi_{\max} = \frac{p+1}{p+7/3} = \frac{\alpha+1}{\alpha+5/3} \tag{6}$$

in the case of a perfectly ordered magnetic field. For typical spectral indices of $p \sim 2$–3, this corresponds to $\sim$70 %–75 % polarization. The observed degree of optical polarization is typically in the range $\Pi_0 \sim$ a few percent—30%, this indicates that the magnetic fields in the optical emission regions must be partially ordered. Changes in the PA then likely indicate a change in the orientation of the magnetic field with respect to our line of sight. This can be either an intrinsic change in a (more or less) straight jet, or it can indicate a change of the jet orientation with respect to the line of sight. Alternatively, PA variations may be the result of stochastic processes in a turbulent jet environment. All of these possibilities will be discussed in more detail below.

Intrinsic magnetic field changes may be caused through magnetic-field compression in a shock. Models based on shock propagation in a jet pervaded by a magnetic field have been advanced, in particular, by [96–98]. Here, the finite transverse light travel time plays a crucial role in producing PA swings. Ref. [97] have used such a model to self-consistently explain SEDs, multi-wavelength light curves, and PA and Π variations, including a $\sim$180° PA rotation, in the FSRQ 3C279, as observed by [58] (see Figure 17). A potential drawback of such a model is that a single shock passage predicts swings of at most 180°. Thus, rotations by multiples of 180° would require a succession of multiple shocks. Furthermore, assuming that the helicity of the magnetic field structure does not change in the same object between different observation periods, such rotations are predicted to occur always in the same direction, which is contrary to observations. Thus, while this model has been very successful in explaining some PA rotations associated with multi-wavelength flares, it can likely not be applied to all.

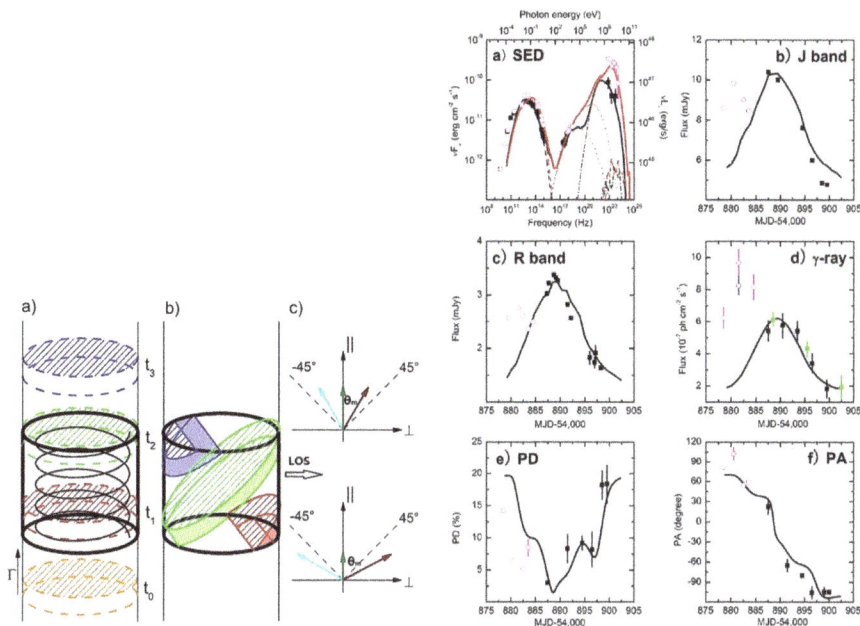

Figure 17. Left: Sketch of the magnetic-field geometry and illustration of light-travel-time effects in a shock-in-jet model with helical magnetic field. Different colors indicate the location of the shock front at different times: In the left-most sketch, equal times in the active galactic nuclei (AGN) rest frame; in the middle sketch: Location of the shock front at equal photon-arrival times at the observer, illustrating how different parts of a helical magnetic fields are "lit-up" by the shock front, as seen by the observer at different times, leading to a gradually changing dominant magnetic-field (and, thus, polarization) direction. **Right**: The resulting (**a**) snap-shot SEDs, (**b–d**) multi-wavelength light curves, (**e**) polarization degree, and (**f**) polarization-angle swing compared to observations of 3C279 by [58]. From [97]. Reproduced with permission from the American Astronomical Society (AAS).

Models along similar lines involve magnetic-field re-structuring and particle energization through the kink instability [132,133] (see Figure 18). This process will also lead to flaring activity, with the strength of flares depending on the initial magnetization, correlated with PA swings. In this model, as with the internal-shock model, the unit of PA rotations is 180°, also expected to occur always in the same direction.

Alternatively, PA swings may result from a change in the jet orientation, such as assumed in the helical-jet model (e.g., [59,126–130,155]). In particular, ref. [126,127] modelled the optical flux and polarization variability of S5 0716+71 and CTA 102, respectively (see Figure 19). Such models have been successful in representing flux and polarization variability in some cases. However, also here the direction of the rotations is expected to be pre-determined by the helicity of the jet (and B-field structure) and thus not changing for a given object. Also, as discussed in Section 3.1, at least in their simplest form, they predict essentially achromatic multi-wavelength variability, which is rarely observed.

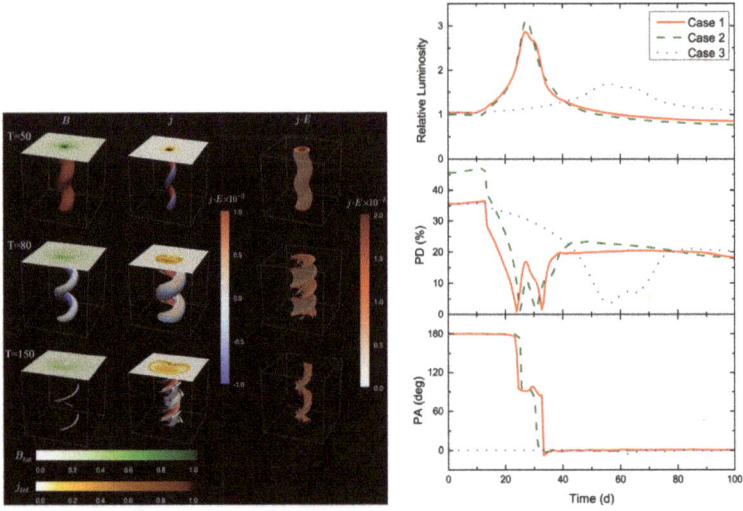

Figure 18. Left: Magneto-Hydrodynamics (MHD) simulations of the evolution of the kink instability in a blazar jet. **Right**: The resulting (top) light curves, (middle) polarization degree, and (bottom) polarization-angle swing. The simulation in the left panel corresponds to Case 1 in the right panel, corresponding to a magnetization of $\sigma_m = 2$. For Case 2, the initial helical B-field is radially more strongly confined; for Case 3, the magnetization is lower ($\sigma_m = 0.2$) compared to Case 1. From [133]. Reproduced with permission from the American Astronomical Society (AAS).

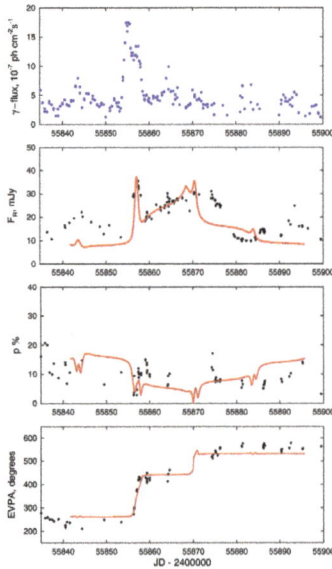

Figure 19. Fits to the optical light curve, polarization degree, and PA of S5 0716+714 in October 2011, using a model of a shock in a helical jet. From [126]. Reproduced with permission from the American Astronomical Society (AAS).

Finally, stochastic polarization variability may result from turbulent environments in the jet. In particular, ref. [156] developed a "Turbulent Extreme Multi-Zone" (TEMZ) model in which different turbulent cells in the jet are characterized by different magnetic-field orientations. The summed radiation of a large number of such turbulent cells will result in a (normally small) residual polarization with stochastically varying direction. This model has successfully reproduced stochastic polarization and flux variations in blazars (see Figure 20) and may occasionally also lead to large-angle PA swings. It naturally accounts for changing directions of PA swings in the same object and has also been used to model circular radio polarization from the inner jet regions of blazars [157]. However, due to the stochasticity of both flux and polarization variations, large-angle PA swings are expected to be very rare, and such swings are not expected to correlate systematically with multi-wavelength flares, as observed by the RoboPol experiment [63].

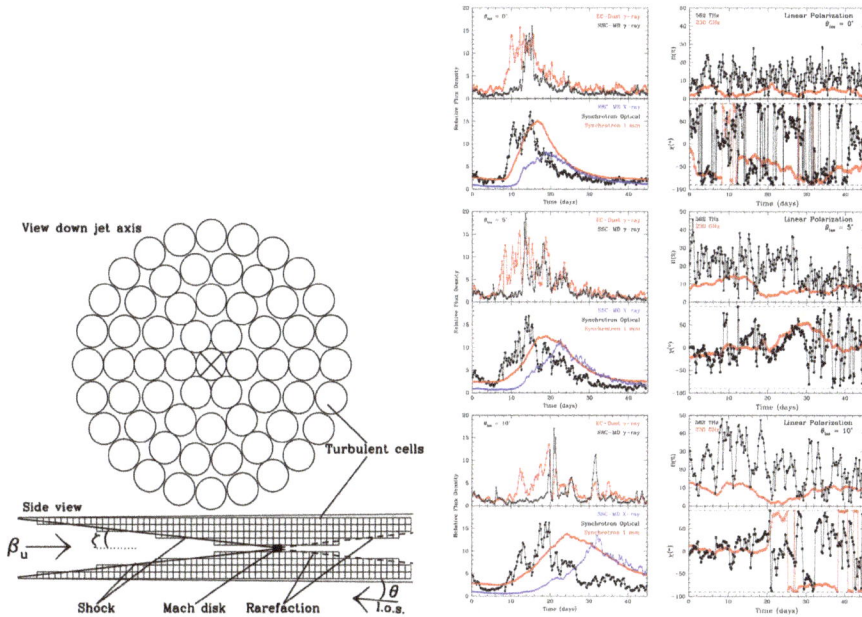

Figure 20. Left: Sketch of the Turbulent Extreme Multi-Zone model. **Right**: Simulated light curves (**left**) and polarization variability (**right**) for a representative test case of the Turbulent Extreme Multi-Zone (TEMZ) model, for three different viewing angles. From [156]. Reproduced with permission from the American Astronomical Society (AAS).

3.2.2. High-Energy Polarization

As will be discussed in more detail below in Section 4, there are currently great prospects for future detections of high-energy (X-ray and γ-ray) polarization from blazars. Thus, it is timely to consider model predictions of such high-energy polarization. In addition to synchrotron radiation, also Compton scattering may induce polarization, but only in the case of scattering off non-relativistic electrons (see, e.g., [158]). Inverse-Compton scattering by relativistic electrons does not induce polarization, but reduces the degree of polarization of a polarized target photon field by at least $\sim 1/2$ (e.g., [159]).

Detailed predictions for the X-ray and γ-ray polarization in blazars have been made by [160] (see Figure 21). In the case of leptonic models, the X-ray emission in blazars is generally dominated by either synchrotron (in HSPs) or SSC emission (in ISPs and LSPs) and thus expected to be polarized. The γ-ray emission in LSP (and ISP) blazars is usually dominated by Compton scattering of external radiation fields by relativistic electrons and, thus, unpolarized, whereas in HSP blazars, it is usually modelled as being due to SSC emission, thus exhibiting a low, but non-zero, degree of polarization. In hadronic models, the high-energy emission is dominated by synchrotron emission of either protons or secondary pairs from photo-pion production and subsequent cascades, and thus expected to be polarized with a similar degree of polarization as the optical. High-energy polarization can thus be used as a diagnostic between leptonic and hadronic models.

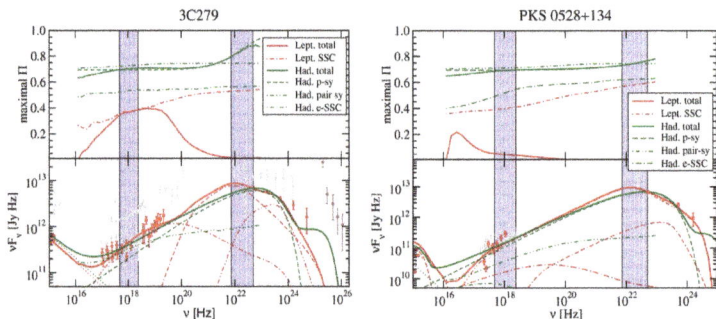

Figure 21. Predictions of X-ray and γ-ray polarization for two FSRQs. **Bottom**: Zoom-in on the UV–γ-ray SED, based on data from [11] and leptonic (red) and hadronic (green) SED fits from [8]. **Top**: Predicted high-energy polarization from the SED models of the bottom panel. The vertical shaded bands indicate the 2–10 keV X-ray and and the 30–200 MeV γ-ray band. From [160]. Reproduced with permission from the American Astronomical Society (AAS).

A first attempt at a time-dependent, multi-zone hadronic model with polarization-dependent radiation transfer has been published by [161]. The model is applicable in a parameter regime in which the high-energy emission is dominated by proton synchrotron radiation, as appears to be preferred, at least for most LSPs when fitted with hadronic models (e.g., [8,162]). Ref. [161] shows that, in such a hadronic scenario, even though optical PA swings may be produced by a shock in a jet with a helical magnetic field, similar PA swings are not expected in the X-ray and γ-ray polarization (see Figure 22). This is because of the much longer radiative cooling time of protons responsible for the X-ray and γ-ray emission, which therefore occupy a much larger active volume than the optical-synchrotron emitting primary electrons, so that the effect of the shock on the magnetic field orientation is less pronounced for the proton synchrotron emission. Consequently, the X-ray and γ-ray PA is expected to remain relatively stable even during a shock-induced flare, which significantly improves the chances of experimental detection of such polarization.

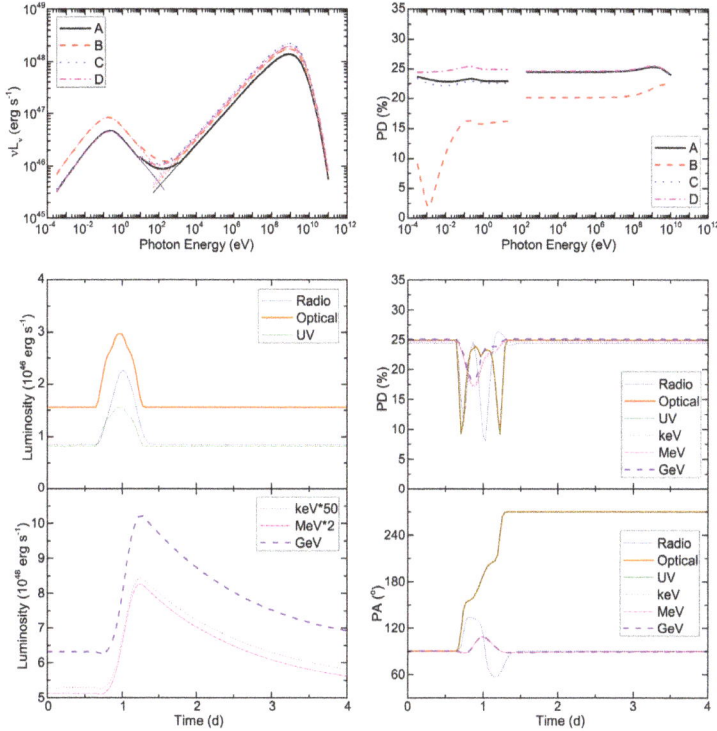

Figure 22. Representative simulations of a hadronic shock emission model with polarization predictions. **Top left**: Snap-shot SEDs, **Middle and bottom left**: Multi-wavelength light curves, **Top right**: Frequency-dependent polarization degree at different times, **Middle right**: Time-dependence of polarization degree at various frequencies, **Bottom right**: Time-dependent PA at various frequencies. The figure illustrates that in a hadronic shock emission model, optical (primary electron synchrotron) polarization-angle swings are not expected to be accompanied by similar PA swings at X-rays and γ-rays. From [161]. Reproduced with permission from the American Astronomical Society (AAS).

3.3. Models of Neutrino Emission from Blazars

Blazars have long been considered a prime candidate for the source of at least part of the VHE neutrinos detected by IceCube (e.g., [21,23,71–75,163–167]). In hadronic models of blazars, protons are accelerated to sufficiently high energies to produce γ-ray emission via proton synchrotron radiation and/or photo-pion production, $p + \gamma \rightarrow p + \pi^0$, $p + \gamma \rightarrow n + \pi^+$, or higher-order processes with multi-pion production, followed by pion decay, $\pi^\pm \rightarrow \mu^\pm + \nu_\mu(\overline{\nu_\mu})$ and muon decay, $\mu^\pm \rightarrow e^\pm + \overline{\nu_\mu}(\nu_\mu) + \nu_e(\overline{\nu_e})$. In typical AGN jet environments, photo-pion production is expected to be significantly more efficient than hadro-nuclear (i.e., proton-proton) interactions. Therefore, almost all works on neutrino production in AGN work on the basis of photo-pion production (see, however, hadro-nuclear emission models for neutrino production in AGN by [168,169]). Usually, $p\gamma$ interactions are strongly dominated by single-pion production through the Δ^+ resonance at an energy of $E_{\Delta^+} = 1232$ MeV, in which case each photo-pion interaction results in the production of 3 neutrinos, each carrying an average energy of ~5% of the proton energy,

$$E'_\nu \approx 0.05\, E'_p \tag{7}$$

where the prime indicates quantities in the rest frame of the emission region, in which the interactions are assumed to take place. Interaction at the Δ^+ resonance energy requires that the photon (E'_γ) and proton energies obey the relation for the center-of-momentum energy squared, s,

$$s \approx E'_\gamma E'_p \approx E^2_{\Delta^+} \tag{8}$$

Hence, combining Equations (7) and (8), one finds that the production of IceCube neutrinos of energies $E_\nu = \delta E'_\nu = 100 E_{14}$ TeV requires protons of energy

$$E'_p \approx 200 E_{14}/\delta_1 \text{ TeV} \tag{9}$$

and target photons of energy

$$E'_\gamma \gtrsim 1.6 \, \delta_1 / E_{14} \text{ keV}. \tag{10}$$

Hence, first, while the sources of IceCube neutrinos must be able to accelerate protons to ~EeV energies, they are not necessarily the sources of ultra-high-energy cosmic rays (UHECRs, with energies $E_{\mathrm{UHECR}} > 10^{19}$ eV). Second, Equation (10) indicates that, for efficient IceCube neutrino production, an intense target photon field at X-ray energies (in the co-moving frame of the emission region) is required. Note that the co-moving primary-electron-synchrotron radiation field in all blazars (especially, LSP and ISP blazars) peaks at much lower energies than required by Equation (10). Therefore, photo-hadronic blazar models utilizing the co-moving synchrotron radiation fields typically produce very hard IceCube neutrino spectra, peaking significantly above the IceCube energy range (see, e.g., [170]).

Further constraints on the photo-hadronic scenario to produce both high-energy γ-rays and PeV neutrinos stem from the fact that the target photon field for photo-pion production also acts to absorb any co-spatially produced γ-rays via $\gamma\gamma$ absorption. The $p\gamma$ cross section is several orders of magnitude smaller than the $\gamma\gamma$ absorption cross section. Efficient $p\gamma$ neutrino (and γ-ray) production requires that the optical depth for relativistic protons to interact with the target photon field, $\tau_{p\gamma} \sim 1$, which then implies that the optical depth of the emission region to $\sim$ GeV photons is $\tau_{\gamma\gamma} \sim 310\,\tau_{p\gamma} \gg 1$ (e.g., [171]). Thus, any γ-rays produced co-spatially with IceCube neutrinos, are expected to be strongly absorbed and initiate electromagnetic cascades, whose energy will ultimately escape the emission region in the optical – UV – X-ray regime.

This led several authors (e.g., [172–174]) to conclude that if blazars are the sources of (at least some) IceCube neutrinos, their γ-ray emission is likely to be dominated by leptonic processes, and no correlation between γ-ray and neutrino emissions is necessarily expected. Instead, the unavoidable cascade emission from photo-hadronic processes in blazar jets is expected to leave an imprint in the X-ray emission from blazars, which may be a better indicator of neutrino-production activity than γ-rays.

This conclusion is further corroborated by detailed studies of electromagnetic cascades initiated by photo-hadronic neutrino-production processes in blazar jet environments by [171]. Ref. [171] investigated possible regimes of electromagnetic cascades based on the energetics requirements to produce the neutrino flux from TXS 0506+056 during the 2014–2015 neutrino flare, and found that a synchrotron-supported cascade regime can be ruled out, as the cascades violate observational constraints (see Figure 23). Further considering the nature of the target photon field, they found that there is only a narrow range of parameters not violating observational constraints, requiring a $p\gamma$ UV–soft X-ray target photon field external to the jet (stationary in the AGN rest frame) and a kinetic luminosity in protons near the Eddington limit of the central black hole in TXS 0506+056. The nature of the required UV–soft X-ray target photon field remains unclear, but could possibly be related to a radially structured jet (spine-sheath, see [175,176]), or a radiatively inefficient accretion flow [177].

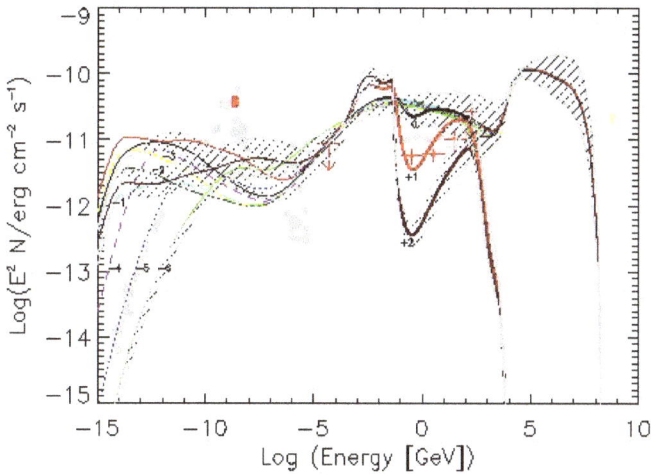

Figure 23. Simulations of synchrotron-supported cascades with target photon and proton spectra appropriate to produce the neutrino flux from TXS 0506+056 during the 2014–2015 neutrino flare, compared to contemporaneous optical, X-ray (upper limit from *Swift*-BAT), and GeV γ-ray (*Fermi*-LAT) data. The different curves (different colors) are labelled by the log of the maximum $\gamma\gamma$ optical depth, $\log(\tau_{\gamma\gamma}^{max})$, and the shaded areas indicate error margins based on the error on the measured IceCube neutrino spectrum. All cases violate either the X-ray or γ-ray constraints. From [171].

To conclude, photo-hadronic PeV neutrino production in blazar jets is not necessarily correlated with γ-ray activity (but rather X-ray activity), and due to the expected high $\gamma\gamma$ opacity of the required $p\gamma$ target photon fields, γ-rays in neutrino-producing blazars are likely to be dominated by leptonic processes, likely not spatially coincident with the site of neutrino production.

4. Future Prospects

The next decade will witness the start of operations of several new ground- and space-based astronomy facilities. Of particular relevance to blazar research will be, in the author's opinion, the Cherenkov Telescope Array (CTA [178]) and the Imaging X-ray Polarimetry Explorer (IXPE, [179]), while ground-based radio and optical flux and polarimetry monitoring projects will continue, and hopefully the *Fermi* Gamma-Ray Space Telescope and the Neil Gehrels *Swift* X-Ray Observatory will continue to provide all-sky GeV γ-ray monitoring and flexible X-ray coverage to blazar observations, respectively, for several years to come. KM3NeT [180,181] and planned upgrades to IceCube (IceCube-Gen2) [182,183]) and the Lake Baikal neutrino detector [184] will greatly improve our view of the neutrino sky and hopefully provide a definitive answer whether blazars are PeV neutrino sources. If selected, future missions such as the All-sky Medium Energy Gamma-ray Observatory (AMEGO, [185]) would also provide a boost to the study of blazars.

4.1. The Cherenkov Telescope Array

CTA[4] will be the next-generation ground-based Cherenkov Telescope facility, consisting of $\sim$100 Cherenkov telescopes of three different sizes at two sites, one in the nothern hemisphere

---

[4] http://www.cta-observatory.org/

(La Palma, Canary Islands, Spain), and one in the southern hemisphere (Paranal/ESO, Chile). It will improve on the sensitivity of current IACT facilities by about an order of magnitude and extend the observable energy range to ~20 GeV–~300 TeV. The current schedule foresees CTA to be operational by 2024. In addition to key science projects conducted by the CTA Consortium, CTA observing time will be open to community through a competitive proposal process. For an in-depth discussion of the science potential and key science projects to be conducted with the CTA, see [186]. Highlights of progress that CTA promises for blazar studies, include:

1. **Short-term variability:** The ~10-fold improved sensitivity of CTA compared to currently operating IACT facilities will allow for the study of blazar variability on short time-scales. In the case of the brightest flares, variability down to sub-minute time-scales could, in principle, be detected, if present. Already the ~5 min. variability observed in a few cases is severely challenging current blazar models, as discussed in Section 3.1. If blazars do indeed show significant variability on sub-minute time scales, this would call for a profound paradigm shift concerning our understanding of γ-ray emission from blazars.

    The detection of minute-scale variability strongly suggests a location of the emission region close to the central black hole, at a distance of the order $r \sim 2\Gamma^2 c\, t_{var} \sim 3.6 \times 10^{14}\, \Gamma_1^2\, t_{var,min}$ cm. Even for bulk Lorentz factors of order $\Gamma \sim 100$, this would normally be within the Broad Line Region (BLR) of an FSRQ. Thus, in the case of FSRQs (such as PKS 1222+216, from which sub-hour variability has been seen), it would be very difficult to avoid $\gamma\gamma$ absorption by the BLR radiation field (e.g., [187–190]). Thus, such detections either require exotic physics to suppress $\gamma\gamma$ absorption within the BLR (e.g., [191,192]), or a mechanism to produce very compact emission regions along the jet at $\sim pc$ distances from the central black hole.

    The improved sensitivity will also enable the detection of short-term variability in fainter γ-ray-detected AGN, including mis-aligned blazars (seen as radio galaxies) and narrow-line Seyfert-1 galaxies, some of which also appear to show blazar-like properties (e.g., [193,194]). Thus, we will be able to study whether sub-hour variability is a property of only a few blazars, or whether it is a common phenomenon among γ-ray emitting radio-loud AGN.

2. **Detailed γ-ray spectral studies:** The improved sensitivity and extended energy range of CTA compared to current IACTs will enable detailed γ-ray spectral studies with substantial overlap in energy with the *Fermi*-LAT. Potential spectral features due to $\gamma\gamma$ absorption in the BLR or dust torus radiation fields of the AGN are expected to arise in the ~10–100 GeV regime (e.g., [190]), i.e., just in the transition region between the energy ranges of *Fermi*-LAT and IACTs. Currently, the study of spectral features in this regime, beyond simple exponential cut-offs or similar smooth spectral shapes (see, e.g., [195]), is complicated by the fact that a high-quality *Fermi*-LAT spectrum, for most blazars, typically requires exposures of (at least) several days, during which the source is likely to show substantial variability. On the other hand, IACTs perform short, pointed observations of at most a few hours per night. Thus, any spectral features in this overlap energy range may well be an artifact of the often vastly different integration times in the GeV and TeV regimes. The substantial energy overlap between *Fermi*-LAT (up to $\lesssim$100 GeV) and CTA ($\gtrsim$20 GeV) will enable to proper flux cross-calibration of the two instruments and allow for a detailed study of spectral features in the overlap region, at least in cases where observations do not indicate significant variability over the course of the joint observations. The identification of BLR $\gamma\gamma$ absorption features would provide a key diagnostic for the location of the γ-ray emission region and, thus, the region where relativistic particles are accelerated to >TeV energies. The location of the particle acceleration region would provide strong clues towards the nature of the acceleration mechanism.

    It has also been suggested (e.g., [196]) that hadronic emission processes might lead to detectable spectral features in the multi-TeV spectra of blazars due to separate spectral components from muon and pion synchrotron emission and photo-pion induced pair cascades. Such features may be detectable, at least for nearby blazars, with the CTA. If they are systematically identified in a sample of blazars, they might provide a "smoking-gun" signature of hadronic processes and point to blazars as sources of high-energy cosmic rays.

4.2. High-Energy Polarimetry

The Imaging X-Ray Polarimetry Explorer (IXPE)[5] has been selected by NASA for launch in (or after) 2020 as the first dedicated high-energy polarimetry mission. It will provide X-ray polarimetry in the 2–8 keV X-ray regime and has the capacity to detect X-ray polarization from bright blazars within a few hours of observations. In the γ-ray regime, the proposed AMEGO[6] mission promises to perform, for the first time, polarimetry in the MeV regime.

X-ray polarimetry of HSP blazars, where the X-ray emission is electron-synchrotron dominated, would probe the degree of ordering and dominant direction of magnetic fields in the high-energy emission region. When compared to simultaneous optical polarimetry, this will allow for test of the co-spatiality of X-ray and optical emission in HSPs and for a comparison of the respective emission region sizes. As the optical emission in HSPs originates from lower-energy electrons than the X-ray emission, one would expect the X-ray emission region to be more confined near the particle acceleration site, likely embedded in a more highly ordered magnetic field than in the, presumably much larger, optical emission site. A significantly higher degree of polarization in X-rays compared to optical would then be expected. Such results will allow us to study the spatial dependence of turbulence in the jet and discriminate between different particle acceleration mechanisms, such as diffusive shock acceleration and magnetic reconnection.

In ISP/LSP blazars, the X-rays are likely produced by synchrotron self-Compton (SSC) emission in leptonic emission scenarios, or by proton synchrotron and cascade synchrotron emission in hadronic scenarios. Thus, if X-ray polarimetry reveals a degree of polarization of the order of the optical polarization in ISPs/LSPs, this would be a strong indication of hadronic emission [160].

Along the same lines, as disussed in Section 3.2, if γ-ray polarimetry by AMEGO shows significant (of the order of the optical degree of polarization) MeV polarization, hadronic emission scenarios will be strongly favoured, thus identifying blazars as (at least) PeV proton accelerators and likely sources of IceCube neutrinos.

As the optical polarization of blazars is known to be variable, at least on daily time scales, both in degree and angle of polarization, one might expect that the same holds true for X-ray and γ-ray polarization. Given that, for the X-ray fainter LSP sources, IXPE might require integration times of days to weeks, PA changes during the exposure might destroy any intrinsic polarization signal, unless data analysis techniques can be developed to account for a changing PA during the exposure, such as using the PA evolution observed in the optical, as a template for the IXPE analysis. However, as shown by [161], if PA rotations are produced by a straight shock-in-jet model with a helical magnetic field, the anticipated high degree of polarization in a hadronic model might be measurable without the expectation of a significant PA change.

4.3. Future Neutrino Detectors

KM3NeT[7] is the next-generation neutrino telescope, to be built at three sites at the bottom of the Mediterranean Sea. It will improve on the sensitivity of IceCube by more than an order of magnitude and might therefore provide the statistics to identify individual sources of neutrinos

---

[5] https://ixpe.msfc.nasa.gov.
[6] https://asd.gsfc.nasa.gov/amego/.
[7] http://www.km3net.org.

confidently. In addition, major upgrades are planned to the existing IceCube detector at the South Pole (IceCube-Gen2)[8] and to the neutrino detector in Lake Baikal[9].

If blazars are identified systematically as a source of neutrinos by future neutrino observatories, it would prove their nature as cosmic-ray proton accelerators (up to at least PeV energies). However, as discussed in Section 3.3, a correlation with GeV–TeV γ-ray emission is not expected if the neutrinos are produced via photo-pion production. As there is no expected constrain on the $\gamma\gamma$ opacity in the case of hadro-nuclear neutrino production, a correlation between neutrino emission and γ-ray activity may therefore hint towards this neutrino production channel. Joint KM3NeT neutrino detections and γ-ray monitoring with *Fermi*-LAT and CTA will allow us to establish or refute such a correlation and thus constrain the mechanism of neutrino production.

5. Summary and Conclusions

The past few years have seen many observational discoveries on blazars, among which this article highlights a few, such as the rapid variability at GeV and TeV energies, down to just a few minutes; large-angle optical polarization-angle rotations, found to be systematically correlated with γ-ray and multi-wavelength flares, and a strong hint of the blazar TXS 0506+056 as a source of IceCube neutrinos.

Minute-scale blazar variability severely challenges existing models for blazar emission, possibly indicating the prominence of small-scale structures due to magnetic reconnection. Polarization-decline seem to hint towards the presence of helical magnetic field structures, possibly, but not necessarily, associated with a changing viewing angle. The possible TXS 0506+056 + IceCube neutrino association provides a hint towards hadronic particle acceleration in blazars. However, a careful study of photo-pion neutrino production suggests that a direct correlation between γ-ray activity and neutrino emission is not necessarily expected, and the cascade emission going in tandem with $p\gamma$ neutrino production is likely to show up more prominently in X-rays than in γ-rays.

Future observations by the CTA will allow for a more sensitive exploration of minute-scale VHE γ-ray variability in blazars, potentially invalidating the current blazar emission paradigm. IXPE and possibly AMEGO promise the first X-ray (and γ-ray) polarimetry results, which will aid in testing the co-spatiality of optical and X-ray emissions and provide further diagnostics for hadronic vs. leptonic emission scenarios. When analysing future IXPE observations of blazars, special care has to be taken to account for possible PA rotations of the X-ray polarization, which might mirror those seen in the optical.

On the multi-messenger side, the future KM3NeT will greatly improve the statistics of astrophysical PeV neutrinos and promises the clear identification of a source class (or multiple source classes), possibly including blazars. This would provide conclusive proof of PeV proton acceleration in AGN jets, but does not necessarily point towards the origin of ultra-high-energy cosmic-rays in AGN.

Author Contributions: M.B. is the sole author of this publication.

Acknowledgments: The work of M.B. is supported by the Department of Science and Technology and National Research Foundation[10] of South Africa through the South African Research Chairs Initiative (SARChI), grant no. 64789.

Conflicts of Interest: The author declares no conflict of interest.

---

[8] https://icecube.wisc.edu/science/beyond.
[9] http://www.inr.ru/eng/ebgnt.html.
[10] Any opinion, finding and conclusion or recommendation expressed in this material is that of the author and the NRF does not accept any liability in this regard.

References

1. Acero, F.; Ackermann, M.; Ajello, M.; Albert, A.; Atwood, W.B.; Axelsson, M.; Baldini, L.; Ballet, J.; Barbiellini, G.; Bastieri, D.; et al. Fermi LAT third source catalog. *Astrophys. J. Suppl. Ser.* **2015**, *218*, 23. [CrossRef]

2. Aharonian, F.; Akhperjanian, A.G.; Bazer-Bachi, A.R.; Behera, B.; Beilicke, M.; Benbow, W.; Berge, D.; Bernlöhr, K.; Boisson, C.; Bolz, O.; et al. An Exceptional Very High Energy Gamma-Ray Flare of PKS 2155-304. *Astrophys. J.* **2007**, *644*, L71. [CrossRef]

3. Albert, J.; Aliu, E.; Anderhub, H.; Antoranz, P.; Armada, A.; Baixeras, C.; Barrio, J.A.; Bartko, H.; Bastieri, D.; Becker, J.K.; et al. Variable Very High Energy γ-Ray Emission from Markarian 501. *Astrophys. J.* **2007**, *669*, 862. [CrossRef]

4. Aleksić, J.; Antonelli, L.A.; Antoranz, P.; Backes, M.; Barrio, J.A.; Bastieri, D.; González, J.B.; Bednarek, W.; Berdyugin, A.; Berger, K.; et al. MAGIC Discovery of Very High Energy Emission from the FSRQ PKS 1221+21. *Astrophys. J.* **2011**, *730*, L8. [CrossRef]

5. Arlen, T.; Aune, T.; Beilicke, M.; Benbow, W.; Bouvier, A.; Buckley, J.H.; Bugaev, V.; Cesarini, A.; Ciupik, L.; Connolly, M.P.; et al. Rapid TeV Gamma-Ray Flaring of BL Lacertae. *Astrophys. J.* **2013**, *762*, 92. [CrossRef]

6. Lister, M.; Aller, M.F.; Aller, H.D.; Hodge, M.A.; Homan, D.C.; Kovalev, Y.Y.; Pushkarev, A.B.; Savolainen, T. MOJAVE XV. VLBA 15 GHz Total Intensity and Polarization Maps of 437 Parsec-scale AGN Jets from 1996 to 2017. *Astrophys. J. Suppl. Ser.* **2018**, *234*, 12. [CrossRef]

7. Blinov, D.; Pavlidou, V.; Papadakis, I.E.; Hovatta, T.; Pearson, T.J.; Liodakis, I.; Panopoulou, G.V.; Angelakis, E.; Baloković, M.; Das, H.; et al. RoboPol: Optical polarization-plane rotations and flaring activity in blazas. *Mon. Not. R. Astron. Soc.* **2016**, *457*, 2252. [CrossRef]

8. Böttcher, M.; Reimer, A.; Sweeney, K.; Prakash, A. Leptonic and Hadronic Modeling of Fermi-Detected Blazars. *Astrophys. J.* **2013**, *768*, 54. [CrossRef]

9. Romero, G.E.; Boettcher, M.; Markoff, S.; Tavecchio, F. Relativistic Jets in Active Galactic Nuclei and Microquasars. *Space Sci. Rev.* **2017**, *207*, 5–61. [CrossRef]

10. Ghisellini, G.; Tavecchio, F.; Foschini, L.; Ghirlanda, G. The transition between BL Lac objects and flat spectrum radio quasars. *Mon. Not. R. Astron. Soc.* **2011**, *414*, 2674. [CrossRef]

11. Abdo, A.A.; Ackermann, M.; Ajello, M.; Axelsson, M.; Baldini, L.; Ballet, J.; Barbiellini, G.; Bastieri, D.; Baughman, B.M.; Bechtol, K.; et al. The Spectral Energy Distribution of Fermi Bright Blazars. *Astrophys. J.* **2010**, *716*, 30. [CrossRef]

12. Maraschi, L.; Ghisellini, G.; Celotti, A. A jet model for the gamma-ray emitting blazar 3C279. *Astrophys. J.* **1992**, *397*, L5. [CrossRef]

13. Dermer, C.D.; Schlickeiser, R. Model for the High-Energy Emission from Blazars. *Astrophys. J.* **1993**, *416*, 458. [CrossRef]

14. Sikora, M.; Begelman, M.C.; Rees, M.J. Comptonization of diffuse ambient radiation by a relativistic jet: The source of gamma-rays from blazars? *Astrophys. J.* **1994**, *421*, 153. [CrossRef]

15. Błażejowski, M.; Sikora, M.; Moderski, R.; Madejski, G.M. Comptonization of Infrared Radiation from Hot Dust by Relativistic Jets in Quasars. *Astrophys. J.* **2000**, *545*, 107.

16. Ghisellini, G.; Tavecchio, F.; Chiaberge, M. Structured jets in TeV BL Lac objects and radiogalaxies. Implications for the observed properties. *Astron. Astrophys.* **2005**, *432*, 401. [CrossRef]

17. Georganopoulos, M.; Kazanas, D. Decelerating Flows in TeV Blazars: A Resolution to the BL Lacertae—FR I Unification Problem. *Astrophys. J.* **2003**, *594*, L27. [CrossRef]

18. Sikora, M.; Madejski, G. On Pair Content and Variability of Subparsec Jets in Quasars. *Astrophys. J.* **2000**, *534*, 109. [CrossRef]

19. Aharonian, F.A. TeV gamma-rays from BL Lac objects due to synchrotron radiation of extremely high energy protons. *New Astron.* **2000**, *5*, 377. [CrossRef]

20. Mücke, A.; Protheroe, R.J. A proton synchrotron blazar model for flaring in Markarian 501. *Astrophys. J.* **2001**, *15*, 121. [CrossRef]

21. Mannheim, K.; Biermann, P.L. Gamma-ray flaring of 3C279—A proton-initiated cascade in the jet? *Astron. Astrophys.* **1992**, *253*, L21.

22. Mannheim, K. The proton blazar. *Astron. Astrophys.* **1993**, *269*, 67.

23. Mücke, A.; Protheroe, R.J.; Engel, R.; Rachen, J.P.; Stanev, T. BL Lac objects in the synchrotron proton blazar model. *Astropart. Phys.* **2003**, *18*, 593. [CrossRef]

24. Aller, M.; Aller, H.; Hughes, P. The University of Michigan Centimeter-Band All Stokes Blazar Monitoring Program: Single-Dish Polarimetry as a Probe of Parsec-Scale Magnetic Fields. *Galaxies* **2017**, *5*, 5. [CrossRef]

25. Gabuzda, D. Parsec-Scale Jets in Active Galactic Nuclei. In *The Formation and Disruption of Black Hole Jets*; Contopoulos, I., Gabuzda, D., Kylafis, N., Eds.; Springer: Cham, Switzerland, 2015; Volume 414, pp. 117–148.

26. Jorstad, S.G.; Marscher, A.P.; Morozova, D.A.; Troitsky, I.S.; Agudo, I.; Casadio, C.; Foord, A.; Gómez, J.L.; MacDonald, N.R.; Molina, S.N.; et al. Kinematics of Parsec-scale Jets of Gamma-Ray Blazars at 43 GHz within the VLBA-BU-BLAZAR Program. *Astrophys. J.* **2017**, *846*, 98. [CrossRef]

27. Lister, M. AGN Jet Kinematics on Parsec-Scales: The MOJAVE Program. *Galaxies* **2016**, *4*, 29. [CrossRef]

28. Ahnen, M.L.; Ansoldi, S.; Antonelli, L.A.; Arcaro, C.; Babić, A.; Banerjee, B.; Bangale, P.; De Almeida, U.B.; Barrio, J.A.; González, J.B.; et al. First multi-wavelength campaign on the gamma-ray loud active galaxy IC 310. *Astron. Astrophys.* **2017**, *603*, A25. [CrossRef]

29. Collaboration, H.E.S.S.; Abdalla, H.; Abramowski, A.; Aharonian, F.; Ait Benkhali, F.; Akhperjanian, A.G.; Andersson, T.; Angüner, E.O.; Arrieta, M.; Aubert, P.; et al. H.E.S.S. discovery of very high energy γ-ray emission from PKS 0625-354. *Mon. Not. R. Astron. Soc.* **2018**, *476*, 4187–4198. [CrossRef]

30. Hayashida, M.; Nalewajko, K.; Madejski, G.M.; Sikora, M.; Itoh, R.; Ajello, M.; Blandford, R.D.; Buson, S.; Chiang, J.; Fukazawa, Y.; et al. Rapid Variability of Blazar 3C 279 during Flaring States in 2013–2014 with Joint Fermi-LAT, NuSTAR, Swift, and Ground-Based Multiwavelength Observations. *Astrophys. J.* **2015**, *807*, 79. [CrossRef]

31. Saito, S.; Tanaka, Y.T.; Takahashi, T.; Madejski, G.; D'Ammando, F. Very Rapid High-amplitude Gamma-Ray Variability in Luminous Blazar PKS 1510-089 Studied with Fermi-LAT. *Astrophys. J.* **2013**, *766*, L11. [CrossRef]

32. Ackermann, M.; Anantua, R.; Asano, K.; Baldini, L.; Barbiellini, G.; Bastieri, D.; Gonzalez, J.B.; Bellazzini, R.; Bissaldi, E.; Blandford, R.D.; et al. Minut-timescale > 100 MeV γ-Ray Variability during the Giant Outburst of Quasar 3C 279 Observed by Fermi-LAT in 2015 June. *Astrophys. J.* **2016**, *824*, L20. [CrossRef]

33. Shukla, A.; Mannheim, K.; Patel, S.R.; Roy, J.; Chitnis, V.R.; Dorner, D.; Rao, A.R.; Anupama, G.C.; Wendel, C. Short-timescale γ-Ray Variability in CTA 102. *Astrophys. J.* **2018**, *854*, L26. [CrossRef]

34. Böttcher, M.; Harris, D.; Krawczynski, H. *Relativistic Jets from Active Galactic Nuclei*; Wiley-VCH: Berlin, Germany, 2012.

35. Dondi, L.; Ghisellini, G. Gamma-ray loud blazars and beaming. *Mon. Not. R. Astron. Soc.* **1995**, *273*, 583. [CrossRef]

36. Begelman, M.C.; Fabian, A.C.; Rees, M.J. Implications of very rapid TeV variability in blazars. *Mon. Not. R. Astron. Soc.* **2008**, *384*, L19. [CrossRef]

37. Hovatta, T.; Valtaoja, E.; Tornikoski, M.; Lähteenmäki, A. Doppler factors, Lorentz factors and viewing angles for quasars, BL Lacertae objects and radio galaxies. *Astron. Astrophys.* **2009**, *494*, 527–537. [CrossRef]

38. Liodakis, I.; Hovatta, T.; Huppenkothen, D.; Kiehlmann, S.; Max-Moerbeck, W.; Readhead, A.C.S. Constraining the Limiting Brightness Temperature and Doppler Factors for the Largest Sample of Radio-Bright Blazars. *Astrophys. J.* **2018**, *866*, 137. [CrossRef]

39. Lyutikov, M.; Lister, M. Resolving the Doppler-factor Crisis in Active Galactic Nuclei: Non-steady Magnetized Outflows. *Astrophys. J.* **2010**, *722*, 197. [CrossRef]

40. Takahashi, T.; Tashiro, M.; Madejski, G.; Kubo, H.; Kamae, T.; Kataoka, J.; Kii, T.; Makino, F.; Makishima, K.; Yamasaki, N. ASCA Observation of an X-Ray/TeV Flare from the BL Lacertae Object Markarian 421. *Astrophys. J.* **1996**, *470*, L89. [CrossRef]

41. Böttcher, M.; Marscher, A.P.; Ravasio, M.; Villata, M.; Raiteri, C.M.; Aller, H.D.; Aller, M.F.; Teräsranta, H.; Mang, O.; Tagliaferri, G.; et al. Coordinated Multiwavelength Observations of BL Lacertae in 2000. *Astrophys. J.* **2003**, *596*, 847. [CrossRef]

42. Błażejowski, M.; Blaylock, G.; Bond, I.H.; Bradbury, S.M.; Buckley, J.H.; Carter-Lewis, D.A.; Celik, O.; Cogan, P.; Cui, W.; Daniel, M.; et al. A Multiwavelength View of the TeV Blazar Markarian 421: Correlated Variability, Flaring, and Spectral Evolution. *Astrophys. J.* **2005**, *630*, 130.

43. Krawczynski, H.; Hughes, S.B.; Horan, D.; Aharonian, F.; Aller, M.F.; Aller, H.; Boltwood, P.; Buckley, J.; Coppi, P.; Fossati, G.; et al. Multiwavelength Observations of Strong Flares from the TeV Blazar 1ES 1959+650. *Astrophys. J.* **2004**, *601*, 151. [CrossRef]

44. Naurois, M.H.E.S.S. Detection of a Strong VHE Activity from the Blazar 3C 279. The Astronomer's Telegram, No. 11239, 2018. Available online: http://www.astronomerstelegram.org/?read=11239 (accessed on 1 November 2018).

45. Capelo, P.R.; Dotti, M.; Volonteri, M.; Mayer, L.; Bellovary, J.M.; Shen, S. A survey of dual active galactic nuclei in simulations of galaxy mergers: Frequency and properties. *Mon. Not. R. Astron. Soc.* **2017**, *469*, 4437. [CrossRef]

46. Draper, A.R.; Ballantyne, D.R. The Merger-triggered Active Galactic Nucleus Contribution to the Ultraluminous Infrared Galaxy Population. *Astrophys. J.* **2012**, *753*, L37. [CrossRef]

47. Fu, H.; Wrobel, J.M.; Myers, A.D.; Djorgovski, S.G.; Yan, L. Binary Active Galactic Nuclei in Stripe 82: Constraints on Synchronized Black Hole Accretion in Major Mergers. *Astrophys. J.* **2015**, *815*, L6. [CrossRef]

48. Shabala, S.S. Delayed triggering of radio active galactic nuclei in gas-rich minor mergers in the local Universe. *Mon. Not. R. Astron. Soc.* **2017**, *464*, 4706. [CrossRef]

49. Lehto, H.J.; Valtonen, M.J. OJ 287 Outburst Structure and a Binary Black Hole Model. *Astrophys. J.* **1996**, *460*, 207. [CrossRef]

50. Valtonen, M.J.; Nilsson, K.; Sillanpää, A.; Takalo, L.O.; Lehto, H.J.; Keel, W.C.; Haque, S.; Cornwall, D.; Mattingly, A. The 2005 November Outburst in OJ 287 and the Binary Black Hole Model. *Astrophys. J.* **2006**, *643*, L9. [CrossRef]

51. Valtonen, M.J.; Lehto, H.J.; Sillanpää, A.; Nilsson, K.; Mikkola, S.; Hudec, R.; Basta, M.; Teräsranta, H.; Haque, S.; Rampadarath, H. Pedicting the Next Outbursts of OJ 287 in 2006 – 2010. *Astrophys. J.* **2006**, *646*, 36. [CrossRef]

52. Ackermann, M.; Ajello, M.; Albert, A.; Atwood, W.B.; Baldini, L.; Ballet, J.; Barbiellini, G.; Bastieri, D.; Gonzalez, J.B.; Bellazzini, R.; et al. Multiwavelength Evidence for Quasi-periodic Modulation in the Gamma-Ray Blazar PG 1553+113. *Astrophys. J.* **2015**, *813*, L41. [CrossRef]

53. Tavani, M.; Cavaliere, A.; Munar-Adrover, P.; Argan, A. The Blazar PG 1553+113 as a Binary System of Supermassive Black Holes. *Astrophys. J.* **2018**, *854*, 11. [CrossRef]

54. Covino, S.; Sandrinelli, A.; Treves, A. Gamma-ray quasio-periodicities in blazars. A cautious approach. *Mon. Not. R. Astron. Soc.* **2018**, *482*, 1270–1274. [CrossRef]

55. Smith, P.S. Extremely High Optical Polarization Observed in the Blazar PKS 1502+106. The Astronomer's Telegram, No. 11047, 2017. Available online: http://www.astronomerstelegram.org/?findmsg (accessed on 1 November 2018).

56. Pavlidou, V.; Angelakis, E.; Myserlis, I.; Blinov, D.; King, O.G.; Papadakis, I.; Tassis, K.; Hovatta, T.; Pazderska, B.; Paleologou, E.; et al. The RoboPol optical polarization survey of gamma-ray loud blazars. *Mon. Not. R. Astron. Soc.* **2014**, *442*, 1693–1705. [CrossRef]

57. Angelakis, E.; Hovatta, T.; Blinov, D.; Pavlidou, V.; Kiehlmann, S.; Myserlis, I.; Böttcher, M.; Mao, P.; Panopoulou, G.V.; Liodakis, I.; et al. RoboPol: The optical polarization of gamma-ray-loud and gamma-ray-quiet blazars. *Mon. Not. R. Astron. Soc.* **2016**, *463*, 3365–3380. [CrossRef]

58. Abdo, A.A.; Ackermann, M.; Ajello, M.; Axelsson, M.; Baldini, L.; Ballet, J.; Barbiellini, G.; Bastieri, D.; Baughman, B.M.; Bechtol, K.; et al. A change in the optical polarization associated with a γ-ray flare in the blazar 3C279. *Nature* **2010**, *463*, 919. [CrossRef] [PubMed]

59. Marscher, A.P.; Jorstad, S.G.; D'Arcangelo, F.D.; Smith, P.S.; Williams, G.G.; Larionov, V.M.; Oh, H.; Olmstead, A.R.; Aller, M.F.; Aller, H.D.; et al. The inner jet of an active galactic nucleus as revealed by a radio-to-γ-ray outburst. *Nature* **2008**, *452*, 7190. [CrossRef] [PubMed]

60. Marscher, P.; Jorstad, S.G.; Larionov, V.M.; Aller, M.F.; Aller, H.D.; Lähteenmäki, A.; Agudo, I.; Smith, P.S.; Gurwell, M.; Hagen-Thorn, V.A.; et al. Probing the Inner Jet of the Quasar PKS 1510-089 with Multi-Waveband Monitoring During Strong Gamma-Ray Activity. *Astrophys. J.* **2010**, *710*, L126. [CrossRef]

61. Blinov, D.; Pavlidou, V.; Papadakis, I.; Kiehlmann, S.; Panopoulou, G.; Liodakis, I.; King, O.G.; Angelakis, E.; Baloković, M.; Das, H.; et al. RoboPol: Fist season rotations of optical polarization plane in blazars. *Mon. Not. R. Astron. Soc.* **2015**, *453*, 1669–1683. [CrossRef]

62. Blinov, D.; Pavlidou, V.; Papadakis, I.; Kiehlmann, S.; Liodakis, I.; Panopoulou, G.V.; Pearson, T.J.; Angelakis, E.; Baloković, M.; Hovatta, T.; et al. RoboPol: Do optical polarization rotations occur in all blazars? *Mon. Not. R. Astron. Soc.* **2016**, *462*, 1775–1785. [CrossRef]

63. Blinov, D.; Pavlidou, V.; Papadakis, I.; Kiehlmann, S.; Liodakis, I.; Panopoulou, G.V.; Angelakis, E.; Baloković, M.; Hovatta, T.; King, O.G.; et al. RoboPol: Connection between optical polarization plane rotations and gamma-ray flares in blazars. *Mon. Not. R. Astron. Soc.* **2018**, *474*, 1296–1306. [CrossRef]

64. Hovatta, T.; Lindfors, E.; Blinov, D.; Pavlidou, V.; Nilsson, K.; Kiehlmann, S.; Angelakis, E.; Ramazani, V.F.; Liodakis, I.; Myserlis, I.; et al. Optical polarization of high-energy BL Lacertae objects. *Astron. Astrophys.* **2016**, *596*, A78. [CrossRef]

65. Abbott, B.P.; Abbott, R.; Abbott, T.D.; Abernathy, M.R.; Acernese, F.; Ackley, K.; Adams, C.; Adams, T.; Addesso, P.; Adhikari, R.X.; et al. Observation of Gravitational Waves from a Binary Black Hole Merger. *Phys. Rev. Lett.* **2016**, *116*, 061102. [CrossRef] [PubMed]

66. Abbott, B.P.; Abbott, R.; Adhikari, R.X.; Ananyeva, A.; Anderson, S.B.; Appert, S.; Arai, K.; Araya, M.C.; Barayoga, J.C.; Barish, B.C.; et al. Multi-messenger Observations of a Binary Neutron Star Merger. *Astrophys. J.* **2017**, *848*, L12. [CrossRef]

67. The IceCube; Fermi-LAT; MAGIC; AGILE; ASAS-SN; HAWC; H.E.S.S.; INTEGRAL; Kanata; Kiso; Kapteyn; Liverpool telescope; Subaru; Swift=NuSTAR; VERITAS; VLA/17B-403 teams. Multimessenger observations of a flaring blazar coincident with high-energy neutrino IceCube-170922A. *Science* **2018**, *361*, 6398.

68. IceCube Collaboration. Neutrino emission from the direction of the blazar TXS 0506+056 prior to the IceCube-170922A alert. *Science* **2018**, *361*, 147.

69. Aartsen, M.G.; Abraham, K.; Ackermann, M.; Adams, J.; Aguilar, J.A.; Ahlers, M.; Ahrens, M.; Altmann, D.; Anderson, T.; Archinger, M.; et al. Evidence for Astrophysical Muon Neutrinos from the Northern Sky with IceCube. *Phys. Rev. Lett.* **2015**, *115*, 081102. [CrossRef] [PubMed]

70. Aartsen, M.G.; Abraham, K.; Ackermann, M.; Adams, J.; Aguilar, J.A.; Ahlers, M.; Ahrens, M.; Altmann, D.; Andeen, K.; Anderson, T.; et al. Observation and Characterization of a Cosmic Muon Neutrino Flux from he Northern Hemisphere Using Six Years of IceCube Data. *Astrophys. J.* **2016**, *833*, 3. [CrossRef]

71. Dimitrakoudis, S.; Mastichiadis, A.; Protheroe, R.J.; Reimer, A. The time-dependent one-zone hadronic model. First principles. *Astron. Astrophys.* **2012**, *546*, A120. [CrossRef]

72. Mastichiadis, A. The Hadronic Model of Active Galactic Nuclei. *Space Sci. Rev.* **1996**, *75*, 317–329. [CrossRef]

73. Murase, K.; Inoue, Y.; Dermer, C.D. Diffuse neutrino intensity from the inner jets of active galactic nuclei: Impacts of external photon fields and the blazar sequence. *Phys. Rev.* **2014**, *90*, 023007. [CrossRef]

74. Protheroe, R.J.; Szabo, A.P. High energy cosmic rays from active galactic nuclei. *Phys. Rev. Lett.* **1992**, *69*, 2885. [CrossRef]

75. Stecker, F.W.; Done, C.; Salamon, M.H.; Sommers, P. High-energy neutrinos from active galactic nuclei. *Phys. Rev. Lett.* **1991**, *66*, 2697. [CrossRef] [PubMed]

76. Aartsen, M.G.; Abraham, K.; Ackermann, M.; Adams, J.; Aguilar, J.A.; Ahlers, M.; Ahrens, M.; Altmann, D.; Andeen, K.; Anderson, T.; et al. All-sky Search for Time-integrated Neutrino Emission from Astrophysical Sources with 7 yr of IceCube Data. *Astrophys. J.* **2017**, *835*, 151. [CrossRef]

77. Adrián-Martínez, S.; Albert, A.; André, M.; Anton, G.; Ardid, M.; Aubert, J.-J.; Baret, B.; Barrios-Martí, J.; Basa, S.; Bertin, V.; et al. The First Combined Search for Neutrino Point-sources in the Southern Hemisphere with the ANTARES and IceCube Neutrino Telescopes. *Astrophys. J.* **2016**, *823*, 65. [CrossRef]

78. Kadler, M.; Krauß, F.; Mannheim, K.; Ojha, R.; Müller, C.; Schulz, R.; Anton, G.; Baumgartner, W.; Beuchert, T.; Buson, S.; et al. Coincidence of a high-fluence blazar outburst with a PeV-energy neutrino event. *Nat. Phys.* **2016**, *12*, 807. [CrossRef]

79. Padovani, P.; Giommi, P.; Resconi, E.; Glauch, T.; Arsioli, B.; Sahakyan, N.; Huber, M. Dissecting the region around IceCube-170922A: The blazar TXS 0506+056 as the first cosmic neutrino source. *Mon. Not. R. Astron. Soc.* **2018**, *480*, 192–203. [CrossRef]

80. Böttcher, M. *Modeling the Emission Processes in Blazars*; Springer: Dordrecht, The Netherlands, 2007; Volume 309, p. 95.

81. Baring, M.G.; Böttcher, M.; Summerlin, E.J. Probing acceleration and turbulence at relativistic shocks in blazar jets. *Mon. Not. R. Astron. Soc.* **2017**, *464*, 4875. [CrossRef]

82. Blandford, R.; Eichler, D. Particle acceleration at astrophysical shocks: A theory of cosmic ray origin. *Phys. Rep.* **1987**, *154*, 1. [CrossRef]

83. Drury, L. An introduction to the theory of diffusive shock acceleration of energetic particles in tenuous plasmas. *Rep. Progr. Phys.* **1983**, *46*, 973. [CrossRef]

84. Jones, F.C.; Ellison, D.C. The plasma physics of shock acceleration. *Space Sci. Rev.* **1991**, *58*, 259. [CrossRef]

85. Summerlin, E.J.; Baring, M.G. Diffusive Acceleration of Particles at Oblique, Relativistic, Magnetohydrodynamic Shocks. *Astrophys. J.* **2012**, *745*, 63. [CrossRef]

86. Marscher, A.P.; Gear, W.K. Models for high-frequency radio outbursts in extragalactic sources, with application to the early 1983 millimeter-to-infrared flare of 3C 273. *Astrophys. J.* **1985**, *298*, 114–127. [CrossRef]

87. Böttcher, M.; Dermer, C.D. Timing Signatures of the Internal-Shock Model for Blazars. *Astrophys. J.* **2010**, *711*, 445. [CrossRef]

88. Chen, X.; Fossati, G.; Liang, E.P.; Böttcher, M. Time-dependent simulations of multi-wavelength variability of the blazar Mrk 421 with a Monte Carlo multizone code. *Mon. Not. R. Astron. Soc.* **2011**, *416*, 2368–2387. [CrossRef]

89. Chen, X.; Fossati, G.; Böttcher, M.; Liang, E.P. Time-dependent simulations of emission from the FSRQ PKS 1510-089: Multiwavelength variability of external Compton and synchrotron-self-Compton models. *Mon. Not. R. Astron. Soc.* **2012**, *424*, 789–799. [CrossRef]

90. Graff, P.B.; Georganopoulos, M.; Perlman, E.S.; Kazanas, D. A Multizone Model for Simulating the High-Energy Variability of TeV Blazars. *Astrophys. J.* **2008**, *689*, 68. [CrossRef]

91. Joshi, M.; Böttcher, M. Time-dependent Radiation Transfer in the Internal Shock Model Scenario for Blazar Jets. *Astrophys. J.* **2011**, *727*, 21. [CrossRef]

92. Joshi, M.; Marscher, A.P.; Böttcher, M. Seed Photon Fields of Blazars in the Internal Shock Scenario. *Astrophys. J.* **2014**, *785*, 132. [CrossRef]

93. Sokolov, A.; Marscher, A.P.; McHardy, I.M. Synchrotron Self-Compton Model for Rapid Nonthermal Flares in Blazars with Frequency-dependent Time Lags. *Astrophys. J.* **2004**, *613*, 725. [CrossRef]

94. Sokolov, A.; Marscher, A.P. External Compton Radiation from Rapid Nonthermal Flares in Blazars. *Astrophys. J.* **2005**, *629*, 52. [CrossRef]

95. Spada, M.; Ghisellini, G.; Lazzati, D.; Celotti, A. Internal shocks in the jets of radio-loud quasars. *Mon. Not. R. Astron. Soc.* **2001**, *325*, 1559–1570. [CrossRef]

96. Zhang, H.; Chen, X.; Böttcher, M. Synchrotron Polarization in Blazars. *Astrophys. J.* **2014**, *789*, 66. [CrossRef]

97. Zhang, H.; Chen, X.; Böttcher, M.; Guo, F.; Li, H. Polarization Swings Reveal Magnetic Energy Dissipation in Blazars. *Astrophys. J.* **2015**, *804*, 58. [CrossRef]

98. Zhang, H.; Deng, W.; Li, H.; Böttcher, M. Polarization Signatures of Relativistic Magnetohydrodynamic Shocks in the Blazar Emission Region. I. Force-free Helical Magnetic Fields. *Astrophys. J.* **2016**, *817*, 63. [CrossRef]

99. Kusunose, M.; Takahara, F. A Structured Leptonic Jet Model for the "Orphan" TeV Gamma-Ray Flares in TeV Blazas. *Astrophys. J.* **2006**, *651*, 113. [CrossRef]

100. Potter, W.J. Modelling blazar flaring using a time-dependent fluid jet emission model—An explanation for orphan flares and radio lags. *Mon. Not. R. Astron. Soc.* **2018**, *473*, 4107. [CrossRef]

101. Weidinger, M.; Spanier, F. A self-consistent and time-dependent hybrid blazar emission model. Properties and application. *Astron. Astrophys.* **2015**, *573*, 7. [CrossRef]

102. Sironi, L.; Keshet, U.; Lemoine, M. Relativistic Shocks: Particle Acceleration and Magnetization. *Space Sci. Rev.* **2015**, *191*, 519. [CrossRef]

103. Ghisellini, G.; Tavecchio, F.; Foschini, L.; Ghirlanda, G.; Maraschi, L.; Celotti, A. General physical properties of bright Fermi blazars. *Mon. Not. R. Astron. Soc.* **2010**, *402*, 497–518. [CrossRef]

104. Guo, F.; Li, X.; Li, H.; Daughton, W.; Zhang, B.; Lloyd-Ronning, N.; Liu, Y.-H.; Zhang, H.; Deng, W. Efficient Production of High-energy Nonthermal Particles during Magnetic Reconnection in a Magnetically Dominated Ion-Electron Plasma. *Astrophys. J.* **2016**, *818*, L9. [CrossRef]

105. Kagan, D.; Sironi, L.; Cerutti, B.; Giannios, D. Relativistic Magnetic Reconnection in Pair Plasmas and Its Astrophysical Applications. *Space Sci. Rev.* **2015**, *191*, 545–573. [CrossRef]

106. Nalewajko, K.; Uzdensky, D.A.; Cerutti, B.; Werner, G.R.; Begelman, M.C. On the Distribution of Particle Acceleration Sites in Plasmoid-dominated Magnetic Reconnection. *Astrophys. J.* **2015**, *815*, 101. [CrossRef]

107. Sironi, L.; Petropoulou, M.; Giannios, D. Relativistic jets shine through shocks or magnetic reconnection? *Mon. Not. R. Astron. Soc.* **2015**, *450*, 183. [CrossRef]

108. Werner, G.R.; Uzdensky, D.A.; Cerutti, B.; Nalewajko, K.; Begelman, M.C. The Extent of Power-law Energy Spectra in Collisionless Relativistic Magnetic Reconnection in Pair Plasmas. *Astrophys. J.* **2016**, *816*, L8. [CrossRef]

109. Guo, F.; Li, H.; Daughton, W.; Liu, Y.-H. Formaton of Hard Power Laws in the Energetic Particle Spectra Resulting from Relativistic Magnetic Reconnection. *Phys. Rev. Lett.* **2014**, *113*, 155005. [CrossRef] [PubMed]

110. Giannios, D.; Uzdensky, D.A.; Begelman, M.C. Fast TeV variability in blazars: Jets in a jet. *Mon. Not. R. Astron. Soc.* **2009**, *395*, L29. [CrossRef]

111. Giannos, D.; Uzdensky, D.A.; Begelman, M.C. Fast TeV variability from misaligned minijets in the jet of M87. *Mon. Not. R. Astron. Soc.* **2010**, *402*, 1649. [CrossRef]

112. Giannios, D. Reconnection-driven plasmoids in blazars: Fast flares on a slow envelope. *Mon. Not. R. Astron. Soc.* **2013**, *431*, 355. [CrossRef]

113. Christie, I.M.; Petropoulou, M.; Sironi, L.; Giannios, D. Radiative signatures of plasmoid-dominated magnetic reconnection in blazar jets. *Mon. Not. R. Astron. Soc.* **2019**, *482*, 65–82. [CrossRef]

114. Ghisellini, G.; Tavecchio, F. Rapid variability in TeV blazars: The case of PKS 2155-304. *Mon. Not. R. Astron. Soc.* **2008**, *386*, L28. [CrossRef]

115. Petropoulou, M.; Giannios, D.; Sironi, L. Blazar flares powered by plasmoids in relativistic reconnection. *Mon. Not. R. Astron. Soc.* **2016**, *462*, 3325. [CrossRef]

116. Araudo, A.T.; Bosch-Ramon, V.; Romero, G.E. Gamma rays from cloud penetration at the base of AGN jets. *Astron. Astrophys.* **2010**, *552*, 97. [CrossRef]

117. Araudo, A.T.; Bosch-Ramon, V.; Romero, G.E. Gamma-ray emission from massive stars interacting with active galactic nuclei jets. *Mon. Not. R. Astron. Soc.* **2013**, *436*, 3626. [CrossRef]

118. Barkov, M.V.; Aharonian, F.A.; Bosch-Ramon, V. Gamma-ray Flares from Red Giant/Jt Interactions in Active Galactic Nuclei. *Astrophys. J.* **2010**, *724*, 1517. [CrossRef]

119. Barkov, M.V.; Aharonian, F.A.; Bogovalov, S.V.; Kelner, S.R.; Khangulyan, D. Rapid TeV Variability in Blazars as a Result of Jet-Star Interaction. *Astrophys. J.* **2012**, *749*, 119. [CrossRef]

120. Khangulyan, D.V.; Barkov, M.V.; Bosch-Ramon, V.; Aharonian, F.A.; Dorodnitsyn, A.V. Star-Jet Interactions and Gamma-Ray Outbursts from 3C454.3. *Astrophys. J.* **2013**, *774*, 113. [CrossRef]

121. Zacharias, M.; Böttcher, M.; Jankowsky, F.; Lenain, J.-P.; Wagner, S.J.; Wierzcholska, A. Cloud Ablation by a Relativistic Jet and the Extended Flare in CTA 102 in 2016 and 2017. *Astrophys. J.* **2017**, *851*, 72. [CrossRef]

122. Böttcher, M. A Hadronic Synchrotron Mirror Model for the "Orphan Flare" of 1ES 1959+650. *Astrophys. J.* **2005**, *621*, 176. [CrossRef]

123. Tavani, M.; Vittorini, V.; Cavaliere, A. An Emerging Class of Gamma-ray Flares from Blazars: Beyond One-zone Models. *Astrophys. J.* **2015**, *814*, 51. [CrossRef]

124. MacDonald, N.R.; Marscher, A.P.; Jorstad, S.G.; Joshi, M. Through the Ring of Fire: Gamma-Ray Variability in Blazars by a Moving Plasmoid Passing a Local Source of Seed Photons. *Astrophys. J.* **2015**, *804*, 111. [CrossRef]

125. MacDonald, N.R.; Jorstad, S.G.; Marscher, A.P. "Orphan" γ-ray Flares and Stationary Sheaths of Blazar Jets. *Astrophys. J.* **2017**, *850*, 87. [CrossRef]

126. Larionov, V.M.; Jorstad, S.G.; Marscher, A.P.; Morozova, D.A.; Blinov, D.A.; Hagen-Thorn, V.A.; Konstantinova, T.S.; Kopatskaya, E.N.; Larionova, L.V.; Larionova, E.G.; et al. The Outburst of the Blazar S5 0716+71 in 2011 October: Shock in a Helical Jet. *Astrophys. J.* **2013**, *768*, 40. [CrossRef]

127. Larionov, V.M.; Villata, M.; Raiteri, C.M.; Jorstad, S.G.; Marscher, A.P.; Agudo, I.; Smith, P.S.; Acosta-Pulido, J.A.; arévalo, M.J.; Arkharov, A.A.; et al. Exceptional outburst of the blazar CTA 102 in 2012: The GASP-WEBT campaign and its extension. *Mon. Not. R. Astron. Soc.* **2016**, *461*, 3047. [CrossRef]

128. Ostorero, L.; Villata, M.; Raiteri, C.M. Helical jets in blazars. Interpretation of the multifrequency long-term variability of AO 0235+16. *Astron. Astrophys.* **2004**, *419*, 913. [CrossRef]

129. Raiteri, C.M.; Villata, M.; Acosta-Pulido, J.A.; Agudo, I.; Arkharov, A.A.; Bachev, R.; Baida, G.V.; Benítez, E.; Borman, G.A.; Boschin, W.; et al. Blazar spectral variability as explained by a twisted inhomogeneous jet. *Nature* **2017**, *552*, 374. [CrossRef] [PubMed]

130. Villata, M.; Raiteri, C.M. Helical jets in blazars. I. The case of MKN 501. *Astron. Astrophys.* **1999**, *347*, 30.

131. Lister, M.L.; Aller, M.F.; Aller, H.D.; Homan, D.C.; Kellermann, K.I.; Kovalev, Y.Y.; Pushkarev, A.B.; Richards, J.L.; Ros, E.; Savolainen, T. MOJAVE. X. Parsec-scale Jet Orientation Variations and Superluminal Motion in Active Galactic Nuclei. *Astron. J.* **2013**, *146*, 120. [CrossRef]

132. Nalewajko, K. A Model of Polarisation Rotations in Blazars from Kink Instabilities in Relativistic Jets. *Galaxies* **2017**, *5*, 64. [CrossRef]

133. Zhang, H.; Li, H.; Guo, F.; Taylor, G. Polarization Signatures of Kink Instabilities in the Blazar Emission Region from Relativistic Magnetohydrodynamic Simulations. *Astrophys. J.* **2017**, *835*, 125. [CrossRef]

134. Liang, E.P.; Fu, W.; Böttcher, M.; Roustazadeh, P. Scaling of Relativistic Shear Flows with the Bulk Lorentz Factor. *Astrophys. J.* **2018**, *854*, 129. [CrossRef]

135. Sobacchi, E.; Lyubarsky, Y.E. On the magnetisation and the radiative efficiency of BL Lac jets. *Mon. Not. R. Astron. Soc.* **2019**, submitted. [CrossRef]

136. Asano, K.; Takahara, F.; Kusunose, M.; Toma, K.; Kakuwa, J. Time-dependent Models for Blazar Emission with the Second-order Fermi Acceleration. *Astrophys. J.* **2014**, *780*, 64. [CrossRef]

137. Böttcher, M.; Chiang, J. X-Ray Spectral Variability Signatures of Flares in BL Lacertae Objects. *Astrophys. J.* **2000**, *581*, 127. [CrossRef]

138. Diltz, C.; Böttcher, M. Time dependent leptonic modeling of Fermi II processes in the jets of flat spectrum radio quasars. *J. High Energy Astrophys.* **2014**, *1*, 63. [CrossRef]

139. Kirk, J.G.; Rieger, F.M.; Mastichiadis, A. Particle acceleration and synchrotron emission in blazar jets. *Astron. Astrophys.* **1998**, *333*, 452.

140. Kirk, J.G.; Mastichiadis, A. Variability patterns of synchrotron and inverse Compton emission in blazars. *Astropart. Phys.* **1999**, *11*, 45. [CrossRef]

141. Kusunose, M.; Takahara, F.; Li, H. Electron Acceleration and Time Variability of High-Energy Emission from Blazars. *Astrophys. J.* **2000**, *536*, 299. [CrossRef]

142. Li, H.; Kusunose, M. Temporal and Spectral Variabilities of High-Energy Emission from Blazars Using Synchrotron Self-Compton Models. *Astrophys. J.* **2000**, *536*, 729. [CrossRef]

143. Mastichiadis, A.; Kirk, J.G. Variability in the synchrotron self-Compton model of blazars. *Astron. Astrophys.* **1997**, *320*, 19.

144. Diltz, C.; Böttcher, M.; Fossati, G. Time Dependent Hadronic Modeling of Flat Spectrum Radio Quasars. *Astrophys. J.* **2015**, *802*, 133. [CrossRef]

145. Petropoulou, M.; Mastichiadis, A. Temporal signatures of leptohadronic feedback mechanisms in compact sources. *Mon. Not. R. Astron. Soc.* **2012**, *421*, 2325. [CrossRef]

146. Chang, J.S.; Cooper, G. A Practical Difference Scheme for Fokker-Planck Equations. *J. Comput. Phys.* **1970**, *6*, 1. [CrossRef]

147. Potter, W.J.; Cotter, G. Synchrotron and inverse-Compton emission from blazar jets—II. An accelerating jet model with a geometry set by observations of M87. *Mon. Not. R. Astron. Soc.* **2012**, *429*, 1189–1205. [CrossRef]

148. Potter, W.J.; Cotter, G. Synchrotron and inverse-Compton emission from blazar jets—I. A uniform conical jet model. *Mon. Not. R. Astron. Soc.* **2012**, *423*, 756. [CrossRef]

149. Richter, S.; Spanier, F. A Numerical Model of Parsec-scale SSC Morphologies and Their Radio Emission. *Astrophys. J.* **2016**, *829*, 56. [CrossRef]

150. Chen, X.; Pohl, M.; Böttcher, M. Particle diffusion and localized acceleration in inhomogeneous AGN Jets—I. Steady-state spectra. *Mon. Not. R. Astron. Soc.* **2015**, *447*, 530. [CrossRef]

151. Finke, J.D.; Becker, P.A. Fourier Analysis of Blazar Variability. *Astrophys. J.* **2014**, *791*, 21. [CrossRef]

152. Finke, J.D.; Becker, P.A. Fourier Analysis of Blazar Variability: Klein-Nishina Effects and the Jet Scattering Environment. *Astrophys. J.* **2015**, *809*, 85. [CrossRef]

153. Lewis, T.R.; Becker, P.A.; Finke, J.D. Time-dependent Electron Acceleration in Blazar Transients: X-ray Time Lags and Spectral Formation. *Astrophys. J.* **2016**, *824*, 108. [CrossRef]

154. Zhang, Y.H. Cross-spectral analysis of the X-ray variability of Markarian 421. *Mon. Not. R. Astron. Soc.* **2002**, *337*, 609–618. [CrossRef]

155. Lyutikov, M.; Kravchenko, E.V. Polarization swings in blazars. *Mon. Not. R. Astron. Soc.* **2017**, *467*, 3876. [CrossRef]

156. Marscher, A.P. Turbulent, Extreme Multi-zone Model for Simulating Flux and Polarization Variability in Blazars. *Astrophys. J.* **2014**, *780*, 87. [CrossRef]

157. MacDonald, N.R.; Marscher, A.P. Faraday Conversion in Turbulent Blazar Jets. *Astrophys. J.* **2018**, *862*, 58. [CrossRef]

158. Krawczynski, H. The Polarization Properties of Inverse Compton Emission and Implications for Blazar Observations with the GEMS X-Ray Polarimeter. *Astrophys. J.* **2012**, *744*, 30. [CrossRef]

159. Bonometto, S.; Saggion, A. Polarization in Inverse Compton Scattering of Synchrotron Radiation. *Astron. Astrophys.* **1974**, *23*, 9.

160. Zhang, H.; Böttcher, M. X-Ray and Gamma-Ray Polarization in Leptonic and Hadronic Jet Models of Blazars. *Astrophys. J.* **2013**, *774*, 18. [CrossRef]

161. Zhang, H.; Diltz, C.; Böttcher, M. Radiation and Polarization signatures of the 3D Multizone Time-dependent Hadronic Blazar Model. *Astrophys. J.* **2016**, *829*, 69. [CrossRef]

162. Petropoulou, M.; Dimitrakoudis, S. Constraints of flat spectrum radio quasars in the hadronic model: The case of 3C 273. *Mon. Not. R. Astron. Soc.* **2015**, *452*, 1303. [CrossRef]

163. Dermer, C.D.; Murase, K.; Inoue, Y. Photopionproduction in black-hole jets and flat-spectrum radio quasars as PeV neutrino sources. *J. High Energy Astrophys.* **2014**, *3*, 29. [CrossRef]

164. Halzen, F. Pionic photons and neutrinos from cosmic ray accelerators. *Astropart. Phys.* **2013**, *43*, 155. [CrossRef]

165. Mannheim, K.; Biermann, P.L. Photomeson production in active galactic nuclei. *Astron. Astrophys.* **1989**, *221*, 211.

166. Reimer, A.; Böttcher, M.; Postnikov, S. Neutrino Emission in the Hadronic Synchrotron Mirror Model: The "Orphan" TeV Flare from 1ES 1959+650. *Astrophys. J.* **2005**, *630*, 186. [CrossRef]

167. Petropoulou, M.; Dimitrakoudis, S.; Padovani, P.; Mastichiadis, A.; Resconi, E. Photohadronic origin of γ-ray BL Lac emission: Implications for IceCube neutrinos. *Mon. Not. R. Astron. Soc.* **2015**, *448*, 2412. [CrossRef]

168. Liu, R.; Murase, K.; Inoue, S.; Ge, C.; Wang, X.-Y. Can Winds Driven by Active Galactic Nuclei Account for the Extragalactic Gamma-Ray and Neutrino Backgrounds? *Astrophys. J.* **2018**, *858*, 9. [CrossRef]

169. Liu, R.-Y.; Wang, K.; Xue, R.; Taylor, A.M.; Wang, X.-Y.; Li, Z.; Yan, H. A hadronuclear interpretation of a high-energy neutrino event coincident with a blazar flare. *arXiv* **2018**, arXiv:1807.05113.

170. Cerruti, M.; Zech, A.; Boisson, C.; Emery, G.; Inoue, S.; Lenain, J.P. Lepto-hadronic single-zone models for the electromagnetic and neutrino emission of TXS 0506+056. *Mon. Not. R. Astron. Soc.* **2018**, submitted.

171. Reimer, A.; Böttcher, M.; Buson, S. Cascading constraints from neutrino emitting blazars: The case of TXS 0506+056. *Astrophys. J.* **2018**, submitted.

172. Gao, S.; Fedynitch, A.; Winter, W.; Pohl, M. Interpretation of the coincident observation of a high energy neutrino and a bright flare. *arXiv* **2018**, arXiv:1807.04275.

173. Keivani, A.; Murase, K.; Petropoulou, M.; Fox, D.B.; Cenko, S.B.; Chaty, S.; Coleiro, A.; DeLaunay, J.J.; Dimitrakoudis, S.; Evans, P.A.; et al. A Mutimessenger Picture of the Flaring Blazar TXS 0506+056: Implications for High-energy Neutrino Emission and Cosmic-Ray Acceleration. *Astrophys. J.* **2018**, *864*, 84. [CrossRef]

174. Murase, K.; Oikonomou, F.; Petropoulou, M. Blazar Flares as an Origin of High-energy Cosmic Neutrinos? *Astrophys. J.* **865**, 124. [CrossRef]

175. Ansoldi, S. The Blazar TXS 0506+056 Associated with a High-energy Neutrino: Insights into Extragalactic Jets and Cosmic-Ray Acceleration. *Astrophys. J.* **2018**, *863*, L10. [CrossRef]

176. Tavecchio, F.; Ghisellini, G. Spine-sheath layer radiative interplay in subparsec-scale jets and the TeV emission from M87. *Mon. Not. R. Astron. Soc.* **2008**, *385*, L98. [CrossRef]

177. Righi, C.; Tavecchio, F.; Inoue, S. Neutrino emission from BL Lac objects: The role of radiatively inefficient accretion flows. *Mon. Not. R. Astron. Soc.* **2018**, submitted. [CrossRef]

178. Acharya, B.S.; Actis, M.; Aghajani, T.; Agnetta, G.; Aguilar, J.; Aharonian, F.; Ajello, M.; Akhperjanian, A.; Alcubierre, M.; Aleksić, J.; et al. Introducing the CTA Concept. *Astropart. Phys.* **2013**, *43*, 3–18. [CrossRef]

179. Weisskopf, M.C.; Ramsey, B.; O'Dell, S.L.; Tennant, A.; Elsner, R.; Soffita, P.; Bellazzini, R.; Costa, E.; Kolodziejczak, J.; Kaspi, V.; et al. The Imaging X-ray Polarimetry Explorer (IXPE). *Results Phys.* **2016**, *6*, 1179–1180. [CrossRef]

180. Margiotta, A. The KM3NeT deep-sea neutrino telescope. *Nucl. Inst. Methods Phys.* **2014**, *766*, 83. [CrossRef]

181. Adrián-Martínez, S.; Ageron, M.; Aharonian, F.; Aiello, S.; Albert, A.; Ameli, F.; Anassontzis, E.; Andre, M.; Androulakis, G.; Anghinolfi, M.; et al. Letter of intent for KM3NeT 2.0. *J. Phys. G* **2016**, *43*, 084001. [CrossRef]

182. Aartsen, M.G.; Ackermann, M.; Adams, J.; Aguilar, J.A.; Ahlers, M.; Ahrens, M.; Al Samarai, I.; Altmann, D.; Andeen, K.; Anderson, T.; et al. The IceCube Neutrino Observatory—Contributions to ICRC 2017 Part VI: IceCube-Gen2, the Next Generation Neutrino Observatory. *arXiv* **2017**, arXiv:1710.01207.

183. Blaufuss, E.; Kopper, C.; Haack, C.; IceCube-Gen2 Collaboration. The IceCube-Gen2 High Energy Array. In Proceedings of the 34th The 34th International Cosmic Ray Conference, The Hague, The Netherlands, 30 July–6 August 2015; Volume 236, p. 1146.

184. Avrorin, A.; Aynutdinov, V.; Belolaptikov, I.; Berezhnev, S.; Bogorodsky, D.; Budnev, N.; Danilchenko, I.; Domogatsky, G.; Doroshenko, A.; Dyachok, A.; et al. The Gigaton Volume Detector in Lake Baikal. *Nucl. Instrum. Methods Phys. Res. Sect. A* **2011**, *639*, 30–32. [CrossRef]

185. Moiseev, A. All-Sky Medium Energy Gamma-Ray Observatory (AMEGO). In Proceedings of the 35th International Cosmic Ray Conference (ICRC2017), Busan, Korea, 12–20 July 2017; Volume 301, p. 798

186. Cherenkov Telescope Array Consortium. Science with the Cherenkov Telescope Array. *World Sci.* **2017**, in press.

187. Böttcher, M.; Els, P. Gamma-Gamma Absorption in the Broad Line Region Radiation Fields of Gamma-Ray Blazars. *Astrophys. J.* **2016**, *821*, 102. [CrossRef]

188. Finke, J.D. External Compton Scattring in Blazar Jets and the Location of the Location of the Gamma-Ray Emitting Region. *Astrophys. J.* **2016**, *830*, 94. [CrossRef]

189. Liu, H.T.; Bai, J.M. Absorption of 10–200 GeV Gamma-Ray sby Radiation from Broad Line Regions in Blazars. *Astrophys. J.* **2006**, *653*, 1089. [CrossRef]

190. Poutanen, J.; Stern, B. GeV Breaks in Blazars as a Result of Gamma-ray Absorption Within the Broad-Line Region. *Astrophys. J.* **2010**, *717*, L118. [CrossRef]

191. Tavecchio, F.; Roncadelli, M.; Galanti, G.; Bonnoli, G. Evidence for axion-like particles from PKS 1222+216? *Phys. Rev. D* **2012**, *86*, 5036. [CrossRef]

192. Tavecchio, F.; Bonnoli, G. On the detectability of Lorentz invariance violation through anomalies in the multi-TeV γ-ray spectra of blazars. *Astron. Astrophys.* **2016**, *585*, 25. [CrossRef]

193. Abdo, A.A.; Ackermann, M.; Ajello, M.; Baldini, L.; Ballet, J.; Barbiellini, G.; Bastieri, D.; Bechtol, K.; Bellazzini, R.; Berenji, B.; et al. Radio-Loud Narrow-Line Seyfert 1 as a New Class of Gamma-Ray Active Galactic Nuclei. *Astrophys. J.* **2009**, *707*, L142. [CrossRef]

194. D'Ammando, F.; Orienti, M.; Finke, J.; Larsson, J.; Giroletti, M.; Raiteri, C.M. A Panchromatic View of Relativistic Jets in Narrow-Line Seyfert 1 Galaxies. *Galaxies* **2016**, *4*, 11. [CrossRef]

195. Şentürk, G.D.; Errando, M.; Böttcher, M.; Mukherjee, R. Gamma-ray Observational Properties of TeV-detected Blazars. *Astrophys. J.* **2013**, *764*, 119. [CrossRef]

196. Zech, A.; Cerruti, M.; Mazin, D. Expected signatures from hadronic emission processes in the TeV spectra of BL Lacertae objects. *Astron. Astrophys.* **2017**, *602*, 25. [CrossRef]

galaxies

MDPI

Article

Plasmas in Gamma-Ray Bursts: Particle Acceleration, Magnetic Fields, Radiative Processes and Environments

Asaf Pe'er

Physics Department, Bar Ilan University, Ramat-Gan 52900, Israel; asaf.peer@biu.ac.il; Tel.: +972-3-5318438

Received: 22 November 2018; Accepted: 7 February 2019; Published: 15 February 2019

Abstract: Being the most extreme explosions in the universe, gamma-ray bursts (GRBs) provide a unique laboratory to study various plasma physics phenomena. The complex light curve and broad-band, non-thermal spectra indicate a very complicated system on the one hand, but, on the other hand, provide a wealth of information to study it. In this chapter, I focus on recent progress in some of the key unsolved physical problems. These include: (1) particle acceleration and magnetic field generation in shock waves; (2) possible role of strong magnetic fields in accelerating the plasmas, and accelerating particles via the magnetic reconnection process; (3) various radiative processes that shape the observed light curve and spectra, both during the prompt and the afterglow phases, and finally (4) GRB environments and their possible observational signature.

Keywords: jets; radiation mechanism: non-thermal; galaxies: active; gamma-ray bursts; TBD

1. Introduction

Gamma-ray bursts (GRBs) are the most extreme explosions known since the big bang, releasing as much as 10^{55} erg (isotropically equivalent) in a few seconds, in the form of gamma rays [1]. Such a huge amount of energy released in such a short time must be accompanied by a relativistic motion of a relativistically expanding plasma. There are two separate arguments for that. First, the existence of photons at energies $\gtrsim$ MeV as are observed in many GRBs necessitates the production of $e^{\pm}$ pairs by photon–photon interactions, as long as the optical depth for such interactions is greater than unity. Indeed, the huge luminosity combined with small system size, as is inferred from light-crossing time arguments ensures that this is indeed the case. Second, as has long been suspected and is well established today, small baryon contamination, originating from the progenitor—being either a single, collapsing star or the merger of binary, degenerate stars (e.g., neutron stars or white dwarfs) implies that some baryon contamination is unavoidable. These baryons must be accelerated for the least by the radiative pressure into relativistic velocities.

This general picture was confirmed already 20 years ago by the detection of afterglow—a continuing radiation that is observed at late times, up to weeks, months and even years after the main GRB, at gradually lower frequencies—from X-ray to radio [2–4]. Lasting for many orders of magnitude longer than the prompt phase, this afterglow radiation is much easier to study. Indeed, it had been extensively studied in the past two decades, after the initial detection enabled by the Dutch-Italian Beppo-SAX satellite.

Fitting the observed spectra shows a clear deviation from a black-body spectra. Instead, the afterglow of many bursts is well fitted by synchrotron radiation from a power law distribution of radiative electrons [5–7]; see Figure 1. The late time decay is well explained by the gradual velocity decay of the expanding plasma as it propagates into the surrounding medium. This decay is well-fitted (at late times) by the Blandford–Mckee self-similar solution [8] of a relativistic explosion (I ignore here the early afterglow phase which typically lasts a few minutes, as during this phase the decay does not follow a simple self-similar law, and is as of yet not fully understood).

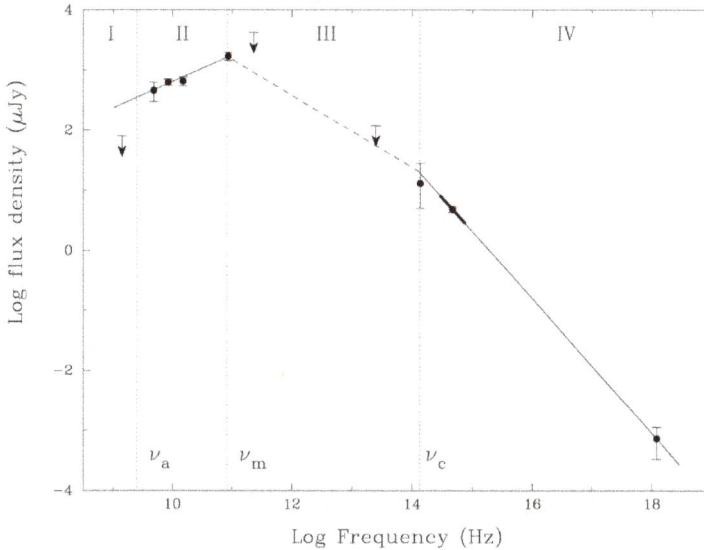

Figure 1. X-ray to radio spectrum of GRB970508 taken 12 days after the event is well fitted by a broken power law, as is expected from a power law distribution of electrons that emit synchrotron radiation. Marked are the transition frequencies: the self absorption frequency ν_a, the peak frequency ν_m and the cooling frequency ν_c. This figure is taken from Galama et al. [9].

At the onset of the afterglow phase, the velocity of the expanding plasma is very close to the speed of light, with initial Lorentz factor of a few hundreds. This is greater than the speed of sound, $c/\sqrt{3}$, and as such necessitates the existence of a highly relativistic shock wave. The combined temporal and spectral analysis thus led to the realization that, at least during the afterglow phase, a relativistic shock wave must exist. This shock wave expands into the circumburst medium, and gradually slows down as it collects and heats material from it.

Interpreting the observed signal during the afterglow phase in the framework of the synchrotron emission model, one finds that the inferred values of the magnetic fields, $\lesssim 1$ G, are about two (and in some cases more) orders of magnitude stronger than the compressed values of the circumburst magnetic field [5,10,11]. This implies that, in order to explain the observed signal, the relativistic shock wave must be able to both (1) accelerate particles to high energies, producing a non-thermal (non-Maxwellian) distribution of particles; and (2) generate a strong magnetic field, which causes the energetic particles to radiate their energy via synchrotron emission.

Studies of the afterglow phase by themselves therefore lead to several very interesting plasma physics phenomena which are not well understood, and are at the forefront of current research. These include (1) the physics of relativistic shock waves, both propagation and stability; (2) particle acceleration to non-thermal distributions; (3) generation of strong magnetic fields; and (4) radiative processes that lead to the observed spectra.

However, the prompt phase of GRB is considered even more challenging. As its name implies, this phenomenon lasts only a short duration of time, typically a few seconds. As opposed to the afterglow phase, this stage is characterized by fluctuative, non-repeating light curve, with no two GRB light curves similar to each other. Furthermore, its spectra does not resemble neither a black body (Planck) spectrum, nor—as has been realized in the past decade—that of a synchrotron emission from a power law distribution of electrons, as in the afterglow phase. Though the large diversity within the bursts prevented, so far, clear conclusions.

Another very challenging aspect is that the origin of the rapid acceleration that results in the relativistic expansion is not yet fully understood. While it was initially thought to occur as a result of the photon's strong radiative pressure (the "fireball" model), in recent years, it has been argued that strong magnetic fields—whose origin may be associated with the progenitor(s), hence external to the outflow, may play a key role in the acceleration process. If this is indeed the case, the plasma must be magnetically dominated, namely $u_B \gg \{u_k, u_{th}\}$, where u_B, u_k and u_{th} are the magnetic, kinetic and thermal energies of the plasma, respectively.

Either way, the plasma in GRBs during its prompt emission phase is characterized by strong interactions accompanied by energy and momentum exchange between the particle and photon fields, and/or the particles and the magnetic fields. Combined with the different conditions during the afterglow phase, one can conclude that GRBs provide a unique laboratory to study various fundamental questions in plasma physics. These are related to the creation of magnetic fields, acceleration of particles, emission of radiation and the interaction between all these three fields. Furthermore, the relativistic expansion can lead to the developments of several instabilities in the expanding plasma, which, in turn, can affect the phenomena previously mentioned.

In this chapter, I highlight the current state of knowledge in these areas. I should stress that I limit the discussion here to plasma physics phenomena only; in recent years, there have been many excellent reviews covering various aspects of GRB phenomenology and physics, and I refer the reader to these reviews for a more comprehensive discussion on the various subjects. A partial list includes Atteia and Boër [12], Gehrels and Mészáros [13], Bucciantini [14], Gehrels and Razzaque [15], Daigne [16], Zhang [17], Berger [18], Meszaros and Rees [19], Pe'er [20], Kumar and Zhang [21], Granot et al. [22], Zhang et al. [23], Toma et al. [24], Pe'er and Ryde [25], Beloborodov and Mészáros [26], Dai et al. [27], van Eerten [28], and Nagataki [29].

Of the many plasma physics effects that exist in GRBs—some of them unique to these objects, I discuss here several fundamental phenomena which emerge directly from GRB studies. Due to the wealth of the subject, I can only discuss each topic briefly. In each section, I refer the reader to (some) relevant literature for further discussion. The topics I cover here include the following: acceleration of particles by relativistic shock waves are discussed in Section 2. Section 3 is devoted to magnetic fields. I discuss generation of magnetic fields by shock waves in Section 3.1, and their possible role during the prompt emission phase in Section 3.2. I briefly discuss the acceleration of particles in magnetically dominated outflow via reconnection of magnetic field lines in Section 3.3. I then discuss the radiation field, which plays a key role in GRBs in Section 4. I first introduce the "classical" radiative processes in Section 4.1 and then introduce the photospheric emission in Section 4.2. Finally, I very briefly consider the different environments into which GRBs may explode and their effects in Section 5 before concluding the paper.

2. Acceleration of Particles in Shock Waves

The idea that shock waves can be the acceleration sites of particles dates back to Enrico Fermi himself [30,31], and had been extensively studied over the years since [32–38]. The key motivation was to explain the observed spectrum and flux of cosmic rays. Fermi's original idea suggests that particles are energized as they bounce back and forth across the shock wave. Its basic details can be found today in many textbooks (e.g., [39]).

In the context of GRBs, it was proposed in the mid 1990s that the relativistic shock waves that exist in GRB plasmas may provide the conditions required for the acceleration of particles to the highest observed energies, $\gtrsim 10^{20}$ eV [40–42]. While this idea is still debatable (e.g., [43]), the observations of >GeV, and up to ∼100 GeV photons [44] during the prompt phase of several GRBs implies that very high energy particles must exist in the emitting region. While these particles can be protons, energetically, it is much less demanding if these are electrons that are accelerated to non-thermal distribution at high energies. This is due to the lighter mass of the electrons, which implies much more efficient coupling to the magnetic and photon fields, and hence much better radiative efficiency. These energetic particles, in turn, radiate their energy in the strong magnetic fields that are believed to exist, as well as Compton scatter the photons to produce the very high energy photons observed.

Fitting the observed spectra of many GRBs in the framework of the synchrotron model (namely, under the assumption that the leading radiative mechanism is synchrotron emission by energetic electrons) strongly suggests that the radiating particles do not follow a (relativistic) Maxwellian distribution. Rather, they follow a power law distribution at high energies, $dn_E/dE \propto E^{-p}$, with power law index $p \approx 2.0 - -2.4$ [5,9,45–47]. This power law distribution is exactly what is expected from acceleration of particles in shock waves within the framework of the Fermi mechanism (e.g., [33–35,48]). Intuitively, the power law shape of the distribution can be understood as there is no characteristic momentum scale that exists during the acceleration process, implying that the rate of momentum gain is proportional to the particle's momentum.

The power law index inferred from observations is close to Fermi's original suggestion of 2.0. This is surprising, given that the shock waves in GRBs both during the prompt (if exist) and afterglow phases must be relativistic, while Fermi's work dealt with ideal, non-relativistic shocks.

In fact, the situation is far more complicated. Despite many decades of research, the Fermi process is still not fully understood from first principles. This is attributed mainly to the highly nonlinear coupling between the accelerated particles and the turbulent magnetic field at the shock front. The magnetic field is both generated by the energetic particles (via the generated currents) and at the same time affects their acceleration. This makes analytical models to be extremely limited in their ability to simultaneously track particle acceleration and magnetic field generation.

Due to this complexity, most analytical and Monte Carlo methods use the "test particle" approximation. According to this approximation, during the acceleration process, the accelerated particles interact with a fixed background magnetic field. These models therefore neglect the contribution of the high energy particles to the magnetic field, which occurs due to the currents they generate. This assumption can be justified as long as the accelerated particles carry only a small fraction of the available energy that can be deposited to the magnetic field. However, as explained above, this assumption is not supported by current observations.

Furthermore, relativistic shocks, as are expected in GRBs, introduce several challenges which do not exist when considering non-relativistic shocks. These include (1) the fact that the distribution of the accelerated particles cannot be considered isotropic; (2) mixing of the electric and magnetic fields when moving between the upstream and downstream shock regions; and (3) the fact that it is more difficult to test the theory (or parts of it), and one has to rely on very limited data, which can often be interpreted in more than a single way.

Very broadly speaking, theoretical works can be divided into three categories. The first is a semi-analytic approach (e.g., [49–54]), in which particles are described in terms of distribution functions, enabling analytic or numerical solutions of the transport equations. Clearly, while this is the fastest method, reliable solutions exist only over a very limited parameter space region, and several considerable simplifications (e.g., about the turbulence, anisotropy, etc.) are needed. The second method is the Monte Carlo approach [55–62]. In this method, the trajectories and properties of representative particles are tracked, assuming some average background magnetic fields. The advantage of this method is that it enables exploring a much larger parameter space region than analytical methods while maintaining fast computational speed. The disadvantages are (a) the simplified treatment of the background magnetic field, which effectively implies that the "test particle" approximation is used; and (b) current Monte Carlo codes use a simplified model to describe the details of the interactions between the particles and magnetic fields. For example, many codes use the "Bohm" diffusion model, which is not well-supported theoretically (see [63]).

The third approach is the Particle-In-Cell (PIC) simulations [64–68]. These codes basically solve simultaneously both particle trajectories and electromagnetic fields in a fully self-consistent way.

They therefore provide the "ultimate answer", namely the entire spectra of the accelerated particles alongside the generated magnetic field. They further provide the details of the generated magnetic turbulence as well as visualize the formation process of collisionless shocks. However, these codes are prohibitively computationally expensive, and are therefore limited to a very small range both in time and space. Modern simulations can compute processes on a length scale of no more than a few thousands of skin depth (c/ω_p). This scale is many orders of magnitudes—typically 7–8 orders of magnitude shorter than the physical length scale of the acceleration region, as is inferred from observations. This is an inherent drawback that cannot be overcome in the nearby future.

To conclude this section, GRB observations provide direct evidence—possibly the most detailed observational evidence for the acceleration of particles in relativistic shock waves. This evidence triggered a huge amount of theoretical work aimed at understanding this phenomenon from first physical principles. Due to its huge complexity, and while a huge progress was made in the past two decades or so mainly due to advances in PIC codes, the problem is still far from being solved.

3. Magnetic Fields in GRBs

In addition to the existence of clear evidence that shock waves in GRBs serve as particle acceleration sites, there is a wealth of evidence for the existence of strong magnetic fields in GRBs. When discussing magnetic fields in GRBs, one has to discriminate between two, very different, scenarios.

First, as already discussed above, fitting the data of GRB afterglow strongly suggests that the main radiative mechanism during this phase is synchrotron emission from energetic electrons. This idea therefore implies that strong magnetic fields must exist in the plasma. Fitting the afterglow provides evidence that the magnetic fields are about two orders of magnitude—and in some cases more—than the values expected from compression of the intergalactic field [5,10,11,69,70]. This provides indirect evidence that the relativistic shock wave that inevitably exists during this phase must generate a strong magnetic field, in parallel to accelerating particles.

Second, while no direct evidence currently exists, it had been proposed that, during the prompt emission phase, the GRB plasma may in fact be Poynting-flux dominated [71–78]. If this idea is correct, then the origin of the magnetic field must be external to the plasma—namely originate at the progenitor. In this scenario, the magnetic field serves as an energy reservoir that is used to both accelerate the plasma and at the same time accelerate particles to high energies.

3.1. Magnetic Field Generation in Shock Waves

As is typical to most (in fact, nearly all) astrophysical plasmas, and certainly in GRBs, the shock waves that exist are collisionless, namely they are not mediated by direct collisions between the particles (as opposed to, e.g., shock waves that occur while a jet plane exceeds the speed of sound in the earth's atmosphere). This can easily be verified for the shock wave in the afterglow phase of GRBs by considering the mean free path for particle interaction, $l = (n\sigma_T)^{-1} \simeq 10^{24}$ cm. Here, $n \simeq 1$ cm^{-3} is the typical interstellar medium (ISM) density, and σ_T is Thompson cross section, which is the typical cross section for particle interaction. This scale is many orders of magnitude longer than the scale of the system, implying that the generated shock waves must be collisionless.

Instead of direct collisions, the shock waves are generated by collective plasma effects, namely the charged particles generate currents. These currents in turn generate magnetic fields that deflect the charged particles trajectories, mixing and randomizing their trajectories, until they isotropize. Thus, the generation of collisionless shock waves must include generation of turbulent magnetic fields. The key questions are therefore related to the details of the process, which are not fully understood. These include: (1) the nature of the instability that generates the turbulent field; (2) the strength—and scale of the generated field; and (3) the interconnection between the particle acceleration process and the magnetic field generation process.

The most widely discussed mechanism by which (weakly magnetized) magnetic fields can be generated is the Weibel instability (e.g., [79–89]). In this model, small fluctuation in the magnetic fields charge separation have opposite charges in the background plasma. These particles then form "filaments" of alternating polarity, which grow with time, as the currents carried by the charged particles positively feed the magnetic fields. This is illustrated in Figure 2, taken from Medvedev and Loeb [80]. Indeed, this instability is routinely observed in many PIC simulations [64–66,90–94], which enable quantifying it. Furthermore, these simulations prove the ultimate connection between the formation of collisionless shock waves and the generation of magnetic fields.

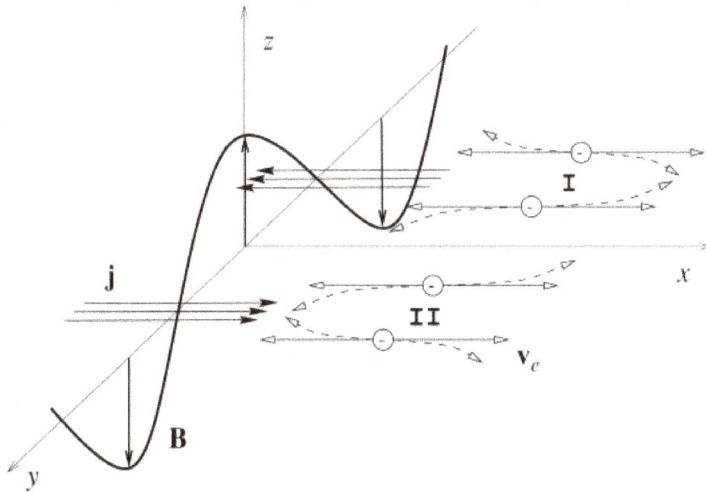

Figure 2. Illustration of the Weibel (also denoted "relativistic two stream") instability, taken from Medvedev and Loeb [80]. A magnetic field perturbation deflects electron motion along the x-axis, and results in current sheets (j) of opposite signs in regions I and II, which in turn amplify the perturbation. The amplified field lies in the plane perpendicular to the original electron motion.

However, these same simulations show that the generated magnetic fields decay over a relatively very short length scale, of few tens—few hundreds of plasma skin depth, as was suggested earlier [95–98]; see Figure 3. This is in sharp contrast to the observed synchrotron signal, which requires that the magnetic field, necessary for the synchrotron emission, will remain substantial over a much larger scale, comparable to the scale of the system.

This drawback, clearly observed in modern PIC simulations, triggered a few alternative suggestions. First, it was suggested that the prompt emission can possibly be generated over a much shorter scale than previously thought [99]. Other works investigate the effect of energetic particles (resulting from the acceleration process) on the evolution of the magnetic fields. It was argued [97,100,101] that strong magnetic fields can last over a substantial range due to other types of instabilities. It was further suggested that the gradual increase in the population of high energy particles that results from the Fermi acceleration process gradually increases the characteristic length scale of the magnetic field [102]. Other suggestions include macroscopic turbulence that is generated by larger scale instabilities that take place as the shock waves propagate through a non homogeneous media. Indeed, inevitable density fluctuations in the ambient medium will trigger several instabilities (e.g., Richtmyer–Meshkov or kink) that can in principle grow over a large scale [103–107]. Another possibility, which is very realistic in a GRB environment, is generation of mgnetic fields by various instabilities (such as kinetic Kelvin–Helmholtz, mushroom or kink instability) that are stimulated if the relativistic jet is propagating into an already magnetized plasma [108,109]. Indeed, helical magnetic fields may be important in jet acceleration and collimation (see the following section), and their existence will stimulate turbulence as the jet propagates through the plasma. This scenario differs than that presented in Figure 3, as it includes both reverse and forward shocks, as well as contact discontinuity [110], all provide possible sites for enhancement of magnetic turbulence.

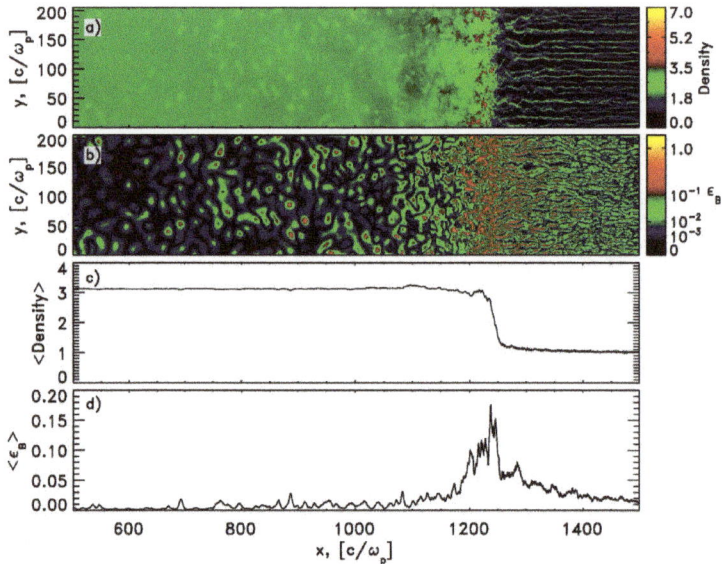

Figure 3. Snapshot of a region from a large 2D relativistic PIC shock simulation, taken from Chang et al. [98]. (**a**) density structure in the simulation plane showing the plasma density enhancements in the foreshock region that steadily grow up to the shock transition region, where the density becomes homogeneous; (**b**) magnetic energy, normalized in terms of upstream energy of the incoming flow. The upstream magnetic filaments, which can be visualized as sheets coming out of the page, that are formed by the Weibel instability reach a peak just before the shock; (**c**) plasma density averaged in the transverse direction as a function of the distance along the flow; (**d**) magnetic energy density averaged in the transverse direction, as a function of distance along the flow. Clearly, strong magnetic fields are generated but quickly decay.

Thus, overall, the origin of the magnetic field as is required to produce the observed (synchrotron) radiation is still an open question. This field remains one of the very active research fields.

3.2. Highly Magnetized Plasma during GRB Prompt Emission?

3.2.1. Motivation

Very early on, it was realized that the extreme luminosity, rapid variability and >MeV photon energies imply that GRB plasma must be moving relativistically during the prompt emission phase; otherwise, the huge optical depth to pair production, $\tau \gtrsim 10^{15}$ would prevent observation of any signal (see, e.g., [20,111,112], for reviews). This idea was confirmed by the late 1990s with the discovery of the afterglow, which proved that the plasma indeed propagates at relativistic speeds.

Thus, two major episodes of energy conversion exist: first, the conversion of the gravitational energy to kinetic energy—namely, the acceleration of plasma that results in the generation of relativistic jet. Second, the huge luminosity suggests that a substantial part of the kinetic energy is dissipated, and used to heat the particles and generate the observed signal.

Originally, it was argued that instabilities within the expanding plasma would generate shock waves, which are internal to the flow ("internal shocks"; see [113–115]). By analogy with the afterglow phase, it was then suggested that a very similar mechanism operates during the prompt phase. Shock waves generated by internal instabilities both generate strong magnetic fields and accelerate particles, which in turn emit the observed prompt radiation [113,116]. In the framework of this model, the internal shocks are therefore the main mechanism of kinetic energy dissipation, and magnetic fields "only" provide the necessary conditions needed for synchrotron radiation.

While this scenario gained popularity by the late 1990s, it was soon realized that it suffers several notable drawbacks. First, the very low efficiency in kinetic energy dissipation, of typically a few % [117–122]. This can be understood, as only the differential kinetic energy between the propagating shells can be dissipated by internal collisions. The only way to overcome this problem is by assuming a very high contrast in the Lorentz factors of the colliding shells [123].

Second, once enough data became available, it became clear that as opposed to the afterglow phase, the simplified version of the synchrotron model does not provide acceptable fits to the vast majority of GRB prompt emission spectra [46,124–127]. Thus, one has to consider alternative emission mechanisms, or, at least, consider ways to modify the synchrotron emission model (see further discussion in Section 4 below).

Third, the details of the initial explosion that triggered the GRB and the mechanism that produce relativistic motion (jet) in the first place remain uncertain. One leading model is the "Collapsar" model [128,129], according to which the core collapse of a massive star triggers the GRB event. In this scenario, the main energy mediators are neutrinos that are copiously produced during the collapse, and transfer the gravitational energy to the outer stellar regions, which are accelerated to relativistic velocities.

An alternative model for the formation of relativistic jets is the mechanism first proposed by Blandford and Znajek [130]. According to this idea, rotational energy and angular momentum are extracted from the created rapidly spinning (Kerr) black hole by strong currents. In this scenario, strong magnetic fields play a key role in the energy extraction process. Thus, the emerging plasma must be Poynting-flux dominated, and the kinetic energy is sub-dominant.

This idea has two great advantages. First, the rotation of a rapidly spinning black hole provides a huge energy reservoir that can in principle be extracted. Second, this mechanism is fairly well understood, and is believed to exist in nature. Furthermore, it does not suffer from the low efficiency problem of the "internal shock" scenario. Indeed, this mechanism gained popularity over the years, and is in wide use for explaining energy extraction in other astronomical objects, such as active galactic nuclei (AGNs; see [131,132]) or X-ray binaries (XRBs; [133]).

I should stress that, as of today, there is no clear evidence that points to which of the two scenarios act in nature to produce the relativistic GRB jets—or possibly a third, as of yet unknown, scenario. However, the possibility that strong magnetic fields may exist motivated studies of the dynamics of highly magnetized plasmas. Under this hypothesis of Poynting-dominated flow, one needs to address two independent questions. The first is the creation of the relativistic jet (namely, the acceleration of the bulk outflow to highly relativistic velocities). The second is the acceleration of (individual) particles to high energies needed to explain the observed radiative signal.

3.2.2. Detailed Models

As opposed to the "internal shock" model, the basic idea in the "Poynting-flux dominated models" is that the strong magnetic fields serve as "energy reservoir". The magnetic energy is converted to kinetic energy and heat (or particle acceleration) by reconnection of the magnetic field lines. In the past few years, many authors considered this possibility. Very crudely speaking, one can divide the models into two categories. The first assumes continuous magnetic energy dissipation (e.g., [72–74,134–141]). These models vary by the different assumptions about the unknown rate of reconnection, outflow parameters, etc. The second type assumes that the magnetic dissipation—hence the acceleration occurs over a finite, short duration [142–149]. The basic idea is that variability in the central engine leads to the ejection of magnetized plasma shells, which expand due to the internal magnetic pressure gradient once they lose causal contact with the source.

A relatively well understood scenario is the "striped wind" model, first proposed in the context of pulsars [150,151]. According to this model, the gravitational collapse that triggers the GRB event leads to a rapidly rotating, highly magnetized neutron star (which can later collapse into a black hole). The rotational axis is misaligned with its dipolar moment, which naturally produces a striped wind above the light-cylinder (see Figure 4). This striped wind consists of cold regions with alternating magnetic fields, separated by hot current sheets. Reconnection of magnetic field lines with opposite polarity is therefore a natural consequence; such reconnection leads to the acceleration of the wind.

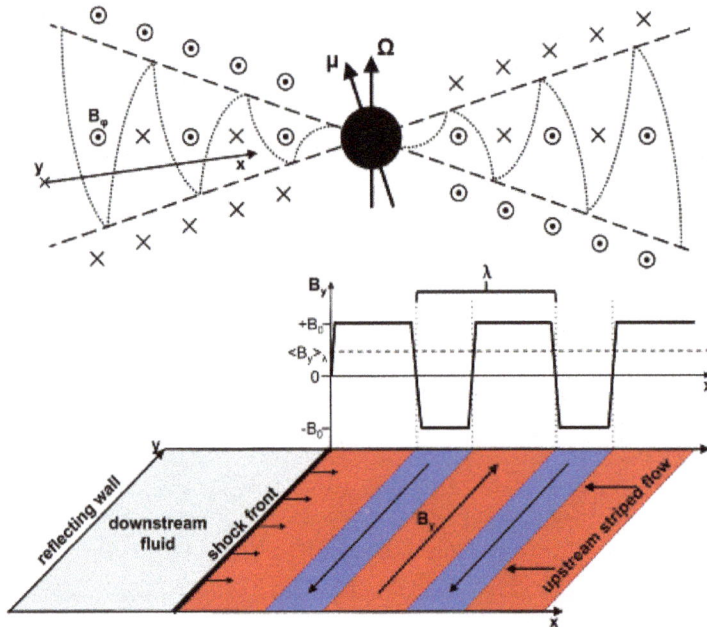

Figure 4. Upper panel: poloidal structure of the striped pulsar wind, according to the solution of Bogovalov [152]. The arrows denote the pulsar rotational axis (along Ω, vertical) and magnetic axis (along μ, inclined). Within the equatorial wedge bounded by the dashed lines, the wind consists of toroidal stripes of alternating polarity, separated by current sheets (dotted lines). Lower panel: 2D PIC simulation setup geometry. The figure is taken from Sironi and Spitkovsky [153].

Evolution of the hydrodynamic quantities in these Poynting-flux dominated outflows within the "striped wind" model was considered by several authors [71–78,154–156]. The scaling laws of the acceleration can be derived under the ideal MHD limit approximation, which is a good approximation due to the high baryon load [71]. Furthermore, in this model, throughout most of the jet evolution the dominated component of the magnetic field is the toroidal component, and so the magnetic field is perpendicular to the outflow direction, $\vec{B} \perp \vec{\beta}$. Under these assumptions, it can be shown that the standard equations of mass, energy and momentum flux conservations combined with the assumption of constant reconnection rate (which is not specified in this model) leads to a well defined scaling law of the Lorentz factor, $\Gamma(r) \propto r^{1/3}$. This is different than the scaling law expected when the acceleration is mediated by photon field, as originally proposed in the classical "fireball" model, $\Gamma(r) \propto r$. Furthermore, these scaling laws lead to testable predictions about the total luminosity that can be achieved in each of the different phases [157]. However, so far, these were not confronted with observations.

The main uncertainty of these models remains the unknown rate in which the reconnection process takes place. This rate is model dependent, and in general depends on the rate of magneto-hydrodynamic (MHD) instabilities that destroy the regular structure of the flow [158–162]. Furthermore, the presence of strong radiative field can affect this rate [156]. It should be noted that several PIC simulations predict a nearly universal reconnection rate, of ∼ 0.1*c* for highly magnetized flows [163–165]. This rate is dictated by the dynamics of the plasmoid instability. However, due to the limitations of existing PIC codes, I think it is fair to claim that this is still an open problem.

3.3. Acceleration of Particles in Highly Magnetized Plasma: Magnetic Reconnection Process

While it is natural to envision a highly magnetized progenitor that results in Poynting-flux dominated outflow in the early stages of GRB evolution, this possibility leads to two basic questions. The first is the details of the reconnection process that dissipates the magnetic energy. The second is the mechanism by which particles are accelerated. Observations of non-thermal emission during the prompt phase necessitates some mechanism that accelerates the particles (for the least, the electrons) to high energies. However, in Poynting-dominated flow, this mechanism needs to be different than the celebrated Fermi process. In this environment of highly magnetized plasmas, both shock formation is limited [166] and particle acceleration by shock waves is suppressed [167].

First, it was shown that the properties of shock waves (if form) and in particular their ability to accelerate particles to high energies are different if these shock waves reside in highly magnetized regions. In this case, the ability of a shock wave to accelerate particles strongly depends on the inclination angle θ between the upstream magnetic field and the shock propagation direction [168–170]. Only if this angle is smaller than a critical angle θ_{crit} can the shock accelerate particle efficiently. At higher angles, charged particles would need to move along the field faster than the speed of light in order to outrun the shock, and therefore cannot be accelerated. I point out, though, that these simulations assumed a simple configuration of the initial magnetic field, and thus the results in GRB jets may differ.

On the other hand, particles can be accelerated to high energies by the reconnection process itself [171–184]; (see [185] for a recent review). The basic idea is that whenever regions of opposite magnetic polarity are present, Maxwell's equations imply that there must be a current sheet in between. In this current layer, magnetic field lines can diffuse across the plasma to reconnect at one or more "x"-lines. When particles cross the current sheet, they are forced back by the reversing magnetic field. This is seen in Figure 5, taken from [186]. The particles can then be accelerated in the direction perpendicular to the plane of reconnection by the generated inductive electric fields [187]. Their energy gain per unit time is therefore $dW/dt = qE \cdot v \sim qEc$ in the relativistic case.

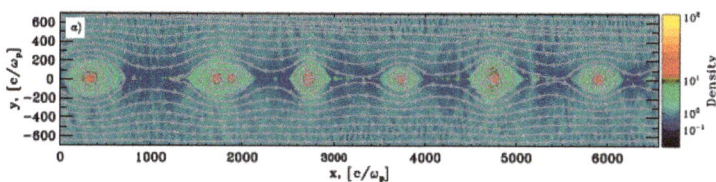

Figure 5. Structure of the particle density in the reconnection layer at $\omega_p t = 3000$, from a 2D simulation of magnetized plasma having magnetization parameter $\sigma = 10$. The figure is taken from [186].

This general idea had been extensively studied over the years by PIC simulations, both 2D [180,181,186–195] and 3D [153,165,169,177,196–199]. These show the generation of hard, non-thermal distribution of energetic particles that are accelerated at (relativistic) reconnection sites. These particles follow a power law distribution, with power law index $p \gtrsim 2$, for strongly magnetized plasma, having magnetization parameter $\sigma \equiv u_B/u_{th} \approx 10$ [186]. Here, u_B, u_{th} are the magnetic field and thermal particle energy densities, respectively. While this index is fairly similar to the one obtained by the Fermi process, it was shown to be sensitive to the exact value of σ [192,194]. Early works suggested that, in a strongly magnetized plasma, the power law index is $p < 2$, implying that most of the energy is carried by the energetic particles. However, recent simulations that were run for longer times and using larger box sizes, showed convergence towards $p \sim 2$ at late times [200].

A heuristic argument for the power law nature of the particle distribution was first suggested by Zenitani and Hoshino [187], and was demonstrated by Sironi and Spitkovsky [186]. More energetic particles have larger Larmor radii, and therefore spend more time near the "x" point of the reconnection layer than particles of lower energies. They suffer less interaction with the reconnected field (in the perpendicular direction), and therefore spend longer time in the accelerated region, where strong electric fields exist. Thus, overall, the gained energy in the acceleration process is proportional to the incoming particle energy, which results in a power law distribution.

Finally, as discussed in the previous section, jet propagation into an already magnetized plasma triggers and enhances several instabilities, such as kinetic Kelvin–Helmholtz or mushroom instabilities. As was recently shown [107,109], these instabilities, which have geometries that are different in nature than the "slab" geometry presented in Figures 4 and 5 above, also serve as acceleration sites of particles [170].

4. Photon Field in GRB Plasmas

Our entire knowledge (or lack thereof) of GRB physics originates from the observed electromagnetic signal. As GRBs are the brightest sources of radiation in the sky, a strong radiation field must exist within the relativistically expanding plasma. This photon field adds to the strong non-thermal particle field and the possible strong magnetic fields that exist.

Similar to the questions outlined above about the sources and role played by magnetic fields (being dominant or sub-dominant), one can divide the basic open questions associated with the photon fields into two categories. The first is understanding the radiative processes that lead to the observed signal. The second is to understand the possible role of the photon field in shaping the dynamics of the GRB outflow.

4.1. Radiative Processes: The Classical Ideas

The most widely discussed model for explaining GRB emission both during the prompt and the afterglow phases is synchrotron emission. This model has several advantages. First, it has been extensively studied since the 1960s [201,202] and its theory is well understood. It is the leading model for interpreting non-thermal emission in many astronomical objects, such as AGNs and XRBs. Second, it is very simple: it requires only two basic ingredients, namely energetic particles and a strong magnetic field. Both are believed to be produced in shock waves or magnetic reconnection process. Third, it is broadband in nature (as opposed, e.g., to the "Planck" spectrum), with a distinctive spectral peak, that could be associated with the observed peak energy. Fourth, it provides a very efficient way of energy transfer, as for the typical parameters, energetic electrons radiate nearly 100% of their energy (during the prompt and parts of the afterglow phases). These properties made synchrotron emission the most widely discussed radiative model in the context of GRB emission (e.g., [5,7,45,203–215], for a very partial list).

Synchrotron emission requires a population of energetic electrons. These electrons, in addition to synchrotron radiation, will inevitably Compton scatter the emitted photons, producing synchrotron-self Compton emission (SSC). This phenomenon is expected to produce high energy photons, that can extend up and beyond the GeV range. Its relative importance depends on the Compton Y parameter, namely the optical depth multiplied by the fractional energy change of each photon. This phenomenon was extensively studied both in the context of the prompt phase [216–219] and the afterglow phase in GRBs [220–225]. Note that the results of the scattering does not only affect the photon field directly, but also indirectly, as the scattering cools the electrons, hence modified the synchrotron emission. Naturally, the importance of this nonlinear effect depends on the Compton Y parameter, and is significantly more pronounced during the prompt phase, where the plasma is much denser and significantly more scattering is expected than during the afterglow phase.

Observations of high energy photons, above the threshold for pair creation, implies that both pair production and pair annihilation can in principle take place. If this happens, then a high energy electromagnetic cascade will occur, namely energetic photons produce $e^{\pm}$ pairs, which lose their energy by synchrotron and SSC, thereby producing another population of energetic photons, etc. These phenomena further modifies the observed spectra in a nonlinear way [216,226,227].

A different suggestion is that the main source of emission is not leptonic, but rather hadronic. This idea lies on the assumptions that the acceleration process, whose details are of yet uncertain, may be more efficient in accelerating protons, rather than electrons to high energies. In this scenario, the main emission mechanism is synchrotron radiation from the accelerated protons [218,221,228–233]. The main drawback of this suggestion is that protons are much less efficient radiators than electrons (as the ratio of proton to electron cross section for synchrotron emission $\sim (m_e/m_p)^2$). Thus, in order to produce the observed luminosity in γ-rays, the energy content of the protons must be very high, with proton luminosity of $\sim 10^{55}$–10^{56} erg s^{-1}. This is at least three orders of magnitude higher than the requirement from leptonic models.

4.2. Photospheric Emission and GRB Dynamics

The idea that photospheric (thermal) emission may play a key role as part of GRB plasma is not new. Already in the very early models of cosmological GRBs, it was realized that the huge energy release, rapid variability that necessitates small emission radii (due to light crossing time argument), and the high $\gtrsim$ MeV photon energy observed, imply the existence of photon-dominated plasma, namely a "fireball" [135,234–236].

Initially, therefore, it was expected that the observed GRB spectra would be thermal. Only with the accumulation of data that showed non-thermal spectra—both during the prompt and afterglow phases—did the synchrotron model gain popularity.

While the synchrotron emission model remains the leading radiative model that can explain the observed signal during the afterglow phase, it was realized by the late 1990s that it fails to explain the low energy part of the prompt emission spectra of many GRBs [46,124–126]. Being well understood, the synchrotron theory provides a robust limit on the maximum low energy spectral slope that can be achieved. As the observed slope in many GRBs was found to be harder than the limiting value, Preece et al. [124] coined the term "synchrotron line of death". Despite two decades of research, this result is still debatable [127,237]. This is due to the different analysis methods chosen. Nonetheless, this observational fact, combined with the fact that the photospheric emission is inherent to GRB fireballs, motivated the study of photospheric emission as a possibly key ingredient in the observed prompt spectra.

Due to the weakness of the observed signal, most of the analysis is done on time integrated signal, as simply not enough photons are observed. However, when analyzing the data of bright GRBs where a time-resolved analysis could be and indeed was done, it was proven that indeed some part (but not all) of the observed prompt spectra can be well fitted with a thermal (Planck) spectrum [238–242]. This was confirmed by several recent observations done with the Fermi satellite [243–249].

From a theoretical perspective, photospheric emission that is combined with other radiative processes was considered as of the early 2000s [250,251]. The key issue is that the thermal photons, similar to the synchrotron photons, can be up-scattered by the energetic electrons. In fact, by definition of the photosphere, they have to be upscattered as the optical depth below the photosphere is >1. This implies both modification of the electron distribution from their initial (accelerated) power law distribution, and modification of the thermal component itself, in a nonlinear way [252,253]. This naturally leads to a broadening of the "Planck" spectrum, which, for a large parameter space region, resembles the observed one [253]. This is demonstrated in Figure 6.

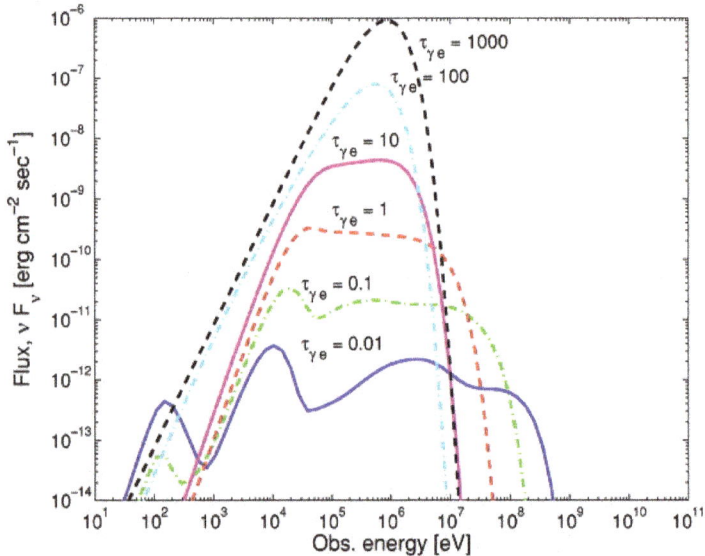

Figure 6. Time averaged broad band spectra expected following kinetic energy dissipation at various optical depths. For low optical depth, the two low energy bumps are due to synchrotron emission and the original thermal component, and the high energy bumps are due to inverse Compton phenomenon. At high optical depth, $\tau \geq 100$, a Wien peak is formed at $\sim 10\,\mathrm{keV}$, and is blue-shifted to the MeV range by the bulk Lorentz factor $\simeq 100$ expected in GRBs. In the intermediate regime, $0.1 < \tau < 100$, a flat energy spectrum above the thermal peak is obtained by multiple Compton scattering. The figure is taken from Pe'er et al. [253].

This idea of modified Planck spectra gained popularity in recent years, as it is capable of capturing the key observational GRB properties in the framework of both photon-dominated and magnetic-dominated flows [254–271].

An underlying assumption here is that a population of energetic particles can exist below the photosphere (or close to it). This is not obvious, as recent works showed that the structure of shock waves, if existing below the photosphere ("sub-photospheric shocks"), does not enable the Fermi acceleration process, at least in its classical form [263,272–274]. Nonetheless, particle heating can still take place below the photosphere via other mechanisms—for example, due to turbulence cascade which passes kinetic fluid energy to photons through scattering [275]. Thus, overall, the question of particle heating below the photosphere and sub-photospheric dissipation is still an open one.

A second, independent way of broadening the "Planck" spectra that enables it to resemble the observed prompt emission spectra of many GRBs is the relativistic "limb darkening" effect, which is geometric in nature. By definition, the photosphere is a region in space in which the optical depth to scattering is >1. In a relativistically expanding plasmas, this surface has a non-trivial shape [276]. Furthermore, as photon scattering is probabilistic in nature, the photospheric region is in a very basic sense "vague"—photons have a finite probability of being scattered anywhere in space in which particles exist [258,276–283]; see Figure 7. The exact shape of this "vague" photosphere depends on the jet geometry, and in particular on the jet velocity profile, namely $\Gamma = \Gamma(r, \theta)$. Under plausible assumptions, this relativistic limb darkening effect can lead to an observed spectra that does not resemble at all a "Planck" spectra, and, in addition, can be very highly polarized—up to 40%, if viewed off the jet axis [279,284,285]. This is demonstrated in Figure 8, taken from Lundman et al. [279].

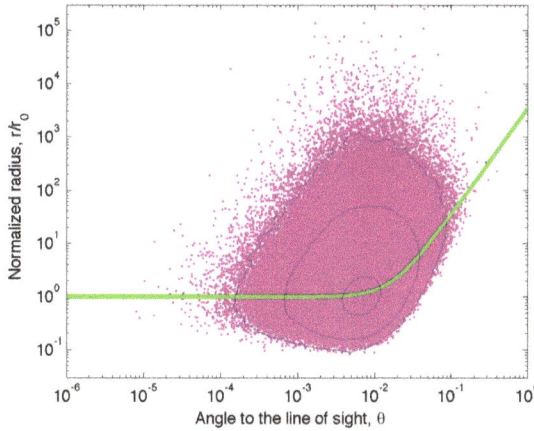

Figure 7. The green line represent the (normalized) photospheric radius r_{ph} as a function of the angle to the line of sight, θ, for spherical explosion. The purple dots represent the last scattering locations of photons emitted in the center of a relativistic expanding "fireball" (using a Monte Carlo simulation). The black lines show contours. Clearly, photons can undergo their last scattering at a range of radii and angles, leading to the concept of "vague photosphere". The observed photospheric signal is therefore smeared both in time and energy. This figure is taken from [276].

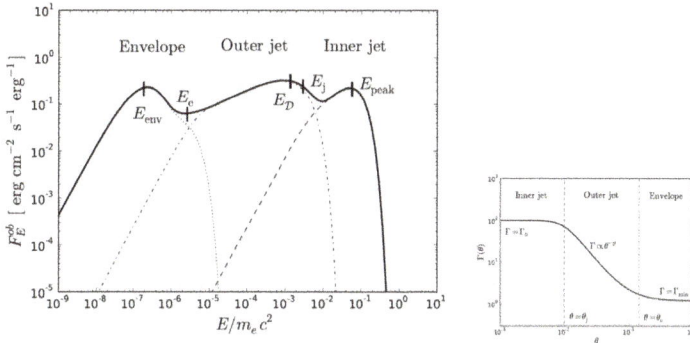

Figure 8. Left: the observed spectrum that emerges from the optically thick regions of an expanding, relativistic jet having a spatial profile, $\Gamma = \Gamma(\theta)$ does not resemble the naively expected "Planck" spectrum. Separate integration of the contributions from the inner jet (where $\Gamma \approx \Gamma_0$), outer jet (where Γ drops with angle) and envelope is shown with dashed, dot dashed and dotted lines, respectively. **Right**: the assumed jet profile. The figure is taken from Lundman et al. [279].

Identification of this thermal emission component has several important implications in understanding the conditions inside the plasma. First, it can be used to directly probe the velocity at the photospheric radius—the innermost region where any electromagnetic signal can reach the observer [286–289]. Second, identification of a photospheric component can be used to constrain the magnetization of the outflow [290–294]. In a highly magnetized outflow, the photospheric component is suppressed, and therefore identifying it can be used to set upper limits on the magnetization. Furthermore, within the context of the "striped wind" model, the existence of strong photon field modifies the rate of reconnection [156]. These identifications led to the suggestion that possibly the GRB outflow is initially strongly magnetized (as is suggested within the Blandford and Znajek [130] mechanism), but that the magnetic field quickly dissipates below the photosphere [293,295].

5. GRB Environments and GRB170817a

One of the key open questions that had been the subject of extensive research over the years is the nature of the GRB progenitors. There are two leading models. The first is the "collapsar" model mentioned above, which involves a core collapse of a massive star, accompanied by accretion into a black hole ([128,129,296–302], and references therein). The second scenario is the merger of two neutron stars (NS–NS), or a black hole and a neutron star (BH–NS). The occurrence rate, as well as the expected energy released, $\sim GM^2/R \sim 10^{53}$ erg (using $M \sim M_\odot$ and $R \gtrsim R_{sch.}$, the Schwarzschild radius of stellar-size black hole), are sufficient for extra-galactic GRBs [303–307].

The association of long GRBs with core collapse supernova, of type Ib/c [6,308–313] serves as a "smoking gun" to confirm that indeed, the long GRB population originates from the "collapsar" scenario. Indeed, in all cases but two (GRB060505 and GRB060614) whenever the GRBs were close enough that evidence for supernovae could be detected, they were indeed observed [314].

While long being suspected, until last year there was only indirect evidence that short GRBs may be associated with the merger scenario. These were mainly based on morphologies of the host galaxies (short GRBs are associated with elliptical galaxies, while long GRBs reside in younger, star-forming galaxies), as well as their position in the sky relative to their host galaxy [315,316]; (For reviews, see, e.g., [18,317,318]). This situation changed with the discovery of the gravitational wave associated with the short GRB170817 [319–321]. This discovery proved that neutron star–neutron star (NS–NS) merger does indeed produce a short GRB, thereby providing the missing "smoking gun". This event, though, was unique in many ways—e.g., the large viewing angle [322], and thus it is not clear whether it is representative of the entire short GRB population. Indeed, a detailed analysis show that the environments of short GRBs do not easily fit this "merger" scenario model [323].

Despite these uncertainties, it is widely believed that these two types of progenitors may end up with very different environments. The merger of binary stars is expected to occur very far from their birthplace, in an environment whose density is roughly constant, and equals the interstellar medium (ISM) density. On the other hand, a massive star (e.g., a Wolf–Rayet type) is likely to emit strong wind prior to its collapse [324], resulting in a "wind" like environment, whose density (for a constant mass ejection rate and constant wind velocity) may vary as $\rho \propto r^{-2}$. I should stress though that this is a very heuristic picture, as the properties of the wind emitted by stars in the last episode before they collapse is highly speculative. Furthermore, even if this is the case, one can predict the existence of a small "bump" in the light curve, resulting from interaction of the GRB blast wave with the wind termination shock [325]. This, though, could be very weak [326], and indeed, no clear evidence for such a jump (whose properties are very uncertain) currently exists.

Nonetheless, as early as a few years after the detection of the first long GRB optical afterglow [3], a split between ISM-like and wind-like environments was observed, with up to 50% of bursts found to be consistent with a homogeneous medium (e.g., [327–329]). In later studies, (e.g., [330,331]), ISM-like environments continued to be found in long GRB afterglows. Further measurements of the spectral and temporal indices for optical [332] and X-ray [70,333–335] afterglows of long GRBs all point to a split in environment types between wind and ISM.

The theoretical analyses that lead to this conclusion are relatively simple, as these are based on measurements of the properties of the late time afterglow. During this stage, the outflow is expected to evolve in a self-similar way. Despite the uncertainty in the detailed of the processes, it is expected that the velocity profile, the particle acceleration and the magnetic field generation all follow well defined scaling laws. These enable the use of the relatively well sampled afterglow data to infer the properties of the environment at late times. From this knowledge, one can hope to constrain the nature of the progenitor, hence the properties of the GRB plasma. The inconsistencies frequently found between the afterglow data of both the long and short GRBs and the simplified environmental models implies that we still have a way to go before understanding the nature of the progenitors, hence the conditions inside the GRB plasmas.

6. Conclusions

GRBs serve as unique laboratories to many plasma physics effects. In fact, GRB observations triggered many basic studies in plasma physics, whose consequences reach far beyond the field of GRBs, and even extend beyond the realm of astrophysics.

GRBs are the only objects known to produce ultra-relativistic shock waves, whose Lorentz factors exceed $\Gamma \gtrsim 100$. As such, they are the only objects that serve as laboratories to study the properties of ultra-relativistic shocks. In Section 2, I highlight a key property that directly follows GRB afterglow observations, that of particle acceleration to high energies in relativistic shock waves. While the existence of cosmic rays implies that such mechanism exists, and although the mechanism by which (non-relativistic) shock waves accelerate particles was discussed in the 1950s by Fermi, the details of the process in two important limits: (1) the relativistic limit, and (2) the "back reaction" of the accelerated particles on the shock structure (i.e., the opposite of the "test particle" limit) are not known. However, from observations in GRB afterglows, it is clear that these two limits are the ones that exist in nature.

Section 3 was devoted to discussing magnetic fields in GRBs. This section was divided into three parts. First, I discussed the current state of knowledge about the generation of strong magnetic fields in shock waves. This again is directly motivated from GRB afterglow observations, which show the existence of strong magnetic fields during this phase. As these fields are several orders of magnitude stronger than the compressed magnetic fields in the ISM, they must be generated by the shock wave itself. It is clear today that the process of magnetic field generation is intimately connected to the process of particle acceleration.

I then discussed energy transfer from magnetic fields by the magnetic reconnection process. This is important in one class of models—the "Poynting flux" dominated models, which assume that early on the main source of energy is magnetic energy. This thus motivates a detailed study of the reconnection process, as a way of transferring this energy to the plasma—both as a way of generating the relativistic jets (accelerating the bulk of the plasma), and as a way of accelerating individual particles to high energies, giving them a non-thermal distribution. This last subject was treated separately in Section 3.3.

In Section 4, I discussed the last ingredient of GRB plasmas, which is the photon field and its interaction with the particle and magnetic fields. The discussion in this section was divided into two parts. I first highlighted the "traditional" radiative processes such as synchrotron emission and Compton scattering that are expected when a population of high energy particles resides in a strongly magnetized region. As far as we know, these are the conditions that exist during the (late time) GRB "afterglow" phase. I then discussed the role of the photosphere in Section 4.2. The photosphere exists in the early stages of GRB evolution, and may be an important ingredient that shapes the prompt emission signal.

However, in addition to shaping the observed prompt emission spectra, the photosphere affects other aspects of the problem as well. As, by definition, the optical depth to scattering below the photosphere is >1, there is a strong coupling between the photon and particle fields. This leads to various effects that modify the structure of sub-photospheric shock waves, affect the dynamics, and can affect the magnetic field–particle interactions, via modification of the magnetic reconnection process. Most importantly, there are several observational consequences that can be tested.

Finally, I briefly discussed in Section 5 our knowledge about the different environments in GRBs. The current picture is very puzzling, and there is no simple way to characterize the environment. The importance of this study lies in the fact that understanding the environment can provide very important clues about the nature of the progenitors, hence on the physical conditions inside the GRB plasmas. Such clues are very difficult to be obtained in any other way.

Thus, overall, GRBs, being the most extreme objects known, provide a unique laboratory to study plasma physics in a unique, relativistic astrophysical environment. While GRB studies triggered and stimulated many plasma physics studies in the laboratory, clearly, unfortunately we cannot mimic the conditions that exist within the GRB environment in the lab. Thus, in the future, by large, we foresee that we will have to continue to rely on GRB observations to provide the necessary input to test the theories.

As I demonstrated here, as of today, there is no consensus on many basic phenomena which are at the forefront of research. However, as the study of GRBs is a very active field—both observationally and theoretically—one can clearly expect a continuous stream of data and ideas that will continue to change this field.

Funding: This research was funded by the European Research Council via the ERC consolidating Grant No. 773062 (acronym O.M.J.).

Acknowledgments: A.P. acknowledges support by the European Research Council via the ERC consolidating Grant No. 773062 (acronym O.M.J.). I wish to thank Antoine Bret for useful discussions.

Conflicts of Interest: The author declares no conflict of interest.

References

1. Abdo, A.A.; Ackermann, M.; Arimoto, M.; Asano, K.; Atwood, W.B.; Axelsson, M.; Baldini, L.; Ballet, J.; Band, D.L.; Barbiellini, G.; et al. Fermi Observations of High-Energy Gamma-Ray Emission from GRB 080916C. *Science* **2009**, *323*, 1688–1693. [CrossRef] [PubMed]
2. Costa, E.; Frontera, F.; Heise, J.; Feroci, M.; in't Zand, J.; Fiore, F.; Cinti, M.N.; Dal Fiume, D.; Nicastro, L.; Orlandini, M.; et al. Discovery of an X-ray afterglow associated with the γ-ray burst of 28 February 1997. *Nature* **1997**, *387*, 783–785. [CrossRef]
3. van Paradijs, J.; Groot, P.J.; Galama, T.; Kouveliotou, C.; Strom, R.G.; Telting, J.; Rutten, R.G.M.; Fishman, G.J.; Meegan, C.A.; Pettini, M.; et al. Transient optical emission from the error box of the γ-ray burst of 28 February 1997. *Nature* **1997**, *386*, 686–689. [CrossRef]
4. Frail, D.A.; Kulkarni, S.R.; Nicastro, L.; Feroci, M.; Taylor, G.B. The radio afterglow from the γ-ray burst of 8 May 1997. *Nature* **1997**, *389*, 261–263. [CrossRef]
5. Wijers, R.A.M.J.; Rees, M.J.; Meszaros, P. Shocked by GRB 970228: The afterglow of a cosmological fireball. *Mon. Notices R. Astron. Soc.* **1997**, *288*, L51–L56. [CrossRef]
6. Galama, T.J.; Vreeswijk, P.M.; van Paradijs, J.; Kouveliotou, C.; Augusteijn, T.; Böhnhardt, H.; Brewer, J.P.; Doublier, V.; Gonzalez, J.F.; Leibundgut, B.; et al. An unusual supernova in the error box of the γ-ray burst of 25 April 1998. *Nature* **1998**, *395*, 670–672. [CrossRef]
7. Wijers, R.A.M.J.; Galama, T.J. Physical Parameters of GRB 970508 and GRB 971214 from Their Afterglow Synchrotron Emission. *Astrophys. J.* **1999**, *523*, 177–186. [CrossRef]
8. Blandford, R.D.; McKee, C.F. Fluid dynamics of relativistic blast waves. *Phys. Fluids* **1976**, *19*, 1130–1138. [CrossRef]
9. Galama, T.J.; Wijers, R.A.M.J.; Bremer, M.; Groot, P.J.; Strom, R.G.; Kouveliotou, C.; van Paradijs, J. The Radio-to-X-ray Spectrum of GRB 970508 on 1997 May 21.0 UT. *Astrophys. J.* **1998**, *500*, L97–L100. [CrossRef]
10. Kumar, P.; Panaitescu, A. Afterglow Emission from Naked Gamma-Ray Bursts. *Astrophys. J.* **2000**, *541*, L51–L54. [CrossRef]
11. Santana, R.; Barniol Duran, R.; Kumar, P. Magnetic Fields in Relativistic Collisionless Shocks. *Astrophys. J.* **2014**, *785*, 29. [CrossRef]
12. Atteia, J.L.; Boër, M. Observing the prompt emission of GRBs. *C. R. Phys.* **2011**, *12*, 255–266. [CrossRef]
13. Gehrels, N.; Mészáros, P. Gamma-Ray Bursts. *Science* **2012**, *337*, 932. [CrossRef] [PubMed]
14. Bucciantini, N. Magnetars and Gamma Ray Bursts. In Proceedings of the IAU Symposium: Death of Massive Stars: Supernovae and Gamma-Ray Bursts, Nikko, Japan, 12–16 March 2012; Volume 279, pp. 289–296. [CrossRef]
15. Gehrels, N.; Razzaque, S. Gamma-ray bursts in the swift-Fermi era. *Front. Phys.* **2013**, *8*, 661–678. [CrossRef]
16. Daigne, F. *GRB Prompt Emission and the Physics of Ultra-Relativistic Outflows*; EAS Publications Series; Castro-Tirado, A.J., Gorosabel, J., Park, I.H., Eds.; Cambridge University Press: Cambridge, UK, 2013; Volume 61, pp. 185–191. [CrossRef]
17. Zhang, B. Gamma-Ray Burst Prompt Emission. *Int. J. Mod. Phys. D* **2014**, *23*, 30002. [CrossRef]
18. Berger, E. Short-Duration Gamma-Ray Bursts. *Annu. Rev. Astron. Astrophys.* **2014**, *52*, 43–105. [CrossRef]
19. Meszaros, P.; Rees, M.J. Gamma-Ray Bursts. *arXiv* **2014**, arXiv:astro-ph.HE/1401.3012.

20. Pe'er, A. Physics of Gamma-Ray Bursts Prompt Emission. *Adv. Astron.* **2015**, *2015*, 907321. [CrossRef]
21. Kumar, P.; Zhang, B. The physics of gamma-ray bursts and relativistic jets. *Phys. Rep.* **2015**, *561*, 1–109. [CrossRef]
22. Granot, J.; Piran, T.; Bromberg, O.; Racusin, J.L.; Daigne, F. Gamma-Ray Bursts as Sources of Strong Magnetic Fields. *Space Sci. Rev.* **2015**, *191*, 471–518. [CrossRef]
23. Zhang, B.; Lü, H.J.; Liang, E.W. GRB Observational Properties. *Space Sci. Rev.* **2016**, *202*, 3–32. [CrossRef]
24. Toma, K.; Yoon, S.C.; Bromm, V. Gamma-Ray Bursts and Population III Stars. *Space Sci. Rev.* **2016**, *202*, 159–180. [CrossRef]
25. Pe'er, A.; Ryde, F. Photospheric emission in gamma-ray bursts. *Int. J. Mod. Phys. D* **2017**, *26*, 1730018–1730296. [CrossRef]
26. Beloborodov, A.M.; Mészáros, P. Photospheric Emission of Gamma-Ray Bursts. *Space Sci. Rev.* **2017**, *207*, 87–110. [CrossRef]
27. Dai, Z.; Daigne, F.; Mészáros, P. The Theory of Gamma-Ray Bursts. *Space Sci. Rev.* **2017**, *212*, 409–427. [CrossRef]
28. van Eerten, H. Gamma-ray burst afterglow blast waves. *Int. J. Mod. Phys. D* **2018**. [CrossRef]
29. Nagataki, S. Theories of central engine for long gamma-ray bursts. *Rep. Prog. Phys.* **2018**, *81*, 026901. [CrossRef]
30. Fermi, E. On the Origin of the Cosmic Radiation. *Phys. Rev.* **1949**, *75*, 1169–1174. [CrossRef]
31. Fermi, E. Galactic Magnetic Fields and the Origin of Cosmic Radiation. *Astrophys. J.* **1954**, *119*, 1. [CrossRef]
32. Axford, W.I.; Leer, E.; Skadron, G. The acceleration of cosmic rays by shock waves. In Proceedings of the International Cosmic Ray Conference, Plovdiv, Bulgaria, 13–26 August 1977; Volume 11, pp. 132–137.
33. Blandford, R.D.; Ostriker, J.P. Particle acceleration by astrophysical shocks. *Astrophys. J.* **1978**, *221*, L29–L32. [CrossRef]
34. Bell, A.R. The acceleration of cosmic rays in shock fronts. I. *Mon. Notices R. Astron. Soc.* **1978**, *182*, 147–156. [CrossRef]
35. Blandford, R.; Eichler, D. Particle acceleration at astrophysical shocks: A theory of cosmic ray origin. *Phys. Rep.* **1987**, *154*, 1–75. [CrossRef]
36. Jones, F.C.; Ellison, D.C. The plasma physics of shock acceleration. *Space Sci. Rev.* **1991**, *58*, 259–346. [CrossRef]
37. Malkov, M.A.; Drury, L.O. Nonlinear theory of diffusive acceleration of particles by shock waves. *Rep. Prog. Phys.* **2001**, *64*, 429–481. [CrossRef]
38. Bell, A.R. Turbulent amplification of magnetic field and diffusive shock acceleration of cosmic rays. *Mon. Notices R. Astron. Soc.* **2004**, *353*, 550–558. [CrossRef]
39. Longair, M.S. *High Energy Astrophysics*; Cambridge University Press: Cambridge, UK, 2011.
40. Milgrom, M.; Usov, V. Possible Association of Ultra–High-Energy Cosmic-Ray Events with Strong Gamma-Ray Bursts. *Astrophys. J.* **1995**, *449*, L37. [CrossRef]
41. Waxman, E. Cosmological Gamma-Ray Bursts and the Highest Energy Cosmic Rays. *Phys. Rev. Lett.* **1995**, *75*, 386–389. [CrossRef]
42. Vietri, M. The Acceleration of Ultra–High-Energy Cosmic Rays in Gamma-Ray Bursts. *Astrophys. J.* **1995**, *453*, 883. [CrossRef]
43. Samuelsson, F.; Bégué, D.; Ryde, F.; Pe'er, A. The Limited Contribution of Low- and High-Luminosity Gamma-Ray Bursts to Ultra-High Energy Cosmic Rays. *arXiv* **2018**, arXiv:astro-ph.HE/1810.06579.
44. Ackermann, M.; Ajello, M.; Asano, K.; Atwood, W.B.; Axelsson, M.; Baldini, L.; Ballet, J.; Barbiellini, G.; Baring, M.G.; Bastieri, D.; et al. Fermi-LAT Observations of the Gamma-Ray Burst GRB 130427A. *Science* **2014**, *343*, 42–47. [CrossRef]
45. Tavani, M. A Shock Emission Model for Gamma-Ray Bursts. II. Spectral Properties. *Astrophys. J.* **1996**, *466*, 768. [CrossRef]
46. Crider, A.; Liang, E.P.; Smith, I.A.; Preece, R.D.; Briggs, M.S.; Pendleton, G.N.; Paciesas, W.S.; Band, D.L.; Matteson, J.L. Evolution of the Low-Energy Photon Spectral in Gamma-Ray Bursts. *Astrophys. J.* **1997**, *479*, L39–L42. [CrossRef]
47. Berger, E.; Kulkarni, S.R.; Frail, D.A. A Standard Kinetic Energy Reservoir in Gamma-Ray Burst Afterglows. *Astrophys. J.* **2003**, *590*, 379–385. [CrossRef]

48. Sironi, L.; Keshet, U.; Lemoine, M. Relativistic Shocks: Particle Acceleration and Magnetization. *Space Sci. Rev.* **2015**, *191*, 519–544. [CrossRef]

49. Kirk, J.G.; Heavens, A.F. Particle acceleration at oblique shock fronts. *Mon. Notices R. Astron. Soc.* **1989**, *239*, 995–1011. [CrossRef]

50. Malkov, M.A. Analytic Solution for Nonlinear Shock Acceleration in the Bohm Limit. *Astrophys. J.* **1997**, *485*, 638–654. [CrossRef]

51. Kirk, J.G.; Guthmann, A.W.; Gallant, Y.A.; Achterberg, A. Particle Acceleration at Ultrarelativistic Shocks: An Eigenfunction Method. *Astrophys. J.* **2000**, *542*, 235–242. [CrossRef]

52. Caprioli, D.; Amato, E.; Blasi, P. Non-linear diffusive shock acceleration with free-escape boundary. *Astropart. Phys.* **2010**, *33*, 307–311. [CrossRef]

53. Keshet, U.; Waxman, E. Energy Spectrum of Particles Accelerated in Relativistic Collisionless Shocks. *Phys. Rev. Lett.* **2005**, *94*, 111102. [CrossRef]

54. Keshet, U. Analytic study of 1D diffusive relativistic shock acceleration. *J. Cosmol. Astropart. Phys.* **2017**, *10*, 025. [CrossRef]

55. Kirk, J.G.; Schneider, P. Particle acceleration at shocks—A Monte Carlo method. *Astrophys. J.* **1987**, *322*, 256–265. [CrossRef]

56. Ellison, D.C.; Reynolds, S.P.; Jones, F.C. First-order Fermi particle acceleration by relativistic shocks. *Astrophys. J.* **1990**, *360*, 702–714. [CrossRef]

57. Achterberg, A.; Gallant, Y.A.; Kirk, J.G.; Guthmann, A.W. Particle acceleration by ultrarelativistic shocks: Theory and simulations. *Mon. Notices R. Astron. Soc.* **2001**, *328*, 393–408. [CrossRef]

58. Lemoine, M.; Pelletier, G. Particle Transport in Tangled Magnetic Fields and Fermi Acceleration at Relativistic Shocks. *Astrophys. J.* **2003**, *589*, L73–L76. [CrossRef]

59. Ellison, D.C.; Double, G.P. Diffusive shock acceleration in unmodified relativistic, oblique shocks. *Astropart. Phys.* **2004**, *22*, 323–338. [CrossRef]

60. Summerlin, E.J.; Baring, M.G. Diffusive Acceleration of Particles at Oblique, Relativistic, Magnetohydrodynamic Shocks. *Astrophys. J.* **2012**, *745*, 63. [CrossRef]

61. Ellison, D.C.; Warren, D.C.; Bykov, A.M. Monte Carlo Simulations of Nonlinear Particle Acceleration in Parallel Trans-relativistic Shocks. *Astrophys. J.* **2013**, *776*, 46. [CrossRef]

62. Bykov, A.M.; Ellison, D.C.; Osipov, S.M. Nonlinear Monte Carlo model of superdiffusive shock acceleration with magnetic field amplification. *Phys. Rev. E* **2017**, *95*, 033207. [CrossRef]

63. Riordan, J.D.; Pe'er, A. Pitch-Angle Diffusion and Bohm-type Approximations in Diffusive Shock Acceleration. *arXiv* **2018**, arXiv:astro-ph.HE/1810.11817.

64. Silva, L.O.; Fonseca, R.A.; Tonge, J.W.; Dawson, J.M.; Mori, W.B.; Medvedev, M.V. Interpenetrating Plasma Shells: Near-equipartition Magnetic Field Generation and Nonthermal Particle Acceleration. *Astrophys. J.* **2003**, *596*, L121–L124. [CrossRef]

65. Frederiksen, J.T.; Hededal, C.B.; Haugbølle, T.; Nordlund, Å. Magnetic Field Generation in Collisionless Shocks: Pattern Growth and Transport. *Astrophys. J.* **2004**, *608*, L13–L16. [CrossRef]

66. Spitkovsky, A. Particle Acceleration in Relativistic Collisionless Shocks: Fermi Process at Last? *Astrophys. J.* **2008**, *682*, L5–L8. [CrossRef]

67. Sironi, L.; Spitkovsky, A. Synthetic Spectra from Particle-In-Cell Simulations of Relativistic Collisionless Shocks. *Astrophys. J.* **2009**, *707*, L92–L96. [CrossRef]

68. Nishikawa, K.I.; Frederiksen, J.T.; Nordlund, Å.; Mizuno, Y.; Hardee, P.E.; Niemiec, J.; Gómez, J.L.; Pe'er, A.; Duţan, I.; Meli, A.; et al. Evolution of Global Relativistic Jets: Collimations and Expansion with kKHI and the Weibel Instability. *Astrophys. J.* **2016**, *820*, 94. [CrossRef]

69. Yost, S.A.; Harrison, F.A.; Sari, R.; Frail, D.A. A Study of the Afterglows of Four Gamma-Ray Bursts: Constraining the Explosion and Fireball Model. *Astrophys. J.* **2003**, *597*, 459–473. [CrossRef]

70. Gompertz, B.P.; Fruchter, A.S.; Pe'er, A. The Environments of the Most Energetic Gamma-Ray Bursts. *Astrophys. J.* **2018**, *866*, 162. [CrossRef]

71. Spruit, H.C.; Daigne, F.; Drenkhahn, G. Large scale magnetic fields and their dissipation in GRB fireballs. *Astron. Astrophys.* **2001**, *369*, 694–705. [CrossRef]

72. Drenkhahn, G. Acceleration of GRB outflows by Poynting flux dissipation. *Astron. Astrophys.* **2002**, *387*, 714–724. [CrossRef]

73. Drenkhahn, G.; Spruit, H.C. Efficient acceleration and radiation in Poynting flux powered GRB outflows. *Astron. Astrophys.* **2002**, *391*, 1141–1153. [CrossRef]

74. Vlahakis, N.; Königl, A. Relativistic Magnetohydrodynamics with Application to Gamma-Ray Burst Outflows. I. Theory and Semianalytic Trans-Alfvénic Solutions. *Astrophys. J.* **2003**, *596*, 1080–1103. [CrossRef]

75. Giannios, D. Spectra of black-hole binaries in the low/hard state: From radio to X-rays. *Astron. Astrophys.* **2005**, *437*, 1007–1015. [CrossRef]

76. Giannios, D. Prompt emission spectra from the photosphere of a GRB. *Astron. Astrophys.* **2006**, *457*, 763–770. [CrossRef]

77. Giannios, D.; Spruit, H.C. Spectra of Poynting-flux powered GRB outflows. *Astron. Astrophys.* **2005**, *430*, 1–7. [CrossRef]

78. Mészáros, P.; Rees, M.J. GeV Emission from Collisional Magnetized Gamma-Ray Bursts. *Astrophys. J.* **2011**, *733*, L40. [CrossRef]

79. Weibel, E.S. Spontaneously Growing Transverse Waves in a Plasma Due to an Anisotropic Velocity Distribution. *Phys. Rev. Lett.* **1959**, *2*, 83–84. [CrossRef]

80. Medvedev, M.V.; Loeb, A. Generation of Magnetic Fields in the Relativistic Shock of Gamma-Ray Burst Sources. *Astrophys. J.* **1999**, *526*, 697–706. [CrossRef]

81. Gruzinov, A.; Waxman, E. Gamma-Ray Burst Afterglow: Polarization and Analytic Light Curves. *Astrophys. J.* **1999**, *511*, 852–861. [CrossRef]

82. Wiersma, J.; Achterberg, A. Magnetic field generation in relativistic shocks. An early end of the exponential Weibel instability in electron-proton plasmas. *Astron. Astrophys.* **2004**, *428*, 365–371. [CrossRef]

83. Lyubarsky, Y.; Eichler, D. Are Gamma-Ray Burst Shocks Mediated by the Weibel Instability? *Astrophys. J.* **2006**, *647*, 1250–1254. [CrossRef]

84. Achterberg, A.; Wiersma, J. The Weibel instability in relativistic plasmas. I. Linear theory. *Astron. Astrophys.* **2007**, *475*, 1–18. [CrossRef]

85. Shaisultanov, R.; Lyubarsky, Y.; Eichler, D. Stream Instabilities in Relativistically Hot Plasma. *Astrophys. J.* **2012**, *744*, 182. [CrossRef]

86. Kumar, R.; Eichler, D.; Gedalin, M. Electron Heating in a Relativistic, Weibel-unstable Plasma. *Astrophys. J.* **2015**, *806*, 165. [CrossRef]

87. Bret, A.; Stockem Novo, A.; Narayan, R.; Ruyer, C.; Dieckmann, M.E.; Silva, L.O. Theory of the formation of a collisionless Weibel shock: Pair vs. electron/proton plasmas. *Laser Part. Beams* **2016**, *34*, 362–367. [CrossRef]

88. Pelletier, G.; Bykov, A.; Ellison, D.; Lemoine, M. Towards Understanding the Physics of Collisionless Relativistic Shocks. Relativistic Collisionless Shocks. *Space Sci. Rev.* **2017**, *207*, 319–360. [CrossRef]

89. Bret, A.; Pe'er, A. On the formation and properties of fluid shocks and collisionless shock waves in astrophysical plasmas. *J. Plasma Phys.* **2018**, *84*, 905840311. [CrossRef]

90. Nishikawa, K.I.; Hardee, P.; Richardson, G.; Preece, R.; Sol, H.; Fishman, G.J. Particle Acceleration and Magnetic Field Generation in Electron-Positron Relativistic Shocks. *Astrophys. J.* **2005**, *622*, 927–937. [CrossRef]

91. Kato, T.N. Saturation mechanism of the Weibel instability in weakly magnetized plasmas. *Phys. Plasmas* **2005**, *12*, 080705. [CrossRef]

92. Ramirez-Ruiz, E.; Nishikawa, K.I.; Hededal, C.B. $e^{+/-}$ Pair Loading and the Origin of the Upstream Magnetic Field in GRB Shocks. *Astrophys. J.* **2007**, *671*, 1877–1885. [CrossRef]

93. Spitkovsky, A. On the Structure of Relativistic Collisionless Shocks in Electron-Ion Plasmas. *Astrophys. J.* **2008**, *673*, L39–L42. [CrossRef]

94. Nishikawa, K.I.; Niemiec, J.; Hardee, P.E.; Medvedev, M.; Sol, H.; Mizuno, Y.; Zhang, B.; Pohl, M.; Oka, M.; Hartmann, D.H. Weibel Instability and Associated Strong Fields in a Fully Three-Dimensional Simulation of a Relativistic Shock. *Astrophys. J.* **2009**, *698*, L10–L13. [CrossRef]

95. Gruzinov, A. Gamma-Ray Burst Phenomenology, Shock Dynamo, and the First Magnetic Fields. *Astrophys. J.* **2001**, *563*, L15–L18. [CrossRef]

96. Spitkovsky, A. Simulations of relativistic collisionless shocks: Shock structure and particle acceleration. In *Astrophysical Sources of High Energy Particles and Radiation*; American Institute of Physics Conference Series; Bulik, T., Rudak, B., Madejski, G., Eds.; American Institute of Physics: College Park, MD, USA, 2005; Volume 801, pp. 345–350. [CrossRef]

97. Milosavljević, M.; Nakar, E. Weibel Filament Decay and Thermalization in Collisionless Shocks and Gamma-Ray Burst Afterglows. *Astrophys. J.* **2006**, *641*, 978–983. [CrossRef]

98. Chang, P.; Spitkovsky, A.; Arons, J. Long-Term Evolution of Magnetic Turbulence in Relativistic Collisionless Shocks: Electron-Positron Plasmas. *Astrophys. J.* **2008**, *674*, 378–387. [CrossRef]

99. Pe'er, A.; Zhang, B. Synchrotron Emission in Small-Scale Magnetic Fields as a Possible Explanation for Prompt Emission Spectra of Gamma-Ray Bursts. *Astrophys. J.* **2006**, *653*, 454–461. [CrossRef]

100. Milosavljević, M.; Nakar, E. The Cosmic-Ray Precursor of Relativistic Collisionless Shocks: A Missing Link in Gamma-Ray Burst Afterglows. *Astrophys. J.* **2006**, *651*, 979–984. [CrossRef]

101. Couch, S.M.; Milosavljević, M.; Nakar, E. Shock Vorticity Generation from Accelerated Ion Streaming in the Precursor of Ultrarelativistic Gamma-Ray Burst External Shocks. *Astrophys. J.* **2008**, *688*, 462–469. [CrossRef]

102. Keshet, U.; Katz, B.; Spitkovsky, A.; Waxman, E. Magnetic Field Evolution in Relativistic Unmagnetized Collisionless Shocks. *Astrophys. J.* **2009**, *693*, L127–L130. [CrossRef]

103. Sironi, L.; Goodman, J. Production of Magnetic Energy by Macroscopic Turbulence in GRB Afterglows. *Astrophys. J.* **2007**, *671*, 1858–1867. [CrossRef]

104. Zhang, W.; MacFadyen, A.; Wang, P. Three-Dimensional Relativistic Magnetohydrodynamic Simulations of the Kelvin-Helmholtz Instability: Magnetic Field Amplification by a Turbulent Dynamo. *Astrophys. J.* **2009**, *692*, L40–L44. [CrossRef]

105. Inoue, T.; Asano, K.; Ioka, K. Three-dimensional Simulations of Magnetohydrodynamic Turbulence Behind Relativistic Shock Waves and Their Implications for Gamma-Ray Bursts. *Astrophys. J.* **2011**, *734*, 77. [CrossRef]

106. Mizuno, Y.; Pohl, M.; Niemiec, J.; Zhang, B.; Nishikawa, K.I.; Hardee, P.E. Magnetic-field Amplification by Turbulence in a Relativistic Shock Propagating Through an Inhomogeneous Medium. *Astrophys. J.* **2011**, *726*, 62. [CrossRef]

107. Mizuno, Y.; Hardee, P.E.; Nishikawa, K.I. Spatial Growth of the Current-driven Instability in Relativistic Jets. *Astrophys. J.* **2014**, *784*, 167. [CrossRef]

108. Mizuno, Y.; Pohl, M.; Niemiec, J.; Zhang, B.; Nishikawa, K.I.; Hardee, P.E. Magnetic field amplification and saturation in turbulence behind a relativistic shock. *Mon. Notices R. Astron. Soc.* **2014**, *439*, 3490–3503. [CrossRef]

109. Nishikawa, K.I.; Mizuno, Y.; Gómez, J.; Duţan, I.; Meli, A.; White, C.; Niemiec, J.; Kobzar, O.; Pohl, M.; Pe'er, A.; et al. Microscopic Processes in Global Relativistic Jets Containing Helical Magnetic Fields: Dependence on Jet Radius. *Galaxies* **2017**, *5*, 58. [CrossRef]

110. Ardaneh, K.; Cai, D.; Nishikawa, K.I. Collisionless Electron-ion Shocks in Relativistic Unmagnetized Jet-ambient Interactions: Non-thermal Electron Injection by Double Layer. *Astrophys. J.* **2016**, *827*, 124. [CrossRef]

111. Piran, T. The physics of gamma-ray bursts. *Rev. Mod. Phys.* **2004**, *76*, 1143–1210. [CrossRef]

112. Mészáros, P. Gamma-ray bursts. *Rep. Prog. Phys.* **2006**, *69*, 2259–2321. [CrossRef]

113. Rees, M.J.; Meszaros, P. Unsteady outflow models for cosmological gamma-ray bursts. *Astrophys. J.* **1994**, *430*, L93–L96. [CrossRef]

114. Sari, R.; Piran, T. Variability in Gamma-Ray Bursts: A Clue. *Astrophys. J.* **1997**, *485*, 270–273. [CrossRef]

115. Ramirez-Ruiz, E.; Fenimore, E.E. Pulse Width Evolution in Gamma-Ray Bursts: Evidence for Internal Shocks. *Astrophys. J.* **2000**, *539*, 712–717. [CrossRef]

116. Kobayashi, S.; Piran, T.; Sari, R. Can Internal Shocks Produce the Variability in Gamma-Ray Bursts? *Astrophys. J.* **1997**, *490*, 92. [CrossRef]

117. Mochkovitch, R.; Maitia, V.; Marques, R. Internal Shocks in a Relativistic Wind as a Source for Gamma-Ray Bursts? *Astrophys. Space Sci.* **1995**, *231*, 441–444. [CrossRef]

118. Panaitescu, A.; Spada, M.; Mészáros, P. Power Density Spectra of Gamma-Ray Bursts in the Internal Shock Model. *Astrophys. J.* **1999**, *522*, L105–L108. [CrossRef]

119. Beloborodov, A.M. On the Efficiency of Internal Shocks in Gamma-Ray Bursts. *Astrophys. J.* **2000**, *539*, L25–L28. [CrossRef]

120. Spada, M.; Panaitescu, A.; Mészáros, P. Analysis of Temporal Features of Gamma-Ray Bursts in the Internal Shock Model. *Astrophys. J.* **2000**, *537*, 824–832. [CrossRef]

121. Guetta, D.; Spada, M.; Waxman, E. Efficiency and Spectrum of Internal Gamma-Ray Burst Shocks. *Astrophys. J.* **2001**, *557*, 399–407. [CrossRef]

122. Pe'er, A.; Long, K.; Casella, P. Dynamical Properties of Internal Shocks Revisited. *Astrophys. J.* **2017**, *846*, 54. [CrossRef]

123. Kobayashi, S.; Sari, R. Ultraefficient Internal Shocks. *Astrophys. J.* **2001**, *551*, 934–939. [CrossRef]

124. Preece, R.D.; Briggs, M.S.; Mallozzi, R.S.; Pendleton, G.N.; Paciesas, W.S.; Band, D.L. The Synchrotron Shock Model Confronts a "Line of Death" in the BATSE Gamma-Ray Burst Data. *Astrophys. J.* **1998**, *506*, L23–L26. [CrossRef]

125. Preece, R.D.; Briggs, M.S.; Giblin, T.W.; Mallozzi, R.S.; Pendleton, G.N.; Paciesas, W.S.; Band, D.L. On the Consistency of Gamma-Ray Burst Spectral Indices with the Synchrotron Shock Model. *Astrophys. J.* **2002**, *581*, 1248–1255. [CrossRef]

126. Ghirlanda, G.; Celotti, A.; Ghisellini, G. Extremely hard GRB spectra prune down the forest of emission models. *Astron. Astrophys.* **2003**, *406*, 879–892. [CrossRef]

127. Axelsson, M.; Borgonovo, L. The width of gamma-ray burst spectra. *Mon. Notices R. Astron. Soc.* **2015**, *447*, 3150–3154. [CrossRef]

128. Woosley, S.E. Gamma-ray bursts from stellar mass accretion disks around black holes. *Astrophys. J.* **1993**, *405*, 273–277. [CrossRef]

129. MacFadyen, A.I.; Woosley, S.E. Collapsars: Gamma-Ray Bursts and Explosions in "Failed Supernovae". *Astrophys. J.* **1999**, *524*, 262–289. [CrossRef]

130. Blandford, R.D.; Znajek, R.L. Electromagnetic extraction of energy from Kerr black holes. *Mon. Notices R. Astron. Soc.* **1977**, *179*, 433–456. [CrossRef]

131. Begelman, M.C.; Blandford, R.D.; Rees, M.J. Theory of extragalactic radio sources. *Rev. Mod. Phys.* **1984**, *56*, 255–351. [CrossRef]

132. Wilson, A.S.; Colbert, E.J.M. The difference between radio-loud and radio-quiet active galaxies. *Astrophys. J.* **1995**, *438*, 62–71. [CrossRef]

133. Narayan, R.; McKinney, J.C.; Farmer, A.J. Self-similar force-free wind from an accretion disc. *Mon. Notices R. Astron. Soc.* **2007**, *375*, 548–566. [CrossRef]

134. Usov, V.V. Millisecond pulsars with extremely strong magnetic fields as a cosmological source of gamma-ray bursts. *Nature* **1992**, *357*, 472–474. [CrossRef]

135. Thompson, C. A Model of Gamma-Ray Bursts. *Mon. Notices R. Astron. Soc.* **1994**, *270*, 480. [CrossRef]

136. Vlahakis, N.; Königl, A. Magnetohydrodynamics of Gamma-Ray Burst Outflows. *Astrophys. J.* **2001**, *563*, L129–L132. [CrossRef]

137. Lyutikov, M.; Blandford, R. Gamma Ray Bursts as Electromagnetic Outflows. *arXiv* **2003**, arXiv:astro-ph/031234.

138. Levinson, A. General Relativistic, Neutrino-assisted Magnetohydrodynamic Winds-Theory and Application to Gamma-Ray Bursts. I. Schwarzschild Geometry. *Astrophys. J.* **2006**, *648*, 510–522. [CrossRef]

139. Giannios, D. Prompt GRB emission from gradual energy dissipation. *Astron. Astrophys.* **2008**, *480*, 305–312. [CrossRef]

140. Komissarov, S.S.; Vlahakis, N.; Königl, A.; Barkov, M.V. Magnetic acceleration of ultrarelativistic jets in gamma-ray burst sources. *Mon. Notices R. Astron. Soc.* **2009**, *394*, 1182–1212. [CrossRef]

141. Beniamini, P.; Giannios, D. Prompt gamma-ray burst emission from gradual magnetic dissipation. *Mon. Notices R. Astron. Soc.* **2017**, *468*, 3202–3211. [CrossRef]

142. Contopoulos, J. A Simple Type of Magnetically Driven Jets: An Astrophysical Plasma Gun. *Astrophys. J.* **1995**, *450*, 616. [CrossRef]

143. Tchekhovskoy, A.; McKinney, J.C.; Narayan, R. Simulations of ultrarelativistic magnetodynamic jets from gamma-ray burst engines. *Mon. Notices R. Astron. Soc.* **2008**, *388*, 551–572. [CrossRef]

144. Tchekhovskoy, A.; Narayan, R.; McKinney, J.C. Magnetohydrodynamic simulations of gamma-ray burst jets: Beyond the progenitor star. *New A.* **2010**, *15*, 749–754. [CrossRef]

145. Komissarov, S.S.; Vlahakis, N.; Königl, A. Rarefaction acceleration of ultrarelativistic magnetized jets in gamma-ray burst sources. *Mon. Notices R. Astron. Soc.* **2010**, *407*, 17–28. [CrossRef]

146. Metzger, B.D.; Giannios, D.; Thompson, T.A.; Bucciantini, N.; Quataert, E. The protomagnetar model for gamma-ray bursts. *Mon. Notices R. Astron. Soc.* **2011**, *413*, 2031–2056. [CrossRef]

147. Granot, J.; Komissarov, S.S.; Spitkovsky, A. Impulsive acceleration of strongly magnetized relativistic flows. *Mon. Notices R. Astron. Soc.* **2011**, *411*, 1323–1353. [CrossRef]

148. Zhang, B.; Yan, H. The Internal-collision-induced Magnetic Reconnection and Turbulence (ICMART) Model of Gamma-ray Bursts. *Astrophys. J.* **2011**, *726*, 90. [CrossRef]

149. Sironi, L.; Petropoulou, M.; Giannios, D. Relativistic jets shine through shocks or magnetic reconnection? *Mon. Notices R. Astron. Soc.* **2015**, *450*, 183–191. [CrossRef]

150. Kennel, C.F.; Coroniti, F.V. Confinement of the Crab pulsar's wind by its supernova remnant. *Astrophys. J.* **1984**, *283*, 694–709. [CrossRef]

151. Coroniti, F.V. Magnetically striped relativistic magnetohydrodynamic winds—The Crab Nebula revisited. *Astrophys. J.* **1990**, *349*, 538–545. [CrossRef]

152. Bogovalov, S.V. On the physics of cold MHD winds from oblique rotators. *Astron. Astrophys.* **1999**, *349*, 1017–1026.

153. Sironi, L.; Spitkovsky, A. Particle-in-cell simulations of shock-driven reconnection in relativistic striped winds. *Comput. Sci. Discov.* **2012**, *5*, 014014. [CrossRef]

154. Lyubarsky, Y.; Kirk, J.G. Reconnection in a Striped Pulsar Wind. *Astrophys. J.* **2001**, *547*, 437–448. [CrossRef]

155. Spruit, H.C.; Drenkhahn, G.D. Magnetically powered prompt radiation and flow acceleration in GRB. In *Gamma-Ray Bursts in the Afterglow Era*; Astronomical Society of the Pacific Conference Series; Feroci, M., Frontera, F., Masetti, N., Piro, L., Eds.; Societá Italiana di Fisica: Bologna, Spain, 2004; Volume 312, p. 357.

156. Bégué, D.; Pe'er, A.; Lyubarsky, Y. Radiative striped wind model for gamma-ray bursts. *Mon. Notices R. Astron. Soc.* **2017**, *467*, 2594–2611. [CrossRef]

157. Pe'er, A. Constraining Magnetization of Gamma-Ray Bursts Outflows Using Prompt Emission Fluence. *Astrophys. J.* **2017**, *850*, 200. [CrossRef]

158. Lyubarskij, Y.E. Energy release in strongly magnetized relativistic winds. *Soviet Astron. Lett.* **1992**, *18*, 356.

159. Begelman, M.C. Instability of Toroidal Magnetic Field in Jets and Plerions. *Astrophys. J.* **1998**, *493*, 291–300. [CrossRef]

160. Giannios, D.; Spruit, H.C. The role of kink instability in Poynting-flux dominated jets. *Astron. Astrophys.* **2006**, *450*, 887–898. [CrossRef]

161. Gill, R.; Granot, J.; Lyubarsky, Y. 2D Relativistic MHD simulations of the Kruskal-Schwarzschild instability in a relativistic striped wind. *Mon. Notices R. Astron. Soc.* **2018**, *474*, 3535–3546. [CrossRef]

162. Sobacchi, E.; Lyubarsky, Y.E. Instability induced by recollimation in highly magnetized outflows. *Mon. Notices R. Astron. Soc.* **2018**, *480*, 4948–4954. [CrossRef]

163. Riquelme, M.A.; Quataert, E.; Sharma, P.; Spitkovsky, A. Local Two-dimensional Particle-in-cell Simulations of the Collisionless Magnetorotational Instability. *Astrophys. J.* **2012**, *755*, 50. [CrossRef]

164. Melzani, M.; Walder, R.; Folini, D.; Winisdoerffer, C.; Favre, J.M. Relativistic magnetic reconnection in collisionless ion-electron plasmas explored with particle-in-cell simulations. *Astron. Astrophys.* **2014**, *570*, A111. [CrossRef]

165. Guo, F.; Liu, Y.H.; Daughton, W.; Li, H. Particle Acceleration and Plasma Dynamics during Magnetic Reconnection in the Magnetically Dominated Regime. *Astrophys. J.* **2015**, *806*, 167. [CrossRef]

166. Bret, A.; Pe'er, A.; Sironi, L.; Sadowski, A.; Narayan, R. Kinetic inhibition of magnetohydrodynamics shocks in the vicinity of a parallel magnetic field. *J. Plasma Phys.* **2017**, *83*, 715830201. [CrossRef]

167. Sironi, L.; Spitkovsky, A.; Arons, J. The Maximum Energy of Accelerated Particles in Relativistic Collisionless Shocks. *arXiv* **2013**, arXiv:astro-ph.HE/1301.5333.

168. Sironi, L.; Spitkovsky, A. Particle Acceleration in Relativistic Magnetized Collisionless Pair Shocks: Dependence of Shock Acceleration on Magnetic Obliquity. *Astrophys. J.* **2009**, *698*, 1523–1549. [CrossRef]

169. Sironi, L.; Spitkovsky, A. Particle Acceleration in Relativistic Magnetized Collisionless Electron-Ion Shocks. *Astrophys. J.* **2011**, *726*, 75. [CrossRef]

170. Matsumoto, Y.; Amano, T.; Kato, T.N.; Hoshino, M. Electron Surfing and Drift Accelerations in a Weibel-Dominated High-Mach-Number Shock. *Phys. Rev. Lett.* **2017**, *119*, 105101. [CrossRef] [PubMed]

171. Romanova, M.M.; Lovelace, R.V.E. Magnetic field, reconnection, and particle acceleration in extragalactic jets. *Astron. Astrophys.* **1992**, *262*, 26–36.

172. Lyutikov, M. Role of reconnection in AGN jets. *New Astron. Rev.* **2003**, *47*, 513–515. [CrossRef]

173. Jaroschek, C.H.; Treumann, R.A.; Lesch, H.; Scholer, M. Fast reconnection in relativistic pair plasmas: Analysis of particle acceleration in self-consistent full particle simulations. *Phys. Plasmas* **2004**, *11*, 1151–1163. [CrossRef]

174. Lyubarsky, Y.E. On the relativistic magnetic reconnection. *Mon. Notices R. Astron. Soc.* **2005**, *358*, 113–119. [CrossRef]

175. Giannios, D. UHECRs from magnetic reconnection in relativistic jets. *Mon. Notices R. Astron. Soc.* **2010**, *408*, L46–L50. [CrossRef]

176. Lazarian, A.; Kowal, G.; Vishniac, E.; de Gouveia Dal Pino, E. Fast magnetic reconnection and energetic particle acceleration. *Planet. Space Sci.* **2011**, *59*, 537–546. [CrossRef]

177. Liu, W.; Li, H.; Yin, L.; Albright, B.J.; Bowers, K.J.; Liang, E.P. Particle energization in 3D magnetic reconnection of relativistic pair plasmas. *Phys. Plasmas* **2011**, *18*, 052105. [CrossRef]

178. Uzdensky, D.A.; McKinney, J.C. Magnetic reconnection with radiative cooling. I. Optically thin regime. *Phys. Plasmas* **2011**, *18*, 042105. [CrossRef]

179. McKinney, J.C.; Uzdensky, D.A. A reconnection switch to trigger gamma-ray burst jet dissipation. *Mon. Notices R. Astron. Soc.* **2012**, *419*, 573–607. [CrossRef]

180. Bessho, N.; Bhattacharjee, A. Fast Magnetic Reconnection and Particle Acceleration in Relativistic Low-density Electron-Positron Plasmas without Guide Field. *Astrophys. J.* **2012**, *750*, 129. [CrossRef]

181. Cerutti, B.; Werner, G.R.; Uzdensky, D.A.; Begelman, M.C. Beaming and Rapid Variability of High-energy Radiation from Relativistic Pair Plasma Reconnection. *Astrophys. J.* **2012**, *754*, L33. [CrossRef]

182. Cerutti, B.; Werner, G.R.; Uzdensky, D.A.; Begelman, M.C. Simulations of Particle Acceleration beyond the Classical Synchrotron Burnoff Limit in Magnetic Reconnection: An Explanation of the Crab Flares. *Astrophys. J.* **2013**, *770*, 147. [CrossRef]

183. Kagan, D.; Milosavljević, M.; Spitkovsky, A. A Flux Rope Network and Particle Acceleration in Three-dimensional Relativistic Magnetic Reconnection. *Astrophys. J.* **2013**, *774*, 41. [CrossRef]

184. Uzdensky, D.A.; Spitkovsky, A. Physical Conditions in the Reconnection Layer in Pulsar Magnetospheres. *Astrophys. J.* **2014**, *780*, 3. [CrossRef]

185. Kagan, D.; Sironi, L.; Cerutti, B.; Giannios, D. Relativistic Magnetic Reconnection in Pair Plasmas and Its Astrophysical Applications. *Space Sci. Rev.* **2015**, *191*, 545–573. [CrossRef]

186. Sironi, L.; Spitkovsky, A. Relativistic Reconnection: An Efficient Source of Non-thermal Particles. *Astrophys. J.* **2014**, *783*, L21. [CrossRef]

187. Zenitani, S.; Hoshino, M. The Generation of Nonthermal Particles in the Relativistic Magnetic Reconnection of Pair Plasmas. *Astrophys. J.* **2001**, *562*, L63–L66. [CrossRef]

188. Bessho, N.; Bhattacharjee, A. Collisionless Reconnection in an Electron-Positron Plasma. *Phys. Rev. Lett.* **2005**, *95*, 245001. [CrossRef] [PubMed]

189. Zenitani, S.; Hoshino, M. Particle Acceleration and Magnetic Dissipation in Relativistic Current Sheet of Pair Plasmas. *Astrophys. J.* **2007**, *670*, 702–726. [CrossRef]

190. Hesse, M.; Zenitani, S. Dissipation in relativistic pair-plasma reconnection. *Phys. Plasmas* **2007**, *14*, 112102. [CrossRef]

191. Lyubarsky, Y.; Liverts, M. Particle Acceleration in the Driven Relativistic Reconnection. *Astrophys. J.* **2008**, *682*, 1436–1442. [CrossRef]

192. Guo, F.; Li, H.; Daughton, W.; Liu, Y.H. Formation of Hard Power Laws in the Energetic Particle Spectra Resulting from Relativistic Magnetic Reconnection. *Phys. Rev. Lett.* **2014**, *113*, 155005. [CrossRef] [PubMed]

193. Nalewajko, K.; Uzdensky, D.A.; Cerutti, B.; Werner, G.R.; Begelman, M.C. On the Distribution of Particle Acceleration Sites in Plasmoid-dominated Relativistic Magnetic Reconnection. *Astrophys. J.* **2015**, *815*, 101. [CrossRef]

194. Werner, G.R.; Uzdensky, D.A.; Cerutti, B.; Nalewajko, K.; Begelman, M.C. The Extent of Power-law Energy Spectra in Collisionless Relativistic Magnetic Reconnection in Pair Plasmas. *Astrophys. J.* **2016**, *816*, L8. [CrossRef]

195. Sironi, L.; Giannios, D.; Petropoulou, M. Plasmoids in relativistic reconnection, from birth to adulthood: First they grow, then they go. *Mon. Notices R. Astron. Soc.* **2016**, *462*, 48–74. [CrossRef]

196. Zenitani, S.; Hoshino, M. The Role of the Guide Field in Relativistic Pair Plasma Reconnection. *Astrophys. J.* **2008**, *677*, 530–544. [CrossRef]

197. Yin, L.; Daughton, W.; Karimabadi, H.; Albright, B.J.; Bowers, K.J.; Margulies, J. Three-Dimensional Dynamics of Collisionless Magnetic Reconnection in Large-Scale Pair Plasmas. *Phys. Rev. Lett.* **2008**, *101*, 125001. [CrossRef]

198. Cerutti, B.; Werner, G.R.; Uzdensky, D.A.; Begelman, M.C. Three-dimensional Relativistic Pair Plasma Reconnection with Radiative Feedback in the Crab Nebula. *Astrophys. J.* **2014**, *782*, 104. [CrossRef]

199. Werner, G.R.; Uzdensky, D.A. Nonthermal Particle Acceleration in 3D Relativistic Magnetic Reconnection in Pair Plasma. *Astrophys. J.* **2017**, *843*, L27. [CrossRef]

200. Petropoulou, M.; Sironi, L. The steady growth of the high-energy spectral cut-off in relativistic magnetic reconnection. *Mon. Notices R. Astron. Soc.* **2018**, *481*, 5687–5701. [CrossRef]

201. Ginzburg, V.L.; Syrovatskii, S.I. Cosmic Magnetobremsstrahlung (synchrotron Radiation). *Annu. Rev. Astron. Astrophys.* **1965**, *3*, 297. [CrossRef]

202. Blumenthal, G.R.; Gould, R.J. Bremsstrahlung, Synchrotron Radiation, and Compton Scattering of High-Energy Electrons Traversing Dilute Gases. *Rev. Mod. Phys.* **1970**, *42*, 237–271. [CrossRef]

203. Rees, M.J.; Meszaros, P. Relativistic fireballs—Energy conversion and time-scales. *Mon. Notices R. Astron. Soc.* **1992**, *258*, 41–43. [CrossRef]

204. Meszaros, P.; Rees, M.J. Relativistic fireballs and their impact on external matter—Models for cosmological gamma-ray bursts. *Astrophys. J.* **1993**, *405*, 278–284. [CrossRef]

205. Meszaros, P.; Laguna, P.; Rees, M.J. Gasdynamics of relativistically expanding gamma-ray burst sources—Kinematics, energetics, magnetic fields, and efficiency. *Astrophys. J.* **1993**, *415*, 181–190. [CrossRef]

206. Mészáros, P.; Rees, M.J.; Papathanassiou, H. Spectral properties of blast-wave models of gamma-ray burst sources. *Astrophys. J.* **1994**, *432*, 181–193. [CrossRef]

207. Paczynski, B.; Xu, G. Neutrino bursts from gamma-ray bursts. *Astrophys. J.* **1994**, *427*, 708–713. [CrossRef]

208. Papathanassiou, H.; Meszaros, P. Spectra of Unsteady Wind Models of Gamma-Ray Bursts. *Astrophys. J.* **1996**, *471*, L91. [CrossRef]

209. Cohen, E.; Katz, J.I.; Piran, T.; Sari, R.; Preece, R.D.; Band, D.L. Possible Evidence for Relativistic Shocks in Gamma-Ray Bursts. *Astrophys. J.* **1997**, *488*, 330. [CrossRef]

210. Sari, R.; Piran, T. Cosmological gamma-ray bursts: Internal versus external shocks. *Mon. Notices R. Astron. Soc.* **1997**, *287*, 110–116. [CrossRef]

211. Sari, R.; Piran, T.; Narayan, R. Spectra and Light Curves of Gamma-Ray Burst Afterglows. *Astrophys. J.* **1998**, *497*, L17. [CrossRef]

212. Pilla, R.P.; Loeb, A. Emission Spectra from Internal Shocks in Gamma-Ray Burst Sources. *Astrophys. J.* **1998**, *494*, L167–L171. [CrossRef]

213. Daigne, F.; Mochkovitch, R. Gamma-ray bursts from internal shocks in a relativistic wind: Temporal and spectral properties. *Mon. Notices R. Astron. Soc.* **1998**, *296*, 275–286. [CrossRef]

214. Granot, J.; Sari, R. The Shape of Spectral Breaks in Gamma-Ray Burst Afterglows. *Astrophys. J.* **2002**, *568*, 820–829. [CrossRef]

215. Gao, H.; Lei, W.H.; Zou, Y.C.; Wu, X.F.; Zhang, B. A complete reference of the analytical synchrotron external shock models of gamma-ray bursts. *New Astron. Rev.* **2013**, *57*, 141–190. [CrossRef]

216. Pe'er, A.; Waxman, E. Prompt Gamma-Ray Burst Spectra: Detailed Calculations and the Effect of Pair Production. *Astrophys. J.* **2004**, *613*, 448–459. [CrossRef]

217. Baring, M.G.; Braby, M.L. A Study of Prompt Emission Mechanisms in Gamma-Ray Bursts. *Astrophys. J.* **2004**, *613*, 460–4767. [CrossRef]

218. Gupta, N.; Zhang, B. Prompt emission of high-energy photons from gamma ray bursts. *Mon. Notices R. Astron. Soc.* **2007**, *380*, 78–92. [CrossRef]

219. Asano, K.; Inoue, S. Prompt GeV-TeV Emission of Gamma-Ray Bursts Due to High-Energy Protons, Muons, and Electron-Positron Pairs. *Astrophys. J.* **2007**, *671*, 645–655. [CrossRef]

220. Sari, R.; Esin, A.A. On the Synchrotron Self-Compton Emission from Relativistic Shocks and Its Implications for Gamma-Ray Burst Afterglows. *Astrophys. J.* **2001**, *548*, 787–799. [CrossRef]

221. Zhang, B.; Mészáros, P. High-Energy Spectral Components in Gamma-Ray Burst Afterglows. *Astrophys. J.* **2001**, *559*, 110–122. [CrossRef]

222. Harrison, F.A.; Yost, S.A.; Sari, R.; Berger, E.; Galama, T.J.; Holtzman, J.; Axelrod, T.; Bloom, J.S.; Chevalier, R.; Costa, E.; et al. Broadband Observations of the Afterglow of GRB 000926: Observing the Effect of Inverse Compton Scattering. *Astrophys. J.* **2001**, *559*, 123–130. [CrossRef]

223. Galli, A.; Piro, L. High energy afterglows and flares from gamma-ray burst by inverse Compton emission. *Astron. Astrophys.* **2007**, *475*, 421–434. [CrossRef]

224. Fan, Y.Z.; Piran, T.; Narayan, R.; Wei, D.M. High-energy afterglow emission from gamma-ray bursts. *Mon. Notices R. Astron. Soc.* **2008**, *384*, 1483–1501. [CrossRef]

225. Fraija, N.; González, M.M.; Lee, W.H. Synchrotron Self-Compton Emission as the Origin of the Gamma-Ray Afterglow Observed in GRB 980923. *Astrophys. J.* **2012**, *751*, 33. [CrossRef]

226. Pe'er, A.; Waxman, E. Time-dependent Numerical Model for the Emission of Radiation from Relativistic Plasma. *Astrophys. J.* **2005**, *628*, 857–866. [CrossRef]

227. Gill, R.; Granot, J. The effect of pair cascades on the high-energy spectral cut-off in gamma-ray bursts. *Mon. Notices R. Astron. Soc.* **2018**, *475*, L1–L5. [CrossRef]

228. Böttcher, M.; Dermer, C.D. High-energy Gamma Rays from Ultra-high-energy Cosmic-Ray Protons in Gamma-Ray Bursts. *Astrophys. J.* **1998**, *499*, L131–L134. [CrossRef]

229. Totani, T. TEV Burst of Gamma-Ray Bursts and Ultra-High-Energy Cosmic Rays. *Astrophys. J.* **1998**, *509*, L81–L84. [CrossRef]

230. Asano, K.; Inoue, S.; Mészáros, P. Prompt High-Energy Emission from Proton-Dominated Gamma-Ray Bursts. *Astrophys. J.* **2009**, *699*, 953–957. [CrossRef]

231. Razzaque, S.; Dermer, C.D.; Finke, J.D. Synchrotron Radiation from Ultra-High Energy Protons and the Fermi Observations of GRB 080916C. *Open Astron. J.* **2010**, *3*, 150–155. [CrossRef]

232. Asano, K.; Mészáros, P. Delayed Onset of High-energy Emissions in Leptonic and Hadronic Models of Gamma-Ray Bursts. *Astrophys. J.* **2012**, *757*, 115. [CrossRef]

233. Crumley, P.; Kumar, P. Hadronic models for Large Area Telescope prompt emission observed in Fermi gamma-ray bursts. *Mon. Notices R. Astron. Soc.* **2013**, *429*, 3238–3251. [CrossRef]

234. Paczynski, B. Gamma-ray bursters at cosmological distances. *Astrophys. J.* **1986**, *308*, L43–L46. [CrossRef]

235. Goodman, J. Are gamma-ray bursts optically thick? *Astrophys. J.* **1986**, *308*, L47–L50. [CrossRef]

236. Shemi, A.; Piran, T. The appearance of cosmic fireballs. *Astrophys. J.* **1990**, *365*, L55–L58. [CrossRef]

237. Burgess, J.M.; Bégué, D.; Bacelj, A.; Giannios, D.; Berlato, F.; Greiner, J. Gamma-ray bursts as cool synchrotron sources. *arXiv* **2018**, arXiv:astro-ph.HE/1810.06965.

238. Ryde, F. The Cooling Behavior of Thermal Pulses in Gamma-Ray Bursts. *Astrophys. J.* **2004**, *614*, 827–846. [CrossRef]

239. Ryde, F. Is Thermal Emission in Gamma-Ray Bursts Ubiquitous? *Astrophys. J.* **2005**, *625*, L95–L98. [CrossRef]

240. Ryde, F.; Pe'er, A. Quasi-blackbody Component and Radiative Efficiency of the Prompt Emission of Gamma-ray Bursts. *Astrophys. J.* **2009**, *702*, 1211–1229. [CrossRef]

241. McGlynn, S.; Foley, S.; McBreen, B.; Hanlon, L.; McBreen, S.; Clark, D.J.; Dean, A.J.; Martin-Carrillo, A.; O'Connor, R. High energy emission and polarisation limits for the INTEGRAL burst GRB 061122. *Astron. Astrophys.* **2009**, *499*, 465–472. [CrossRef]

242. Larsson, J.; Ryde, F.; Lundman, C.; McGlynn, S.; Larsson, S.; Ohno, M.; Yamaoka, K. Spectral components in the bright, long GRB 061007: Properties of the photosphere and the nature of the outflow. *Mon. Notices R. Astron. Soc.* **2011**, *414*, 2642–2649. [CrossRef]

243. Ackermann, M.; Asano, K.; Atwood, W.B.; Axelsson, M.; Baldini, L.; Ballet, J.; Barbiellini, G.; Baring, M.; et al. Fermi Observations of GRB 090510: A Short-Hard Gamma-ray Burst with an Additional, Hard Power-law Component from 10 keV TO GeV Energies. *Astrophys. J.* **2010**, *716*, 1178–1190. [CrossRef]

244. Abdo, A.A.; Ackermann, M.; Ajello, M.; Asano, K.; Atwood, W.B.; Axelsson, M.; Baldini, L.; Ballet, J.; Barbiellini, G.; Baring, M.G.; et al. Fermi Observations of GRB 090902B: A Distinct Spectral Component in the Prompt and Delayed Emission. *Astrophys. J.* **2009**, *706*, L138–L144. [CrossRef]

245. Ryde, F.; Axelsson, M.; Zhang, B.B.; McGlynn, S.; Pe'er, A.; Lundman, C.; Larsson, S.; Battelino, M.; Zhang, B.; Bissaldi, E.; et al. Identification and Properties of the Photospheric Emission in GRB090902B. *Astrophys. J.* **2010**, *709*, L172–L177. [CrossRef]

246. Ryde, F.; Pe'er, A.; Nymark, T.; Axelsson, M.; Moretti, E.; Lundman, C.; Battelino, M.; Bissaldi, E.; Chiang, J.; Jackson, M.S.; et al. Observational evidence of dissipative photospheres in gamma-ray bursts. *Mon. Notices R. Astron. Soc.* **2011**, *415*, 3693–3705. [CrossRef]

247. Guiriec, S.; Connaughton, V.; Briggs, M.S.; Burgess, M.; Ryde, F.; Daigne, F.; Mészáros, P.; Goldstein, A.; McEnery, J.; Omodei, N.; et al. Detection of a Thermal Spectral Component in the Prompt Emission of GRB 100724B. *Astrophys. J.* **2011**, *727*, L33. [CrossRef]

248. Iyyani, S.; Ryde, F.; Axelsson, M.; Burgess, J.M.; Guiriec, S.; Larsson, J.; Lundman, C.; Moretti, E.; McGlynn, S.; Nymark, T.; Rosquist, K. Variable jet properties in GRB 110721A: Time resolved observations of the jet photosphere. *Mon. Notices R. Astron. Soc.* **2013**, *433*, 2739–2748. [CrossRef]

249. Guiriec, S.; Daigne, F.; Hascoët, R.; Vianello, G.; Ryde, F.; Mochkovitch, R.; Kouveliotou, C.; Xiong, S.; Bhat, P.N.; Foley, S.; et al. Evidence for a Photospheric Component in the Prompt Emission of the Short GRB 120323A and Its Effects on the GRB Hardness-Luminosity Relation. *Astrophys. J.* **2013**, *770*, 32. [CrossRef]

250. Mészáros, P.; Rees, M.J. Steep Slopes and Preferred Breaks in Gamma-Ray Burst Spectra: The Role of Photospheres and Comptonization. *Astrophys. J.* **2000**, *530*, 292–298. [CrossRef]

251. Mészáros, P.; Ramirez-Ruiz, E.; Rees, M.J.; Zhang, B. X-ray-rich Gamma-Ray Bursts, Photospheres, and Variability. *Astrophys. J.* **2002**, *578*, 812–817. [CrossRef]

252. Pe'er, A.; Mészáros, P.; Rees, M.J. Peak Energy Clustering and Efficiency in Compact Objects. *Astrophys. J.* **2005**, *635*, 476–480. [CrossRef]

253. Pe'er, A.; Mészáros, P.; Rees, M.J. The Observable Effects of a Photospheric Component on GRB and XRF Prompt Emission Spectrum. *Astrophys. J.* **2006**, *642*, 995–1003. [CrossRef]

254. Ioka, K.; Murase, K.; Toma, K.; Nagataki, S.; Nakamura, T. Unstable GRB Photospheres and $e^{+/-}$ Annihilation Lines. *Astrophys. J.* **2007**, *670*, L77–L80. [CrossRef]

255. Thompson, C.; Mészáros, P.; Rees, M.J. Thermalization in Relativistic Outflows and the Correlation between Spectral Hardness and Apparent Luminosity in Gamma-Ray Bursts. *Astrophys. J.* **2007**, *666*, 1012–1023. [CrossRef]

256. Lazzati, D.; Morsony, B.J.; Begelman, M.C. Very High Efficiency Photospheric Emission in Long-Duration γ-Ray Bursts. *Astrophys. J.* **2009**, *700*, L47–L50. [CrossRef]

257. Lazzati, D.; Begelman, M.C. Non-thermal Emission from the Photospheres of Gamma-ray Burst Outflows. I. High-Frequency Tails. *Astrophys. J.* **2010**, *725*, 1137–1145. [CrossRef]

258. Beloborodov, A.M. Collisional mechanism for gamma-ray burst emission. *Mon. Notices R. Astron. Soc.* **2010**, *407*, 1033–1047. [CrossRef]

259. Mizuta, A.; Nagataki, S.; Aoi, J. Thermal Radiation from Gamma-ray Burst Jets. *Astrophys. J.* **2011**, *732*, 26. [CrossRef]

260. Lazzati, D.; Morsony, B.J.; Begelman, M.C. High-efficiency Photospheric Emission of Long-duration Gamma-ray Burst Jets: The Effect of the Viewing Angle. *Astrophys. J.* **2011**, *732*, 34. [CrossRef]

261. Toma, K.; Wu, X.F.; Mészáros, P. Photosphere-internal shock model of gamma-ray bursts: Case studies of Fermi/LAT bursts. *Mon. Notices R. Astron. Soc.* **2011**, *415*, 1663–1680. [CrossRef]

262. Bromberg, O.; Mikolitzky, Z.; Levinson, A. Sub-photospheric Emission from Relativistic Radiation Mediated Shocks in GRBs. *Astrophys. J.* **2011**, *733*, 85. [CrossRef]

263. Levinson, A. Observational Signatures of Sub-photospheric Radiation-mediated Shocks in the Prompt Phase of Gamma-Ray Bursts. *Astrophys. J.* **2012**, *756*, 174. [CrossRef]

264. Veres, P.; Zhang, B.B.; Mészáros, P. The Extremely High Peak Energy of GRB 110721A in the Context of a Dissipative Photosphere Synchrotron Emission Model. *Astrophys. J.* **2012**, *761*, L18. [CrossRef]

265. Vurm, I.; Lyubarsky, Y.; Piran, T. On Thermalization in Gamma-Ray Burst Jets and the Peak Energies of Photospheric Spectra. *Astrophys. J.* **2013**, *764*, 143. [CrossRef]

266. Beloborodov, A.M. Regulation of the Spectral Peak in Gamma-Ray Bursts. *Astrophys. J.* **2013**, *764*, 157. [CrossRef]

267. Hascoët, R.; Daigne, F.; Mochkovitch, R. Prompt thermal emission in gamma-ray bursts. *Astron. Astrophys.* **2013**, *551*, A124. [CrossRef]

268. Lazzati, D.; Morsony, B.J.; Margutti, R.; Begelman, M.C. Photospheric Emission as the Dominant Radiation Mechanism in Long-duration Gamma-ray Bursts. *Astrophys. J.* **2013**, *765*, 103. [CrossRef]

269. Asano, K.; Mészáros, P. Photon and neutrino spectra of time-dependent photospheric models of gamma-ray bursts. *J. Cosmol. Astropart. Phys.* **2013**, *9*, 8. [CrossRef]

270. Deng, W.; Zhang, B. Low Energy Spectral Index and E_p Evolution of Quasi-thermal Photosphere Emission of Gamma-Ray Bursts. *Astrophys. J.* **2014**, *785*, 112. [CrossRef]

271. Cuesta-Martínez, C.; Aloy, M.A.; Mimica, P.; Thöne, C.; de Ugarte Postigo, A. Numerical models of blackbody-dominated gamma-ray bursts—II. Emission properties. *Mon. Notices R. Astron. Soc.* **2015**, *446*, 1737–1749. [CrossRef]

272. Beloborodov, A.M. Sub-photospheric Shocks in Relativistic Explosions. *Astrophys. J.* **2017**, *838*, 125. [CrossRef]

273. Lundman, C.; Beloborodov, A.M.; Vurm, I. Radiation-mediated Shocks in Gamma-Ray Bursts: Pair Creation. *Astrophys. J.* **2018**, *858*, 7. [CrossRef]

274. Lundman, C.; Beloborodov, A. Radiation mediated shocks in gamma-ray bursts: Subshock photon production. *arXiv* **2018**, arXiv:astro-ph.HE/1804.03053.

275. Zrake, J.; Beloborodov, A.M.; Lundman, C. Sub-photospheric turbulence as a heating mechanism in gamma-ray bursts. *arXiv* **2018**, arXiv:astro-ph.HE/1810.02228.

276. Pe'er, A. Temporal Evolution of Thermal Emission from Relativistically Expanding Plasma. *Astrophys. J.* **2008**, *682*, 463–473. [CrossRef]

277. Beloborodov, A.M. Radiative Transfer in Ultrarelativistic Outflows. *Astrophys. J.* **2011**, *737*, 68. [CrossRef]

278. Pe'er, A.; Ryde, F. A Theory of Multicolor Blackbody Emission from Relativistically Expanding Plasmas. *Astrophys. J.* **2011**, *732*, 49. [CrossRef]

279. Lundman, C.; Pe'er, A.; Ryde, F. A theory of photospheric emission from relativistic, collimated outflows. *Mon. Notices R. Astron. Soc.* **2013**, *428*, 2430–2442. [CrossRef]

280. Ruffini, R.; Siutsou, I.A.; Vereshchagin, G.V. A Theory of Photospheric Emission from Relativistic Outflows. *Astrophys. J.* **2013**, *772*, 11. [CrossRef]

281. Aksenov, A.G.; Ruffini, R.; Vereshchagin, G.V. Comptonization of photons near the photosphere of relativistic outflows. *Mon. Notices R. Astron. Soc.* **2013**, *436*, L54–L58. [CrossRef]

282. Ito, H.; Nagataki, S.; Ono, M.; Lee, S.H.; Mao, J.; Yamada, S.; Pe'er, A.; Mizuta, A.; Harikae, S. Photospheric Emission from Stratified Jets. *Astrophys. J.* **2013**, *777*, 62. [CrossRef]

283. Vereshchagin, G.V. Physics of Nondissipative Ultrarelativistic Photospheres. *Int. J. Mod. Phys. D* **2014**, *23*, 30003. [CrossRef]

284. Lundman, C.; Pe'er, A.; Ryde, F. Polarization properties of photospheric emission from relativistic, collimated outflows. *Mon. Notices R. Astron. Soc.* **2014**, *440*, 3292–3308. [CrossRef]

285. Chang, Z.; Lin, H.N.; Jiang, Y. Gamma-Ray Burst Polarization via Compton Scattering Process. *Astrophys. J.* **2014**, *783*, 30. [CrossRef]

286. Pe'er, A.; Ryde, F.; Wijers, R.A.M.J.; Mészáros, P.; Rees, M.J. A New Method of Determining the Initial Size and Lorentz Factor of Gamma-Ray Burst Fireballs Using a Thermal Emission Component. *Astrophys. J.* **2007**, *664*, L1–L4. [CrossRef]

287. Larsson, J.; Racusin, J.L.; Burgess, J.M. Evidence for Jet Launching Close to the Black Hole in GRB 101219b—A Fermi GRB Dominated by Thermal Emission. *Astrophys. J.* **2015**, *800*, L34. [CrossRef]

288. Pe'er, A.; Barlow, H.; O'Mahony, S.; Margutti, R.; Ryde, F.; Larsson, J.; Lazzati, D.; Livio, M. Hydrodynamic Properties of Gamma-Ray Burst Outflows Deduced from the Thermal Component. *Astrophys. J.* **2015**, *813*, 127. [CrossRef]

289. Wang, Y.Z.; Wang, H.; Zhang, S.; Liang, Y.F.; Jin, Z.P.; He, H.N.; Liao, N.H.; Fan, Y.Z.; Wei, D.M. Evaluating the Bulk Lorentz Factors of Outflow Material: Lessons Learned from the Extremely Energetic Outburst GRB 160625B. *Astrophys. J.* **2017**, *836*, 81. [CrossRef]

290. Zhang, B.; Mészáros, P. An Analysis of Gamma-Ray Burst Spectral Break Models. *Astrophys. J.* **2002**, *581*, 1236–1247. [CrossRef]

291. Daigne, F.; Mochkovitch, R. The expected thermal precursors of gamma-ray bursts in the internal shock model. *Mon. Notices R. Astron. Soc.* **2002**, *336*, 1271–1280. [CrossRef]

292. Zhang, B.; Pe'er, A. Evidence of an Initially Magnetically Dominated Outflow in GRB 080916C. *Astrophys. J.* **2009**, *700*, L65–L68. [CrossRef]

293. Beniamini, P.; Piran, T. The emission mechanism in magnetically dominated gamma-ray burst outflows. *Mon. Notices R. Astron. Soc.* **2014**, *445*, 3892–3907. [CrossRef]

294. Bégué, D.; Pe'er, A. Poynting-flux-dominated Jets Challenged by their Photospheric Emission. *Astrophys. J.* **2015**, *802*, 134. [CrossRef]

295. Meng, Y.Z.; Geng, J.J.; Zhang, B.B.; Wei, J.J.; Xiao, D.; Liu, L.D.; Gao, H.; Wu, X.F.; Liang, E.W.; Huang, Y.F.; Dai, Z.G.; Zhang, B. The Origin of the Prompt Emission for Short GRB 170817A: Photosphere Emission or Synchrotron Emission? *Astrophys. J.* **2018**, *860*, 72. [CrossRef]

296. Paczyński, B. Are Gamma-Ray Bursts in Star-Forming Regions? *Astrophys. J.* **1998**, *494*, L45–L48. [CrossRef]

297. Paczyński, B. Gamma-ray bursts as hypernovae. In *Gamma-Ray Bursts, Proceedings of the 4th Hunstville Symposium*; Meegan, C.A., Preece, R.D., Koshut, T.M., Eds.; American Institute of Physics: College Park, MD, USA, 1998; Volume 428, pp. 783–787. [CrossRef]

298. Fryer, C.L.; Woosley, S.E.; Hartmann, D.H. Formation Rates of Black Hole Accretion Disk Gamma-Ray Bursts. *Astrophys. J.* **1999**, *526*, 152–177. [CrossRef]

299. Popham, R.; Woosley, S.E.; Fryer, C. Hyperaccreting Black Holes and Gamma-Ray Bursts. *Astrophys. J.* **1999**, *518*, 356–374. [CrossRef]

300. MacFadyen, A.I.; Woosley, S.E.; Heger, A. Supernovae, Jets, and Collapsars. *Astrophys. J.* **2001**, *550*, 410–425. [CrossRef]

301. Woosley, S.E.; Bloom, J.S. The Supernova Gamma-Ray Burst Connection. *Annu. Rev. Astron. Astrophys.* **2006**, *44*, 507–556. [CrossRef]

302. Sobacchi, E.; Granot, J.; Bromberg, O.; Sormani, M.C. A common central engine for long gamma-ray bursts and Type Ib/c supernovae. *Mon. Notices R. Astron. Soc.* **2017**, *472*, 616–627. [CrossRef]

303. Eichler, D.; Livio, M.; Piran, T.; Schramm, D.N. Nucleosynthesis, neutrino bursts and gamma-rays from coalescing neutron stars. *Nature* **1989**, *340*, 126–128. [CrossRef]

304. Paczynski, B. Super-Eddington winds from neutron stars. *Astrophys. J.* **1990**, *363*, 218–226. [CrossRef]

305. Narayan, R.; Piran, T.; Shemi, A. Neutron star and black hole binaries in the Galaxy. *Astrophys. J.* **1991**, *379*, L17–L20. [CrossRef]

306. Meszaros, P.; Rees, M.J. Tidal heating and mass loss in neutron star binaries—Implications for gamma-ray burst models. *Astrophys. J.* **1992**, *397*, 570–575. [CrossRef]

307. Narayan, R.; Paczynski, B.; Piran, T. Gamma-ray bursts as the death throes of massive binary stars. *Astrophys. J.* **1992**, *395*, L83–L86. [CrossRef]

308. Hjorth, J.; Sollerman, J.; Møller, P.; Fynbo, J.P.U.; Woosley, S.E.; Kouveliotou, C.; Tanvir, N.R.; Greiner, J.; Andersen, M.I.; Castro-Tirado, A.J.; et al. A very energetic supernova associated with the γ-ray burst of 29 March 2003. *Nature* **2003**, *423*, 847–850. [CrossRef]

309. Stanek, K.Z.; Matheson, T.; Garnavich, P.M.; Martini, P.; Berlind, P.; Caldwell, N.; Challis, P.; Brown, W.R.; Schild, R.; Krisciunas, K.; et al. Spectroscopic Discovery of the Supernova 2003dh Associated with GRB 030329. *Astrophys. J.* **2003**, *591*, L17–L20. [CrossRef]

310. Campana, S.; Mangano, V.; Blustin, A.J.; Brown, P.; Burrows, D.N.; Chincarini, G.; Cummings, J.R.; Cusumano, G.; Della Valle, M.; Malesani, D.; et al. The association of GRB 060218 with a supernova and the evolution of the shock wave. *Nature* **2006**, *442*, 1008–1010. [CrossRef] [PubMed]

311. Pian, E.; Mazzali, P.A.; Masetti, N.; Ferrero, P.; Klose, S.; Palazzi, E.; Ramirez-Ruiz, E.; Woosley, S.E.; Kouveliotou, C.; Deng, J.; et al. An optical supernova associated with the X-ray flash XRF 060218. *Nature* **2006**, *442*, 1011–1013. [CrossRef] [PubMed]

312. Cobb, B.E.; Bloom, J.S.; Perley, D.A.; Morgan, A.N.; Cenko, S.B.; Filippenko, A.V. Discovery of SN 2009nz Associated with GRB 091127. *Astrophys. J.* **2010**, *718*, L150–L155. [CrossRef]

313. Starling, R.L.C.; Wiersema, K.; Levan, A.J.; Sakamoto, T.; Bersier, D.; Goldoni, P.; Oates, S.R.; Rowlinson, A.; Campana, S.; Sollerman, J.; et al. Discovery of the nearby long, soft GRB 100316D with an associated supernova. *Mon. Notices R. Astron. Soc.* **2011**, *411*, 2792–2803. [CrossRef]

314. Cano, Z.; Wang, S.Q.; Dai, Z.G.; Wu, X.F. The Observer's Guide to the Gamma-Ray Burst Supernova Connection. *Adv. Astron.* **2017**, *2017*, 8929054. [CrossRef]

315. Gehrels, N.; Sarazin, C.L.; O'Brien, P.T.; Zhang, B.; Barbier, L.; Barthelmy, S.D.; Blustin, A.; Burrows, D.N.; Cannizzo, J.; Cummings, J.R.; et al. A short γ-ray burst apparently associated with an elliptical galaxy at redshift z = 0.225. *Nature* **2005**, *437*, 851–854. [CrossRef]

316. Fruchter, A.S.; Levan, A.J.; Strolger, L.; Vreeswijk, P.M.; Thorsett, S.E.; Bersier, D.; Burud, I.; Castro Cerón, J.M.; Castro-Tirado, A.J.; Conselice, C.; et al. Long γ-ray bursts and core-collapse supernovae have different environments. *Nature* **2006**, *441*, 463–468. [CrossRef]

317. Nakar, E. Short-hard gamma-ray bursts. *Phys. Rep.* **2007**, *442*, 166–236. [CrossRef]

318. Gehrels, N.; Ramirez-Ruiz, E.; Fox, D.B. Gamma-Ray Bursts in the Swift Era. *Annu. Rev. Astron. Astrophys.* **2009**, *47*, 567–617. [CrossRef]

319. Abbott, B.P.; Abbott, R.; Abbott, T.D.; Acernese, F.; Ackley, K.; Adams, C.; Adams, T.; Addesso, P.; Adhikari, R.X.; Adya, V.B.; et al. GW170817: Observation of Gravitational Waves from a Binary Neutron Star Inspiral. *Phys. Rev. Lett.* **2017**, *119*, 161101. [CrossRef] [PubMed]

320. Abbott, B.P.; Abbott, R.; Abbott, T.D.; Acernese, F.; Ackley, K.; Adams, C.; Adams, T.; Addesso, P.; Adhikari, R.X.; Adya, V.B.; et al. Multi-messenger Observations of a Binary Neutron Star Merger. *Astrophys. J.* **2017**, *848*, L12. [CrossRef]

321. Goldstein, A.; Veres, P.; Burns, E.; Briggs, M.S.; Hamburg, R.; Kocevski, D.; Wilson-Hodge, C.A.; Preece, R.D.; Poolakkil, S.; Roberts, O.J.; et al. An Ordinary Short Gamma-Ray Burst with Extraordinary Implications: Fermi-GBM Detection of GRB 170817A. *Astrophys. J.* **2017**, *848*, L14. [CrossRef]

322. Alexander, K.D.; Berger, E.; Fong, W.; Williams, P.K.G.; Guidorzi, C.; Margutti, R.; Metzger, B.D.; Annis, J.; Blanchard, P.K.; Brout, D.; et al. The Electromagnetic Counterpart of the Binary Neutron Star Merger LIGO/Virgo GW170817. VI. Radio Constraints on a Relativistic Jet and Predictions for Late-time Emission from the Kilonova Ejecta. *Astrophys. J.* **2017**, *848*, L21. [CrossRef]

323. Nysewander, M.; Fruchter, A.S.; Pe'er, A. A Comparison of the Afterglows of Short- and Long-duration Gamma-ray Bursts. *Astrophys. J.* **2009**, *701*, 824–836. [CrossRef]

324. Weaver, R.; McCray, R.; Castor, J.; Shapiro, P.; Moore, R. Interstellar bubbles. II—Structure and evolution. *Astrophys. J.* **1977**, *218*, 377–395. [CrossRef]

325. Pe'er, A.; Wijers, R.A.M.J. The Signature of a Wind Reverse Shock in Gamma-Ray Burst Afterglows. *Astrophys. J.* **2006**, *643*, 1036–1046. [CrossRef]

326. van Eerten, H.J.; Meliani, Z.; Wijers, R.A.M.J.; Keppens, R. No visible optical variability from a relativistic blast wave encountering a wind termination shock. *Mon. Notices R. Astron. Soc.* **2009**, *398*, L63–L67. [CrossRef]

327. Chevalier, R.A.; Li, Z.Y. Gamma-Ray Burst Environments and Progenitors. *Astrophys. J.* **1999**, *520*, L29–L32. [CrossRef]

328. Panaitescu, A.; Kumar, P. Jet Energy and Other Parameters for the Afterglows of GRB 980703, GRB 990123, GRB 990510, and GRB 991216 Determined from Modeling of Multifrequency Data. *Astrophys. J.* **2001**, *554*, 667–677. [CrossRef]

329. Panaitescu, A.; Kumar, P. Properties of Relativistic Jets in Gamma-Ray Burst Afterglows. *Astrophys. J.* **2002**, *571*, 779–789. [CrossRef]

330. Starling, R.L.C; van der Horst, A.J.; Rol, E.; Wijers, R.A.M.J.; Kouveliotou, C.; Wiersema, K.; Curran, P.A.; Weltevrede, P. Gamma-Ray Burst Afterglows as Probes of Environment and Blast Wave Physics. II. The Distribution of p and Structure of the Circumburst Medium. *Astrophys. J.* **2008**, *672*, 433–442. [CrossRef]

331. Curran, P.A.; Starling, R.L.C.; van der Horst, A.J.; Wijers, R.A.M.J. Testing the blast wave model with Swift GRBs. *Mon. Notices R. Astron. Soc.* **2009**, *395*, 580–592. [CrossRef]

332. Oates, S.R.; Page, M.J.; De Pasquale, M.; Schady, P.; Breeveld, A.A.; Holland, S.T.; Kuin, N.P.M.; Marshall, F.E. A correlation between the intrinsic brightness and average decay rate of Swift/UVOT gamma-ray burst optical/ultraviolet light curves. *Mon. Notices R. Astron. Soc.* **2012**, *426*, L86–L90. [CrossRef]

333. Oates, S.R.; Racusin, J.L.; De Pasquale, M.; Page, M.J.; Castro-Tirado, A.J.; Gorosabel, J.; Smith, P.J.; Breeveld, A.A.; Kuin, N.P.M. Exploring the canonical behaviour of long gamma-ray bursts using an intrinsic multiwavelength afterglow correlation. *Mon. Notices R. Astron. Soc.* **2015**, *453*, 4121–4135. [CrossRef]

334. Li, L.; Wu, X.F.; Huang, Y.F.; Wang, X.G.; Tang, Q.W.; Liang, Y.F.; Zhang, B.B.; Wang, Y.; Geng, J.J.; Liang, E.W.; et al. A Correlated Study of Optical and X-ray Afterglows of GRBs. *Astrophys. J.* **2015**, *805*, 13. [CrossRef]

335. Racusin, J.L.; Oates, S.R.; de Pasquale, M.; Kocevski, D. A Correlation between the Intrinsic Brightness and Average Decay Rate of Gamma-Ray Burst X-ray Afterglow Light Curves. *Astrophys. J.* **2016**, *826*, 45. [CrossRef]

galaxies

MDPI

Review

The Origin of the Most Energetic Galactic Cosmic Rays: Supernova Explosions into Massive Star Plasma Winds

Peter L. Biermann [1,2,3,4,*,†], Philipp P. Kronberg [5,6,†], Michael L. Allen [7,†], Athina Meli [8,9,†] and Eun-Suk Seo [10,†]

[1] MPI for Radioastronomy, 53121 Bonn, Germany
[2] Department of Physics, Karlsruhe Institut für Technologie, 76344 Karlsruhe, Germany
[3] Department of Physics & Astronomy, University of Alabama, Tuscaloosa, AL 35487, USA
[4] Department of Physics & Astronomy, University of Bonn, 53115 Bonn, Germany
[5] Department of Physics, University of Toronto, Toronto, ON M5S 1A7, Canada; philkronberg@gmail.com
[6] Theoretical Division, Los Alamos National Laboratory, Los Alamos, NM 87545, USA
[7] Department of Physics & Astronomy, Washington State University, Pullman, WA 99164, USA; mlfa@wsu.edu
[8] AGO Department, University of Liège, B-4000 Liège, Belgium; ameli@uliege.be
[9] Department of Physics, University of Gent, 9000 Gent, Belgium
[10] Department of Physics, University of Maryland, College Park, MD 20742, USA; Seo@Umd.edu
* Correspondence: plbiermann@mpifr-bonn.mpg.de; Tel.: +1-205-348-3797
† These authors contributed equally to this work.

Received: 24 January 2019; Accepted: 25 March 2019; Published: 14 April 2019

Abstract: We propose that the high energy Cosmic Ray particles up to the upturn commonly called the *ankle*, from around the spectral turn-down commonly called the *knee*, mostly come from Blue Supergiant star explosions. At the upturn, i.e., the *ankle*, Cosmic Rays probably switch to another source class, most likely extragalactic sources. To show this we recently compiled a set of Radio Supernova data where we compute the magnetic field, shock speed and shock radius. This list included both Blue and Red Supergiant star explosions; both data show the same magnetic field strength for these two classes of stars despite very different wind densities and velocities. Using particle acceleration theory at shocks, those numbers can be transformed into characteristic *ankle* and *knee* energies. Without adjusting any free parameters both of these observed energies are directly indicated by the supernova data. In the next step in the argument, we use the Supernova Remnant data of the starburst galaxy M82. We apply this analysis to Blue Supergiant star explosions: The shock will race to their outer edge with a magnetic field that is observed to follow over several orders of magnitude $B(r) \times r \sim const.$, with in fact the same magnetic field strength for such stellar explosions in our Galaxy, and other galaxies including M82. The speed is observed to be $\sim$0.1 c out to about 10^{16} cm radius in the plasma wind. The Supernova shock can run through the entire magnetic plasma wind region at full speed all the way out to the wind-shell, which is of order parsec scale in M82. We compare and identify the Cosmic Ray spectrum in other galaxies, in the starburst galaxy M82 and in our Galaxy with each other; we suggest how Blue Supergiant star explosions can provide the Cosmic Ray particles across the *knee* and up to the *ankle* energy range. The data from the ISS-CREAM (Cosmic Ray Energetics and Mass Experiment at the International Space Station) mission will test this cosmic ray concept which is reasonably well grounded in two independent radio supernova data sets. The next step in developing our understanding will be to obtain future more accurate Cosmic Ray data near to the *knee*, and to use unstable isotopes of Cosmic Ray nuclei at high energy to probe the "piston" driving the explosion. We plan to incorporate these data with the physics of the budding black hole which is probably forming in each of these stars.

Keywords: cosmic rays; massive star supernovae; cosmic ray *knee* and *ankle*

1. Introduction

Energetic particles far above thermal levels, called Cosmic Rays (CRs), were discovered by Hess (1912) [1] and Kohlhörster (1913) [2]; their energies are now known to range up to about 10^{20} eV, and at that energy scale first seen by Linsley (1963) [3]. It was proposed very early (Baade and Zwicky 1934 [4]) that Supernovae (SNe) could easily provide the energies required for Galactic Cosmic Rays (GCRs). Ginzburg and Syrovatskij (1963) [5] suggested that radio galaxies could provide the highest energy component. The mechanism of acceleration was originally proposed by Fermi (1949, 1954) [6,7] as repeated reflections between interstellar clouds with magnetic field bottle configurations which are approaching each other. This was later generalized to today's concept of energetic particles being reflected back and forth across a shock wave, and separating converging flows (Axford et al., 1977; Krymskii 1977; Bell 1978a,b; Blandford and Ostriker 1978; [8–12]), with early general overviews by Ginzburg and Ptuskin (1976) [13] and Drury (1983) [14]. It is now recognized that such particles permeate interstellar space, as well as much of intergalactic space. Particles with energies above order GeV are found throughout the Solar system. The early arguments on the possible origin of CRs were all made on the grounds of sufficient energy supply, and they still stand. However we now know of many other processes and events around stars and black holes that clearly produce energetic particles. Energetic particles are also produced on the Sun, and in the Solar wind, teaching us basic plasma physical concepts.

In addition to the *supernova* origin of accelerated CRs and outflow (the theme of this paper), is energy outflow and particle acceleration due to central supermassive black holes (SMBHs), and their jets and lobes. They produce CRs, magnetic fields and energy. They have been discussed by, e.g., Ginzburg and Syrovatskij (1963) [5], Lovelace (1976) [15], Biermann and Strittmatter (1987) [16], Kronberg et al. (2001) [17], Kronberg (2002) [18], Kronberg et al., 2004 [19], Colgate (2004) [20], Colgate et al. (2014) [21], Biermann et al. (2016) [22], and in other papers. These phenomena clearly can produce much higher particle energies, and might explain the highest energies detected above the *ankle*. We note that, since this paper confirms that massive star explosions as well as SN Ia explosions are just two sources of magnetic fields in galaxies and, via galactic winds, in the cosmos. Bigger outflows, such as from super-massive black hole activity via, e.g., radio galaxies, are an additional source.

Some recent relevant reviews and books are Aharonian (2004) [23], Stanev (2010) [24], Kotera and Olinto (2011) [25], Letessier-Selvon and Stanev (2011) [26], Bykov et al. (2012) [27], Diehl (2013, 2017) [28,29], Blasi (2013) [30], Gaisser et al. (2016) [31], Kronberg (2016) [32], Amato and Blasi (2018) [33], Biermann et al. (2018), cited as ASR18 [34], Aartsen et al. (2018c) [35], these latter articles all part of a collection of reviews on CR physics, edited by Moskalenko and Seo 2018 [36]. Many further tests can now be done with newer results from the Pierre Auger Observatory, the Telescope Array, IceCube, AMS, HAWC, H.E.S.S., MAGIC, PAMELA, Fermi-LAT, CALET, DAMPE, and VERITAS, further balloon flights, such as TIGER, and other experiments in space, on balloons, and on the ground. ISS-CREAM is the next major space experiment with new data yet to come, following on AMS (Seo et al., 2014, Seo 2018, [37,38]).

In Section 2 we develop the theory of massive star plasma winds with the goal of deriving characteristic *ankle* and *knee* energies. In Section 3 we compare the energetics of the resolved radio super nova remnants (SNRs) in M82 and demonstrate consistency with our predictions. In Section 4 we discuss a wide variety of concerns involving generation and transport, both theoretical and observational, in both a Galactic and extragalactic context, including alternate acceleration mechanisms found in the literature. We stress that our goal in this paper is to demonstrate one path to understand the CR spectrum, without excluding other mechanisms.

2. Massive Star Plasma Winds

Massive stars explode as Supernovae; and they have magnetic plasma winds. The shock pushed out by the Supernova (SN) explosion can accelerate CR particles. When this shock races through a stellar magnetic plasma wind, accelerated CR particle energies can go as high a few 10^{18} eV

(Stanev et al., 1993, referred to as CR-IV, [39]). Massive stars with magnetized plasma winds come in two varieties, Red Supergiant (RSG) stars above about 25 M$_\odot$ Zero Age Main Sequence (ZAMS) mass, and Blue Supergiant (BSG) stars above about 33 M$_\odot$ ZAMS mass. All of these stars are in binary systems, but not all of them are in tight orbits. RSG star magnetic plasma winds have a low velocity and high density, whereas BSG star magnetic plasma winds have a very high velocity and low density. During the time preceding the SN explosion, this plasma wind experiences short episodes during which the star may enhance its mass loss and its magnetic fields (as summarized with many references, e.g., in ASR18 [34]). A fair number of both RSG star explosions as well as BSG explosions have now been observed, at many wavelengths and spatial resolutions. Some data are collected in ASR18 [34]. The observation that the strength of the magnetic field in the SN shock is the same for RSG explosions as for BSG explosions is not consistent with all mechanisms that enhance the magnetic field just in the SN shock (Bell and Lucek 2001) [40]. In such a case the magnetic field should be an order of magnitude stronger in RSG star explosions, since the implied magnetic field energy density would scale with the ram pressure in the SN-driven shock. (For a detailed discussion including the possibility of a small contribution from the Bell-Lucek mechanism, see ASR18, [34]). We note that the review by Helder et al. (2012) [41] on SNRs and CRs was written before most of these detailed data became available. A newer corresponding article on CRs in stellar-wind bubbles by Zirakashvili and Ptuskin (2018) [42] does not use any of these data either. The arguments in Jokipii (1982, 1986, [43,44]) and Zirakashvili and Ptuskin (2018, [42]) all refer quite explicitly to weak particle scattering, whereas Jokipii (1987, [45]) refers to the strong scattering limit, used here as well as in earlier work as cited. The plasma wind terminates in a shell, where it encounters the interstellar medium (ISM) and the molecular cloud (MC) out of which the star formed, and/or the wind-bubble environment of preceding SN explosions in its immediate environment. As the ram pressure of RSG star plasma winds is two orders of magnitude lower than for BSG star plasma winds, the wind shell may have a radial scale about an order of magnitude smaller than BSG star plasma winds. It follows that the life time of activity for RSG star explosions is an order of magnitude shorter than for BSG star explosions. This provides a strong counter selection effect in identifying such sources. We note that BSG star explosions are very rare in our Galaxy, and yet often enough to be able to contribute significantly to the observed CRs. When the SN shock does hit the wind shell, it slows down considerably; for radio supernovae (RSNe) the SN-shock speed is typically 0.1 c, while for the Supernova Remnants in M82 (Kronberg et al., 1981 [46], 1985, hereafter referred to as KBS85 [47]) the currently observed velocities of expansion (e.g., Fenech et al., 2008, 2010 [48,49]) are all far below this value; this observation matches the expectation that when encountering the wind shell the SN plasma shock reverts to environment-dominated expansion (e.g., Cox 1972 [50]). That reversion causes the shocked region to be about 1/4 of the spherical volume encompassed by the shock, rather than 3/4, as in the shocked region in a wind (Biermann and Cassinelli 1993 [51], Biermann and Strom 1993 [52], referred to as CR-II and CR-III; this uses a strong shock, see Landau and Lifshitz 1959 [53], and Drury 1983 [14]). When a SN shock races through a stellar plasma wind, it produces radio emission at resolutions reachable with interferometry (i.e., sub-arcsecond radio imaging). This permits better specification of the magnetic field $B(r)$, the stellar mass loss $\dot{M}_\star$, its shock radius r, and velocity V_{SN}/c. We note that V_{SN} is the velocity of the shock in the medium through which it propagates, here the wind of the star. As these shocks are observed to show about 0.1 c, we can neglect the prior velocity of the wind. From these three quantities, scale r, magnetic field B, and shock speed V_{SN}/c we can derive two critical energies for the maximal energies particles can reach in shock acceleration, E_{ankle} and E_{knee}. We first show the maximal energy E_{ankle} of a particle fitting into the space available. For a plasma simulation see Meli and Biermann 2006, [54], and for an analytical derivation see, e.g., Biermann 1993, referred to as CR-I, [55], or ASR18 [34].

$$E_{ankle} = \frac{1}{8} Z e B_0 r_0,$$ (1)

where e is the elementary charge, Z is the charge of the particle under consideration (cgs units). We note here that a Parker configuration of a magnetic field in a wind implies that any SN-shock racing through the wind encounters a perpendicular magnetic field (Parker 1958 [56]). We emphasize that this interpretation requires and predicts that all participating stars have about the same magnetic field strength B_0 in their SN-shock racing through the wind (see CR-I [55]). B_0 is the magnetic field strength at a nominal radius of r_0, in observations of radio supernovae, usually taken to be 10^{16} cm. This can be understood as the Larmor motion fitting into the space available. We have used here this plasma wind property which is (Parker (1958) [56])

$$B_\varphi(r) \times r = B_0\, r_0 \sim const. \tag{2}$$

This is consistent with the observations discussed here. Here B_φ is the dominant component of the magnetic field (Parker 1958, Weber and Davis 1967 [56,57]). We interpret the associated energy as that turnup in the spectrum of Galactic CRs (GCRs) which is called the *ankle*. This corresponds to energies reached in perpendicular shocks (see Jokipii 1987, [58], CR-I, CR-IV, Meli and Biermann 2006, [39,54,55], and ASR18 [34]). The argument is simply that the Jokipii limit suggests $\kappa \sim r_g\, V_{SN}$ for the scattering coefficient, where r_g is the Larmor radius, and V_{SN} is again the shock speed. Using shock acceleration theory requires the time scale of acceleration $\sim \{\kappa/V_{SN}^2\} \sim \{r/V_{SN}\}$, which derives from the fact, that in a perpendicular shock particles can only escape via strong turbulence, driven by the shock speed. This then cancels the factor containing the shock speed, and gives the spatial or Hillas limit (Hillas 1984 [59]) limit. In parallel shock acceleration the same argument gives then an additional factor of $(V_{SN}/c)^2$, since in the shock region the escape time scale is then r/c. This is because CRs can stream out along the field lines in the highly non-stationary and un-relaxed post-shock region, and the scattering coefficient does not carry the shock speed anymore (for another derivation of this expression see CR-I [55], using the Parker magnetic field configuration, see Parker 1958 [56]). It is obvious that this *ankle* energy is independent of radius r as long as the Parker configuration is maintained. That is one step in the argument which this paper proposes to check. The lower break energy, given by a parallel shock configuration, corresponds then to the *knee*, where the spectrum turns down:

$$E_{knee} = \frac{1}{8}\, Z\, e\, B_0\, r_0 \left(\frac{V_{SN}}{c}\right)^2. \tag{3}$$

This energy is then also constant with radius r, as long as the SN-shock maintains its velocity V_{SN}. For the observed numbers of RSNe (ASR18 [34]) the energies derived match the energies for *ankle* and *knee* directly to within the errors, without using any parameter adjustment. This expression follows for plane-parallel shock configurations (Drury 1983 [14], Völk and Biermann 1988 [60]). This energy is the *ankle* energy multiplied by $(V_{SN}/c)^2$. Some small fraction of the surface of the SN-shock can be expected to have a parallel configuration of the magnetic field in the form of temporary islands due to turbulence. The Parker solution also has a polar cap in its magnetic field configuration, where the radial magnetic field becomes locally dominant $B_r > B_\varphi$ (see, e.g., Biermann et al., 2009 [61]). We note that to obtain these numbers we take the observed data and average the products $\langle\, r \times B(r)\,\rangle$ and $\langle\, r \times B(r) \times (V_{SN}/c)^2\,\rangle$. The numbers show that the ratio of these two averages is $10^{-1.6}$ between them, instead of 10^{-2}, as would be expected by multiplying separate averages, suggesting that these numbers might be weakly correlated.

However, as these observations show the situation typically around $r \simeq 10^{16}$ cm, due to opacity effects, we need to ascertain what happens at the larger scale, when the SN-shock actually hits the wind shell or other environment. We need to know whether the compact radio sources in M82 can be understood as the late stage development of the same kind of supernova explosions as those observed directly in other galaxies. This is possible to find out with the data from M82:

We propose to test this concept with the observations of the SNRs in M82 (KBS85 [47]; Bartel et al., 1987 [62]; Muxlow et al., 1994 [63]; Golla et al., 1996 [64]; Allen and Kronberg 1998 [65]; Allen 1999 [66]; Kronberg et al., 2000 [67]; McDonald et al., 2002 [68]; Muxlow et al., 2005 [69]; Fenech et al., 2008, 2010 [48,49]; Gendre et al. 2013 [70]).

3. The Supernova Remnants in the Starburst Galaxy M82

Here we show the data on all supernova remnants (SNRs) in M82 that have published derived estimated magnetic field strengths, and for which the size was clearly established (Allen and Kronberg 1998, Allen 1999 [65,66] with many further references therein):

The flux density values listed in Table 1 for the non-thermal emission are referenced to a frequency of 1 GHz, and were obtained by a fit across very many frequencies, from 408 MHz all the way to 23.46 GHz, using a screen model absorption and separating thermal from non-thermal emission. The spectral index given is that of the non-thermal straight power-law component. We note that this magnetic field strength determination posits that the combined energy density of CR electrons and heavier CR particles and the magnetic field strength is a minimum (see, e.g., Miley 1980 [71]). This corresponds roughly to equipartition between energetic particles and the magnetic field, and can be understood on the basis of stability arguments (such as in Parker 1966 [72] and Ames 1973 [73]). The magnetic field thus derived, in Allen and Kronberg (1998) [65], has to be scaled with the factor $\{(1 + k)/\Phi\}^{2/7}$, which is difficult to measure; here k is the ratio of the CR proton energy density to that of CR electrons, and Φ is the volume filling factor. At lower particle energy than a few GeV this ratio is expected to be smaller, since protons do not contribute any extra energy below their rest mass energy, but electrons continue to contribute due to their smaller rest mass. From above we adopt the uncertain filling factor $\Phi = 1/4$, since clearly the SN-shock has encountered the outer shell of the wind, is observed to have slowed down, and so is not filling the entire spherical volume inside the radial scale given. We list here the magnetic field strength without any such correction.

Table 1. Supernova remnants (SNRs) in starburst galaxy M82, based on Allen and Kronberg (their Table 5; 1998) [65], in turn based on Allen (1999) [66].

Coordinate Name	Size 2 r in pc	Flux Density in mJy	sp. Index	est. Magnetic Field Strength B in mGauss	log($B r$) in Gauss × cm
40.68 + 550	3.72	17.9	−0.52	1.80	16.0
41.31 + 596	2.17	8.59	−0.54	2.32	15.9
41.96 + 574	0.33	122.8	−0.72	26.4	16.1
42.53 + 619	1.71	30.9	−1.84	11.7	16.5
42.67 + 556	3.02	4.44	−0.61	1.46	15.8
43.19 + 583	1.16	15.3	−0.67	4.79	15.9
43.31 + 591	3.02	30.3	−0.64	2.54	16.1
44.01 + 595	0.78	62.0	−0.51	9.83	16.1
44.52 + 581	3.72	7.2	−0.61	1.40	15.9
45.18 + 612	3.49	24.1	−0.68	2.13	16.1
45.86 + 640	1.09	4.10	−0.53	3.39	15.8
46.52 + 638	3.88	9.71	−0.73	1.53	16.0
46.70 + 670	3.41	5.22	−0.57	1.37	15.9

3.1. The CR Proton/Electron Ratio k

To determine the magnetic field strength based on radio data it is important to have an estimate of the proton/electron factor k. For the CR proton/electron ratio k, observed to be about 100 above a few GeV, we can make the following argument, based on the observations of young RSNe: The first process that happens in an electron-proton shock is the thermalization of these particles (see also Bell 1978b [10] and Drury 1983 [14] for analogous descriptions, developed before the new radio spectral data were known, and before the importance of drifts was recognized, see Jokipii 1987, [58]). Thermalization has to work at all radii, from below 10^{16} cm out to parsec scale, rendering the density lower by a factor of 10^5 at least. The wind is magnetic with the basic configuration of the magnetic field a Parker type spiral (Parker 1958 [56]), so essentially yielding a perpendicular shock configuration. This process has been discussed in Spitzer (1962) [74], but his treatment is non-relativistic and uses isotropic phase space distributions; then simple electron-proton scattering is far too slow at high temperature. When the protons go through the shock they fill, in phase space, a torus with the radius in momentum $m_p V_{SN}$, and with the relative width of the inverse of the Mach-number, so corresponding to the upstream speed of sound. That entails that this anisotropic momentum distribution causes turbulence at all the scales from the maximum on down, over a range of the Mach-number. However, because the Mach-number is close to m_p/m_e the smallest scales in this range correspond to the waves matching the scales of the electrons with the maximal momentum $m_e V_{SN}$. This means that at all these scales the protons and electrons are scattered in momentum phase space. We have argued in ASR18 [34] that the spectrum of irregularities in the direct shock region is $I(k) k \sim 1$ and so the power per log bin of wave-number is equal at all scales. It follows that thermalization occurs. Due to all this turbulence the time scale of thermalization is of order a small multiple of the thermal Larmor radius divided by the Alfvén speed, so highly adequate, by a factor of $2 E_{knee}/(m_p V_{SN}^2)$, and *a fortiori* $2 E_{ankle}/(m_p V_{SN}^2)$. The time scale for electrons might be m_p/m_e times slower due to the anisotropy of the scattering, but that is still quite adequate by many powers of ten. The time scale scales with the radius, and works similarly at all relevant radii, the condition noted above.

Thermalization yields a first characteristic electron energy $\gamma_{e,sh} c^2 m_e$. Beyond this energy, our interpretation of the radio observations then says that the electrons gain energy only by drifts (CR-I [55]); shock acceleration does not work yet, because the electrons do not see the shock. In ASR18 [34] we noted that the electrons only get *Stochastic Shock Drift Acceleration* (StSDA), giving them a spectrum of −3.06, from observations, say about −3 (this line was reasoning was also used for ions by Caprioli 2012 [75]). This drift-dominated spectrum extends to the momentum at which the energetic electrons reach a Larmor radius that equals that of the shock thermal protons, $\gamma_{sh,inj} c m_e$, (if another CR ion is dominant, corresponding to the Larmor radius of that ion, with mass number A_{dom} and charge at injection Z_{dom}). From this second momentum electrons "see" the shock, and have the normal CR accelerated spectrum, say −7/3 for the most common CR component of CRs accelerated in a wind (see papers CR-I, CR-II, CR-III, CR-IV, and ASR18 [34,39,51,52,55], and the earlier references cited there). Protons attain the maximal energy content only at a few GeV (i.e., just above their rest mass energy). This applies to any particle number spectrum slightly steeper than −2. That implies that electrons have a higher energy content at that momentum where they still are just above thermal in the post-shocked region. We assume that at this stage electrons are in equipartition with the protons (or dominant ions) at the start, due to charge neutrality. We posit that in the initial post-shock thermal phase electric neutrality holds.

This means that

$$\gamma_{e,sh} c^2 m_e = \frac{1}{2} A_{dom} m_p c^2 \beta_{sh}^2 \tag{4}$$

and

$$\gamma_{sh,inj} c^2 m_e = \frac{A_{dom}}{Z_{dom}} m_p c^2 \beta_{sh} \tag{5}$$

So between these two energies the energy density of CR electrons runs down by a factor of

$$\frac{2}{\beta_{sh}} \frac{1}{Z_{dom}} \tag{6}$$

in the first step. Here we use the observations of a spectral index of about E^{-3} implying that the energy content above a given energy runs as E^{-1}, giving a factor of about 20, with the observed $\beta_{sh} = 0.1$. We interpret this as due to *Stochastic Shock Drift Acceleration* (StSDA, first used in CR-I [55]).

There is an observational check on this, as the electron energies required to explain the radio observations of RSNe need to be above that first energy: We observe the emission at a frequency of 5 GHz at a magnetic field strength of order Gauss, requiring a Lorentz factor of $\gamma_e \simeq 30$, while $\gamma_{e,sh}$ is about 10, so well below this. However, this argument implies that the emission should have a cutoff on the low frequency side, perhaps visible somewhat early in the emission phase and at a lower frequency, such as 400 MHz. It also follows that at that stage of the emission there are constraints on what ions could possibly dominate to explain the emission, implying that in this numerical example $A_{dom} \lesssim 3$, eliminating basically all common elements other than Hydrogen (We used $A_{dom} = 1$ and $Z_{dom} = 1$ in our numerical examples).

In the second step, using a spectrum of $p^{-7/3}$ the electrons reduce their energy in that range where the proton contribution is not significant. They are below relativistic, so up to the energy of the proton rest mass (or dominant ion again) the factor is

$$\left(\frac{Z_{dom}}{\beta_{sh}} \right)^{+1/3} \tag{7}$$

which is about 2, using the same data again, with protons. So the CR electron energy density is reduced by a combined factor of $k \simeq 40$ when their momentum reaches that of the relativistic protons, where this factor is determined by observations. Already Allen and Kronberg (1998) [65] had mentioned 40 as a possible and indeed plausible value for k.

The argument makes it obvious that other sources, for which β_{sh} is smaller, contribute even less. On the other hand, if the shock speed is larger, i.e. the factor can be smaller and electrons can contribute more, consistent with many other well established arguments (e.g., Bell 1978b [10], Drury 1983 [14]). However, usually a parallel shock configuration is used (where the magnetic field is parallel to the shock normal). Here we emphasize perpendicular shocks and *Stochastic Shock Drift Acceleration* (StSDA), in a standard magnetic field MHD configuration in a SN-shock racing through a magnetic stellar wind as shown by Parker (1958) [56] (see also, e.g., CR-II, McClements et al., 1997 [76], and Dieckmann et al., 2000 [77]). That in turn implies that the combination of various sources contributing to electrons and protons can reduce the electron net contribution even more, down to the observed factor of 1 in 100. This line of reasoning starts with the energetic electrons having exactly the same energy density as the dominant ions in the shock, due to charge conservation at thermal energies in the thermal post-shock phase. To the degree that protons constitute a strong contribution from ISM-SN-CRs, which have a slightly steeper spectrum than the most common wind-SN-CR component as well as slower shocks, this suggests that at moderate energies the CR electrons have a spectrum slightly flatter than protons.

3.2. The Magnetic Field in the SN-Shocked Wind, and Implications

All this suggests that the proton/electron ratio $k = 40$ is appropriate as a fiducial parameter to derive the magnetic field strength from radio observation via the minimum energy condition. This leads to a correction factor of 4.29 over the magnetic field strength values listed in Allen and Kronberg (1998) [65], and repeated above. However the shock must have slowed down significantly from its free expansion phase through the wind. Using the ISM-Sedov volume fraction of the shocked shell of 1/4, we need to recognize that the magnetic field is likely to have been squashed by a new reverse shock pushed back through the material. This gives the magnetic field strength an extra factor of 4, effectively cancelling the factor of 4.29 mentioned above. The magnetic field strength which we wanted to test was the value before any additional enhancement, over the pure wind-shock mode. Therefore we should reduce the magnetic field strength obtained by a factor of 4, coming back to the original numbers by Allen and Kronberg (1998) [65]. This supports the argument by Allen and Kronberg (1998) [65] to give their magnetic field strength value without such an uncertain correction.

We emphasize, that the values of $r \times B(r)$ in this table show a rather narrow distribution (Allen 1999 [66]), strongly suggesting that this product is not only the same for various massive star explosions, but also that this product is maintained in their evolution. The average of the last column is 16.0: Using the calibration from Allen and Kronberg (1998) [65] and their flux density values we can also estimate the corresponding value of $< \log\{B \times r\} >$ from the independent table of SNRs in the starburst galaxy M82 in Fenech et al. (2010) [49], who did not include a separation of thermal and non-thermal emission, nor any absorption. However, to within the errors, we obtain the same number with a larger scatter confirming the analysis done by Allen and Kronberg (1998) [65]. The uncertainty in $< \log\{B \times r\} >$ due to the different distances to M82 used by Gendre et al. (2013) [70] and Allen and Kronberg (1998) [65] is 0.05. Just using the formula given for the magnetic field estimate in Miley (1980) [71] implies that $\{B \times r\} \sim D^{5/7}$, where D is the distance used. Using the specific numbers here, a factor of $10^{0.1}$ higher than in ASR18 [34], gives for the *ankle* energy

$$E_{ankle} = Z \, 10^{17.6 \pm 0.2} \text{ eV}, \tag{8}$$

now using eV for energy as is custom in the CR community. This specific number for the error was derived from the data given in paper ASR18 [34].

Thus we have two independent estimates of the product $< \log\{B \times r\} >$, one in exploded RSG stars as well as BSG stars, at a typical radius of 10^{16} cm in various other galaxies, and one at a typical scale here of order pc, 3×10^{18} cm in the starburst galaxy M82. The fact that these two numbers coincide allows the interpretation that the population of SNRs in the starburst galaxy M82 is just the later stage of the same explosions (see Allen 1999 [66]). We note that this behavior differs significantly both from a standard Sedov expansion into the ISM (see, e.g., CR-III [52]) as well as from an expansion model of a relativistic blob for compact radio source expansion (see, e.g., van der Laan 1966 [78] and Shklovskii 1960 [79]), for which $B \times r \sim r^{+1}$ or $\sim r^{-1}$, respectively. This is for the radial scale differences here a factor of either $10^{+2.5}$ or $10^{-2.5}$. So the Parker solution is not only maintained across many powers of ten in radius (Allen 1999 [66]), but these explosions have the same magnetic field strength in the SN-shock region racing through the stellar wind of stellar explosions in our Galaxy, other galaxies, and in M82. It seems as we show below only BSG star winds can reach such a radius of parsec scale in the environment of M82. This view is by far the most straightforward, and simplest interpretation of the data. One might then ask why we do not observe such explosions in our Galaxy. These are very rare in our Galaxy. It is not all surprising that we have not seen one during the observations with modern tools. As noted size selection effects make the more common RSG star explosions very difficult to detect; BSG star explosions are quite rare compared to the time scale of modern observations. However, there is a possibility that the recent PeV source seen by H.E.S.S. (High Energy Stereoscopic System) is just such an explosion (Abramowski et al. (2016) [80]). We note that BSG explosions can contribute significantly during the CR storage time, which is much larger than the historical observation time scale, even at the highest energy.

Observations of the CR spectrum, mass composition and arrival direction anisotropy will be important from below the *knee*, such as with IceCube, HAWC and ISS-CREAM, between *knee* and *ankle*, and to higher energies above the *ankle* (see, e.g., Aab et al., 2017a, b, 2018 [81–83], Abbasi et al. 2018a, b, c [84–86] and Abeysekara et al. 2018a, c [87,88]): This will allow us to discern the origins of GCRs and the transition to extragalactic CRs with the Pierre Auger Observatory and the Telescope Array.

Two immediate questions remain: These are (i) whether the plasma wind maintains its magnetic profile $B_\varphi \sim r^{-1}$ (Parker 1958, Weber and Davis 1967, [56,57]), and (ii) whether the SN-shock slows down (see Kronberg et al., 2000 [67] for variability). Simple estimates of the ram-pressure in RSG star winds and BSG star winds show that a RSG wind shell may be an order of magnitude smaller than a BSG star wind shell. That difference entails that in BSG star winds the SN-shock has the chance to race all the way to the wind shell before converting into a Sedov expansion (Cox 1972 [50]). In RSG star explosions this seems to be a more remote possibility, depending on the outside environment; a Sedov expansion implies conservation of energy (Sedov 1958 [89]). The sparse data on velocities (e.g., Fenech et al., 2008, 2010 [48,49]) show that all measured velocities of the SNR expansion in M82 are below 0.1 c, usually far below. In the data listed above in Table 1 there is no hint of a correlation between size and spectral index, possibly suggestive of a changing shock speed and Mach number. Just from simple mass function arguments, the ratio of RSG star explosions to BSG star explosions is expected to be of order 1 to 2. However, in RSG star wind shells the density is far higher than in BSG star wind shells. This makes for more efficient cooling. It implies that we have four development phases: (i) First we have the racing phase, when the SN-shock goes through the plasma wind at constant velocity. (ii) For RSG star winds we may have an intermediate wind-Sedov phase, when the shock velocity slows down. (iii) Then we have the phase when the shock hits the outer shell and becomes a ISM-Sedov expansion. (iv) Finally the shock region transforms to a cooling shock. For RSG star explosions these stages are much shorter than for BSG star explosions. Consistent with such an expectation, the typical radius is about 1 *pc*, with only one source at about 0.1 *pc*, which may be a RSG star explosion, or a BSG star explosion in a very high pressure environment. During the ISM-Sedov phase the radio luminosity is constant, deviating as soon as cooling sets in. The observations appear consistent with this (Kronberg et al., 2000 [67]). The only exception with a well defined light-curve is 41.9 + 58 (Bartel et al., 1987 [62], Kronberg et al., 2000 [67]). It is the source suspected to be a recent mis-aligned GRB (Muxlow et al., 2005 [69]; see also below), and/or possibly a binary BH merger (ASR18 [34]).

We have also derived the required piston mass needed to keep driving the shock; the number we obtained from γ-ray line observations was about 0.1 $M_\odot$ (ASR18 [34]). One question is whether this could be the ^{56}Ni mass derived from optical observations (Maeda et al., 2010; Hamuy 2003; Utrobin et al., 2017; Nakar et al., 2016; Lusk and Baron 2017, [90–94]).

The computed mass accumulated in a BSG star wind can be written as

$$\Delta M_{BSG,wind} = 10^{-2.3} M_\odot \frac{\dot{M}_{\star-5}}{V_{W,8.3}} \frac{r}{\mathrm{pc}}, \tag{9}$$

where $\dot{M}_{\star-5}$ is the mass loss of the star prior to the explosion in units of $10^{-5} M_\odot \, \mathrm{yr}^{-1}$, and $V_{W,8.3}$ is its wind velocity in units of 2000 km/s. The mass loss is directly determined from modelling the radio observations of the radio supernovae (the references in Tables 1 and 2 of ASR18 [34]). The wind velocity is typical for such stars. This suggests that a piston of 0.1 $M_\odot$ could drive the SN-shock unimpeded to scales of order 10 parsec as long as the ram pressure of the wind is sufficient to overcome the environment.

Another consistency test of our concept and its numbers is that the kinetic energy should significantly exceed the magnetic field energy. A piston of 0.1 $M_\odot$ at 0.1 c corresponds to 10^{51} erg, while the typical magnetic field strength of about $\simeq 5\,\mathrm{mGauss}$ corresponds to $10^{49.5}$ erg $\ll 10^{51}$ erg, using the volume fraction of 1/4 adopted above.

For the ram pressure we need to consider first the wind ram pressure against the outside medium, and then the SN-shock ram pressure. The wind ram pressure condition for matching the ISM pressure is

$$r < 30\,\mathrm{pc}\,\dot{M}_{\star-5}^{1/2}\,V_{W,8.3}^{1/2}\,P_{ISM,-12}^{-1/2} \tag{10}$$

where $P_{ISM,-12}$ is the outside pressure in units of $10^{-12}\,\mathrm{dyn\,cm^{-2}}$, a typical number for the outside pressure in our Galaxy. The higher pressure in M82 would allow smaller sizes for the wind shell, but of course the local pressure, where the SNRs are seen, may be lower than in the central region. A pressure 10^3 times higher would still allow the size scales observed for the SNRs in M82. The region containing the SNRs has been described as a ring of about 500 pc radius (Weliachew et al. 1984, Kronberg and Wilkinson 1975, and Kronberg et al. 1981, [46,95,96]). If the over-pressure inside that ring were two orders of magnitude higher, for instance as argued in KBS85 [47], then the maximal size would only be about 3 pc, near the upper end of the range of the SNR radii actually seen.

The corresponding SN-shock ram pressure condition gives

$$r < 10^{2.7}\,\mathrm{pc}\,\dot{M}_{\star-5}^{1/2}\,V_{W,8.3}^{-1/2}\,V_{SN,-1}\,P_{ISM,-12}^{-1/2} \tag{11}$$

where $V_{SN,-1}$ is the shock speed in units of 0.1 c. Another check is when the mass accumulated outside within the extra radial range Δr slows down the shock's progression. That is given by

$$\left(\frac{\Delta r}{r}\right) = 10^{-0.7} M_{pist,-1}\,r_{pc}^{-3}\,n^{-1} \tag{12}$$

where $M_{pist,-1}$ is the mass of the piston in units of 0.1 $M_\odot$, r_{pc} is the radius in pc when the SN-shock hits the outer medium, and n is the outside medium density in units of cm^{-3}. That implies that once the SN-shock hits the wind shell and outside medium at full speed, the shock slows almost immediately to the Sedov stage, and only shortly later to the cooling phase. Adding in the mass accumulated in the wind does not modify this conclusion significantly. In a BSG star wind nothing earlier slows down the SN-shock.

The *knee* energy predicted here is

$$E_{knee} = 10^{16.0\pm0.2} \, Z \, \text{eV}. \tag{13}$$

We conclude that, within the errors these numbers of the product of radial scale and magnetic field strength (Allen 1999 [66], here in Table 1 and in Tables 1 and 2 in ASR18 [34]) are the same. This result supports our argument that in these examples the SN-shock pushed out without losing significant strength, kept following the Parker regime (Parker 1958 [56], Allen 1999 [66]), and so maintained these energies. All or most of these examples in M82 are consistent with BSG star explosions. We note again that we identify massive exploding stars in many other galaxies, the starburst galaxy M82, and in our Galaxy as all being essentially the same, and assume that they produce the same CR spectrum—the data shown here and earlier certainly allow such an approach. All of the SNRs seen in M82 are consistent with being in the slow-down phase, which is after the SN-shock has hit the wind shell or other environment. We note that other physical concepts may also yield similar values for the quantity $\{B \times r\}$, for instance for SN Ia explosions, where the transition from free expansion to Sedov expansion readily yields the same value, thereafter with a slow diminishing over time in the expansion.

4. The Origin of High Energy Galactic Cosmic Rays

As discussed above: RSG star explosions produce High Energy (HE) GCRs, i.e., energies near to, but with a kink down below the *knee*, and with a final cutoff below the *ankle*, with near normal abundances but not extending to the highest energies. BSG star explosions produce HE GCRs with all heavy elements enhanced up to the highest energies for GCRs (see, e.g., Todero Peixoto et al. 2015 [97], Thoudam et al. 2016 [98]). If our simple picture as deduced above, is correct, that RSG star explosions just terminate earlier at a radius about 1/10 of that of BSG star explosions. The anti-proton data (ASR18 [34]) in our Galaxy suggest that the RSG star explosions run their shock down in velocity, whereas the BSG star explosions hit their respective wind shell at full speed. We note that this particular distinction may be different in the galaxy M82 due to the higher environmental over-pressure (see Allen 1999 [66]). Therefore the characteristic energies, E_{ankle} and E_{knee} are maintained only for the BSG star explosions. It ensues that at the highest CR energies, RSG star explosions contribute far less than the BSG star explosions. Since in our model these energies scale with nuclear charge number Z, at a down-turn of the spectrum, the heavier elements go into dominance near the *knee* (CR-IV [39]); we note that this concept of 1993 required and predicted that the product $\{B \times r\}$ has a similar value in all relevant explosions of very massive stars exploding into their wind. At the final cutoff, at the *ankle* energy, protons will disappear very much earlier than Carbon, Oxygen or heavier elements such as Iron (see Figure 6 in CR-IV, as well as Figure 1 in ASR18 [34,39]). We emphasize again that the observed characteristic energies E_{knee} and E_{ankle} are directly reproduced by two independent data sets on RSNe and SNRs, as both are proportional to $\{B \times r\}$: This seems confirmed by a third independent data set (Fenech et al., 2010 [49]). It is based on both RSNe in other galaxies, and SNRs in the starburst galaxies M82, without assigning any special parameter a particular value. We do assume that the CR population and its characteristics are the same in our Galaxy as in other galaxies, and in the starburst galaxy M82. The two data sets agree well within the errors. The spectrum both below and above the *knee* can be explained with a combination of *Diffusive Shock Acceleration* (DSA) and *Stochastic Shock Drift Acceleration* (StSDA), as shown first in CR-I, CR-II, CR-III, and CR-IV, rederived in ASR18 [34,39,51,52,55], all based on earlier work by Drury (1983), Jokipii (1987), Völk and Biermann (1988), [14,58,60] and others. One consequence is that most interaction happens in the wind-shell, when the SN-driven shock hits (e.g., Biermann et al., 2001 [99]). This concept can be tested with the spectra of secondary particles. Here we note that the concept requires that the γ-ray -spectra of the Galaxy should reflect this early interaction, showing evidence for a spectrum of mostly $E^{-7/3}$ from the 4π component in wind-SN-CRs, and a weaker polar cap component of E^{-2}. The Fermi data suggest two such spectral components (de Boer et al., 2017 [100]). HAWC and H.E.S.S. data are fully consistent with this prediction of a dominant 4π component (e.g., Nayerhoda et al., 2018 [101]). At higher γ-ray energies the resulting spectrum is predicted to become flatter due to the emergence of the polar-cap component with E^{-2} in wind-SN-CRs (CR-IV [39]). Other new data can also be used for further tests, e.g., from AMS (Alpha Magnetic Spectrometer on the ISS) or HAWC, such as the spectra of various elements and secondary and primary contributions to Ni in cosmic rays (ASR18, Alfaro et al. 2017, Aguilar et al. 2017, 2018a,b [34,102–105]).

Nonetheless several unanswered questions remain in this simple picture. We list some of them in the following. All the processes discussed up to here may happen as argued. However, they do not have to dominate, as there are many alternatives, especially considering SN Ia, as shown below.

(1) Very rarely, at a rate of order 10^{-2} of massive stars, we do have Gamma Ray Bursts (GBRs) in any star forming galaxy, i.e. for order 100 very massive BSG star formed, there is one GRB. There is no question that they accelerate particles to high energy. Where is the contribution from GRBs in our own Galaxy? Could they also contribute below the *ankle*? Or is this already part of the *extragalactic contribution*, as argued in many papers, such as Milgrom and Usov (1995, 1996) [106,107]; Vietri (1995, 1996, 1998) [108–110]; Waxman (1995), Miralda-Escudé and Waxman (1996), Waxman and Bahcall (1997) [111–113]; Zhang and Mészáros (2004) [114]; Piran (2004) [115]? Based on the radio maps showing some evidence of a jet- and BH-spin-flip, we proposed in ASR18 [34] that the compact radio source 41.9 + 58 in the starburst galaxy M82 (KBS85 [47]) is actually the result of a recent merger of two stellar black holes (BHs). This could have happened concurrent with a GRB (see Muxlow et al., 2005 [69], who first argued for a misaligned GRB for this source). The argument then proceeds to propose that this BH merger could explain the Ultra High Energy CR (UHECR) particles detected by the Telescope Array (TA) detector (Abbasi et al., 2014, 2018a [84,116]). Many arguments have been made (e.g., Meli et al., 2008, Abbasi et al., 2012, Albert et al., 2017a, [117–119]) that GRBs do not contribute significantly to the UHECR population. All this depends on the model assumptions, and so may not be in contradiction.

(2) There are many arguments in support of UHECRs coming from radio galaxies (e.g., Ginzburg and Syrovatskij 1963, Hillas 1984, Biermann and Strittmatter 1987, Rachen and Biermann 1993a, Rachen et al., 1993b, Meli et al., 2008, [5,16,59,119–121]). A special interest is in the case when their relativistic jets are pointed at us, visible in the form of flat-spectrum-radio quasars (FSRQs) or blazars (Kadler et al., 2016; Kun et al., 2017, 2019; Aartsen et al., 2018a,b [122–126]): There are now a total of nine candidate identifications for high energy neutrino events with FSRQs (see also the lectures of F. Halzen at Erice 2018 [127]), many of them suggest an ongoing merger of two super-massive black holes (Gergely and Biermann 2009, Caramete and Biermann 2010, Kun et al., 2017, 2019, [125,126,128,129]). One of these candidates, TXS0506+056 is the most convincing, since it shows evidence for several neutrino events. To make HE neutrinos requires the prior acceleration of HE and UHECR particles first, so this tentative finding allows the speculation that mergers of super-massive black holes may lead to ubiquitous production of UHECR particles. This entails the notion that UHECR production is highly episodic. Do these sources also contribute below the *ankle*?

(3) BSG explosions can yield the energy, but can the CRs produced actually reach us? This is especially an issue for electrons and positrons: we have argued that we need electrons up to about 30 TeV in the source region; H.E.S.S. has now detected electrons at 20 TeV, published on its website [130], and in conference proceedings. The CR electron spectrum shown on the H.E.S.S. website is consistent with our 2009 prediction (Biermann et al. 2009 [61]) of at lower energies first $E^{-10/3}$, then E^{-3}, and last, at the highest energies, E^{-4}. These are all based on the idea that many sources contribute under several loss mechanisms (Kardashev 1962 [131]). At the highest energies the observed spectrum appears to

be slightly flatter than E^{-4}, possibly due to local and/or recent nearby sources. We note that the synchrotron loss time of electrons at 20 TeV is of order 3×10^4 yrs (using a magnetic field strength of order 5 μGauss: Beck et al., 1996 [132]; a crude estimate of 5 μGauss was first made in the 1940s, then using equipartition with CRs as necessary for isotropy), implying a source rather close by (as has been argued by many, e.g., Yüksel et al., 2009, 2012 [133,134]), or a very anisotropic transport to allow larger distances. An alternate way of phrasing this speculation could be that the magnetic fields are extremely structured, allowing some transport over large distances.

(4) The relative contribution of CR particles from RSG star and BSG star explosions at low energy is about 4:1, following Binns et al. (e.g., Murphy et al., 2016 [135]). What is it at high energy? The anti-proton data suggest that the RSG star explosions shocks run out of steam, and so that RSG star explosions in our Galaxy contribute much less at the maximal energies than the BSG star explosions. This may imply that the piston mass in RSG star explosions is smaller than in BSG star explosions, consistent with a naive interpretation of the ^{56}Ni data:

(5) Can the ^{56}Ni mass derived from optical observations (Hamuy et al., 2003, Maeda et al., 2010, Nakar et al., 2016, Lusk and Baron, 2017, Utrobin et al., 2017, [90–94]) be identified with the piston (see also, e.g., Biermann et al., 1992 [136] and Wesson et al., 2015 [137])? The numbers seem to match rather well, but have not been derived for BSG star explosions, and rely mostly on rather simple 1D models. It is apparent that we do not actually know the explosion mechanism for very massive stars, be it due to neutrinos (e.g., Bethe 1990 [138]), magnetic fields (e.g., Bisnovatyi-Kogan 1970, Bisnovatyi-Kogan et al., 2008 [139,140]), some combination of the two mechanisms (à la Seemann and Biermann 1997 [141]), or—another suggestion—quark deconfinement (Fischer et al., 2018 [142]).

(6) Do SN Ia explosions contribute anything to the observed CR particle population? The D^6 scenario seems currently to be favored (Shen et al., 2018a [143]; D^6 here stands for *Dynamically Driven Double-Degenerate Double-Detonation*): This argument has been strengthened by identifying some dwarf stars in the Gaia data as remaining from SN Ia explosions (Shen et al., 2018b, Raddi et al., 2018 [144,145]). The radio data (e.g., Dickel et al., 1991 [146]) and the interpretation of the radio and X-ray data (e.g., Chevalier 1982 [147], Archambault et al. 2017 [148]) demonstrate unequivocally that these explosions do accelerate energetic particles (see also Bykov et al., 2018 [149]).

Assuming a fraction of $0.1\,\epsilon_{B,-1}$ of the kinetic energy of the ejecta in SN Ia $E_{SN,51}\,10^{51}$ erg (e.g., Dwarkadas and Chevalier 1998; Mazzali et al., 2007; Townsley et al., 2009: [150–152]) goes into the amplification of magnetic fields following Lucek and Bell (2000), Bell and Lucek (2001), Bell (2004, 2008) [40,153–155], we obtain, in a Sedov expansion (Cox 1972) [50] at radius $10^{18.5}$ cm r_{pc} (this is scaled to a parsec, typically what is observed in M82)

$$B \times r \simeq 10^{16.2}\,\text{Gauss cm}\,\epsilon_{B,-1}^{1/2}\,E_{SN,51}^{1/2}\,r_{pc}^{-1/2} \tag{14}$$

Obviously, in this mechanism there is a corresponding energy density of the energetic particles. The mass M_{ISM} reached by the SN shock can be reached by a typical ejected mass of about $M_{ej} \simeq 1\,M_\odot$ (e.g., Dwarkadas and Chevalier 1998; Mazzali et al., 2007; Townsley et al., 2009; [150–152]). The radius implied is only weakly dependent on ISM density n_{ISM} (in cm^{-3}), as

$$r_{pc} \simeq 2\left(\frac{M_{ej}}{M_\odot}\frac{1}{n_{ISM}}\right)^{1/3} \tag{15}$$

and can readily be reached in free expansion. At larger radii Sedov expansion implies a slow lowering of the quantity $\{B \times r\}$. Therefore the quantity $\{B \times r\}$ at first rises linearly, as long as the shock is in free expansion, and so the shock velocity constant, to rise to its maximum:

$$(B \times r)_{max} \simeq 10^{16.0}\,\text{Gauss cm}\,\epsilon_{B,-1}^{1/2}\,E_{SN,51}^{1/2}\left(\frac{M_\odot}{M_{ej}}n_{ISM}\right)^{1/6} \tag{16}$$

That is remarkably close to what we observe in M82, and is only weakly dependent on any single given parameter. Thereafter this quantity declines as $r^{-1/2}$, until the Sedov expansion shock reverts to a cooling shock, at a larger radius (see, e.g., Cox (1972) [50]).

Thus explosions of SN Ia would seem to be a viable alternative to explain the observations of M82. However, SN Ia explosions are not enhanced in their rate immediately in a starburst (Scannapieco and Bildsten 2005; Raskin et al. 2009; [156,157]). Only very massive stars explode nearly simultaneously, as only their evolutionary time scale is shorter than a typical starburst time of just a few million years (e.g., Biermann and Fricke 1977; Huchra 1977; Huchra et al., 1983; Alonso-Herrero et al., 2001; [158–161]). Very massive stars show the same magnetic field behavior, i.e., the derived values of $\{B \times r\}$ are consistent with the same number over several powers of ten in radius, from near 10^{16} cm to parsec scale. How do we then exclude the mechanism proposed by Lucek and Bell (2000), Bell and Lucek (2001), Bell (2004, 2008), [40,153–155]? Since the SN shock is argued to run at undiminished velocity over the radial scale from 10^{16} cm to parsec scale, the Bell-Lucek mechanism would in fact also produce a constant value of $\{B \times r\}$ across this radial range, matching all these observations; a condition for this would be a SN shock expansion into a Parker limit wind (Parker 1958, [56]), fulfilled for both RSG and BSG star explosions into their winds. However, that is contradicted by the interpretation of the observations, as SN shocks racing through the dense winds of Red Super Giant (RSG) stars showing the same magnetic field as those racing through the tenuous winds of Blue Super Giant (BSG) stars (see Bell 2004, 2008; ASR18; [34,153,154]). A straightforward conclusion is that all these SNRs in M82 are produced by BSG star explosions, which come from the most massive normal stars. In ASR18 [34] some options are discussed for the Bell-Lucek mechanism to contribute to the observed magnetic fields. This would allow a possible understanding that both RSG and BSG stars show a similar magnetic field behavior. However, this line of reasoning assumes that all such SNRs and RSNe derive from the same class of stars, the most massive stars: The quantity $\{B \times r\}$ turns out to be similar (i) at its peak value for SN Ia explosions, using the Bell-Lucek mechanism, and (ii) for the most massive stars generally. Therefore the option is viable to assume that CRs derive in our Galaxy from SN Ia explosions and in M82 and other such galaxies from the most massive stars: However, SN Ia explosions only approximately give the same energy for *knee* and *ankle* due to the time dependence of $\{B \times r\}$. Thus using many such explosions would imply considerable smearing at the *knee* and *ankle* energies, as well as a somewhat different number for the *knee* energy. This may be ruled out already (e.g., Thoudam et al., 2016, [98]), but should be carefully checked out to confirm or refute this option. What is the spread of individual *knee* energies among the population of contributing explosions?

(7) Pulsars and pulsar wind nebulae are directly observed to accelerate particles to high energy (see, e.g., for a hint of neutrinos the TeVPA 2018 IceCube lectures; e.g., Aliu et al., 2014 [162]). What is their contribution? The high energy gamma data data clearly demonstrate that they produce energetic particles, but they cannot contribute easily to the observed positrons (Yüksel et al., 2009; Abeysekara et al., 2017 [133,163]).

(8) Relativistic SNe: this is amply presented in Soderberg et al. (2010) [164], Bykov et al. (2018) [149], in ASR18 [34], and in earlier papers cited therein. They surely contribute, since they are a variant of GRBs; however, is their contribution mostly at the very highest energies?

(9) Active single and binary stars: Any massive star with a powerful wind contributes (Seo et al. 2018 [165]). Micro-quasars are binary stars, as are all massive stars (Chini et al. 2012, 2013 [166,167]). Micro-quasars surely produce energetic particles and cosmic rays (e.g., Heinz and Sunyaev 2002, Mirabel 2004, 2011 [168–170]). In sum the final evolutionary stages will contribute to energetic particles (e.g., Vulic et al., 2018, Abeysekara et al., 2018b [171,172]). Is their contribution discernible in the CR data?

(10) Neutron star and stellar mass BH mergers: This may lead to a GRB. As observed they likely contribute to the observed CRs (Abbott et al., 2017a,b; Albert et al., 2017b [173–175]). However, if such CRs are produced in other galaxies, what path do they take through the combined magnetic fields of intergalactic space with its filaments, halo, wind and disk of our Galaxy (see, e.g., Pshirkov et al., 2011 [176], Mao et al., 2012 [177], Yüksel et al., 2012 [134], Kronberg 2013 [178])?

(11) Transport out of the Galaxy? Radio data suggest that this CR transport is typically convective, i.e., in a galactic plasma wind (Heesen et al., 2018 [179], Krause et al. (2018) [180] and Miskolczi et al. (2018) [181]). They give a threshold condition in the form of star formation rate per area (see also Rossa and Dettmar 2003 [182]). This confirms some very old ideas, going back to Weber and Davis (1967) [57], Parker (1958) [56] and even earlier notions in the late 1940s.

(12) A CR contribution from the wind-shock of the Galactic plasma wind? This has been explored repeatedly (Jokipii and Morfill 1987 [45]), most recently by Merten et al. (2018) [183].

(13) Neutrino production in the Galaxy? The IceCube and Antares observations have begun to constrain the models (Albert et al., 2018 [184]). The γ-ray data already suggest plausible models, all consistent with the concept that interaction is dominant when the SN-shocks hit the wind-shell (Biermann et al., 2001, 2009 [61,99]). It suggests that interactions during CR transport through the ISM of the galaxy are less important (Nath et al., 2012 [185]), as had been suspected for a long time. Analogously to our comments on diffuse γ-ray emission there should be two neutrino spectra, one at $E^{-7/3}$ from the 4π component, and at higher energy a lower E^{-2} from the polar-cap component (CR-IV and de Boer et al. 2017, [39,100]). The corresponding neutrino emission spectra have a transition at an energy which is poorly determined right now. At TeV energies we cannot easily separate the RSG and BSG contributions where the latter may or may not dominate the neutrino emission. The stronger interaction in the RSG star explosions (leading to anti-protons in our proposed model) may push the transition energy to a higher value in the observations when summed over a line of sight; we note that this approach may clarify the anti-proton spectrum.

The overall diffuse Galactic neutrino emission is predicted to show a sharp turn-off of both components near $1/20$ of the *knee* energy (for protons), so here

$$E_{voff} = 10^{14.7 \pm 0.2} \text{ eV}, \tag{17}$$

IceCube should be able to test this prediction in the future.

(14) What is the explosion mechanism, and do we obtain a BH every time, most of the time, or only occasionally when BSG stars explode? What is the nature of the piston? The innermost layers that get ejected may come from a zone just outside the budding black hole. Is this process the source of the strong magnetic fields observed, coming out with the piston? Are there some stars that produce a black hole without any significant explosion (see Mirabel 2017a,b,c [186–188])? Do we have a chance to learn about BH physics from these explosions?

5. Summary

The main task that we have set ourselves was to check whether the characteristic energies implied by the numbers for radius, magnetic field strength, and shock velocity deduced from observed Radio Supernova (RSN) data (listed in ASR18 [34]) can be confirmed with the observations of the compact sources, interpreted as Supernova Remnants (SNRs) in the starburst galaxy M82 (KBS85, Allen and Kronberg 1998, Allen 1999, [47,65,66] as well as earlier and later references): The answer seems to be positive. In fact we can confirm not only that the Parker magnetic field solution ($B \times r = const$) is maintained over many powers of ten in radial range of the SN-shock racing through the stellar wind for massive stars, but that they all have the same magnetic field strength, in our Galaxy, other galaxies and in starburst galaxies such as M82. The energies for the downturn, called the *knee*, and the upturn, called the *ankle*, seen in the Cosmic Ray (CR) spectrum can be derived directly from the data, again without any free parameter. That supports the interpretation that the observed CRs in the energy range from below the *knee* all the way to the *ankle* may derive directly from Blue Supergiant (BSG) star explosions. It also suggests that the general CR spectrum is the same in our Galaxy, other galaxies and in starburst galaxies such as M82. Given that BSG star explosions are suspected to produce black holes, and to drive the piston which gives rise to CRs, can CRs and their detailed properties teach us about BH formation?

Author Contributions: This work is based on research by M.L.A. with P.P.K.; A.M. covered the plasma physics aspects, and E.-S.S. the space aspects. P.L.B. wrote it up with lots of input.

Funding: This research received no external funding.

Acknowledgments: P.L.B. wishes to thank N. Barghouty, J. Becker Tjus, W.R. Binns,W. de Boer, R. Buta, L. Caramete, S. Casanova, A. Chieffi, R. Chini, R. Diehl, R. Engel, I. Gebauer, L.Á. Gergely, E. Haug, A. Jones, E. Kun, B.B. Nath, V. Ptuskin, F. Sardei, P. Sokolsky, T. Stanev, D. Townsley, and many others for intense discussions of these topics; P.L. Biermann is a member of the Pierre Auger Coll., and the JEM-EUSO-Coll.; E.-S. Seo is a member of the AMS coll., the CREAM-Coll. and the ISS-CREAM Coll. Some of the essential ideas about acceleration in shocks expounded here were developed and tested in discussions between J.R. Jokipii and P.L.B. over many months in 1991 as well as later, as acknowledged in CR-I and other papers.

Conflicts of Interest: The authors declare no conflict of interest.

References

1. Hess, V.F. Die Beobachtungen der durchdringenden Strahlung bei sieben Freiballonfahrten. *Phys. Z.* **1912**, *13*, 1084.

2. Kohlhörster, W. Messungen der durchdringenden Strahlung im Freiballon in größeren Höhen. *Phys. Z.* **1913**, *14*, 1153.

3. Linsley, J. Evidence for a Primary Cosmic-Ray Particle with Energy 10^{20} eV. *Phys. Rev. Lett.* **1963**, *10*, 146–148. [CrossRef]

4. Baade, W.; Zwicky, F. Cosmic rays from supernovae. *Proc. Nat. Acad. Sci. USA* **1934**, *20*, 259. [CrossRef]

5. Ginzburg, V.L.; Syrovatskii, S.I. Cosmic Rays in Metagalactic Space. *Astron. Zh.* **1963**, *40*, 466; translation in *Sov. Astron. A. J.* **1963**, *7*, 357.

6. Fermi, E. On the Origin of the Cosmic Radiation. *Phys. Rev.* **1949**, *75*, 1169–1174. [CrossRef]

7. Fermi, E. Galactic Magnetic Fields and the Origin of Cosmic Radiation. *Astrophys. J.* **1954**, *119*, 1–6. [CrossRef]

8. Axford, W.I.; Leer, E.; Skadron, G. The acceleration of cosmic rays by shock waves. In Proceedings of the 15th International Cosmic Ray Conference, Plovdiv, Bulgaria, 13–26 August 1977; Volume 11, pp. 132–137.

9. Bell, A.R. The acceleration of cosmic rays in shock fronts. I. *Mon. Not. R. Astron. Soc.* **1978**, *182*, 147–156. [CrossRef]

10. Bell, A.R. The acceleration of cosmic rays in shock fronts. II. *Mon. Not. R. Astron. Soc.* **1978**, *182*, 443–455. [CrossRef]

11. Blandford, R.D.; Ostriker, J.P. Particle acceleration by astrophysical shocks. *Astrophys. J. L* **1978**, *221*, L29–L32. [CrossRef]

12. Krymskii, G.F. A regular mechanism for the acceleration of charged particles on the front of a shock wave. *Akad. Nauk Dokl.* **1977**, *234*, 1306–1308; translation in *Sov. Phys. Dokl.* **1977**, *22*, 327–328.

13. Ginzburg, V.L.; Ptuskin, V.S. On the origin of cosmic rays: some problems in high-energy astrophysics. *Rev. Mod. Phys.* **1976**, *48*, 161–189. [CrossRef]

14. Drury, L.O'C. An introduction to the theory of diffusive shock acceleration of energetic particles in tenuous plasmas. *Rep. Pro. Phys.* **1983**, *46*, 973–1027. [CrossRef]

15. Lovelace, R.V.E. Dynamo model of double radio sources. *Nature* **1976**, *262*, 649–652. [CrossRef]

16. Biermann, P.L.; Strittmatter, P.A. Synchrotron emission from shockwaves in active galactic nuclei. *Astrophys. J.* **1987**, *322*, 643–649. [CrossRef]

17. Kronberg, P.P.; Dufton, Q.W.; Li, H.; Colgate, S.A. Magnetic energy of the Intergalactic Medium from galactic black holes. *Astrophys. J.* **2001**, *560*, 178–186. [CrossRef]

18. Kronberg, P.P. Intergalactic Magnetic Fields. *Phys. Today* **2002**, *55*, 40–46. [CrossRef]

19. Kronberg, P.P.; Colgate, S.A.; Li, H.; Dufton, Q.W. Giant Radio Galaxies and Cosmic-Ray Acceleration. *Astrophys. J.* **2004**, *604*, L77–L80. [CrossRef]

20. Colgate, S.A. Acceleration mechanisms 2: Force-free reconnection. *Comptes Rendus Phys.* **2004**, *5*, 431–440. [CrossRef]

21. Colgate, S.A.; Fowler, T.K.; Li, H.; Pino, J. Quasi-Static Model of Collimated Jets I. Accretion Disk and Jets. *Astrophys. J.* **2014**, *789*, 144. [CrossRef]

22. Biermann, P.L.; Caramete, L.I.; Fraschetti, F.; Gergely, L.Á.; Harms, B.C.; Kun, E.; Lundquist, J.P.; Meli, A.; Nath, B.B.; Seo, E.S.; et al. The Nature and Origin of Ultra-High Energy Cosmic Ray Particles, Review at the Vulcano Meeting, Vulcano Island, May 2016. Available online: http://www.lnf.infn.it/sis/frascatiseries/Volume64/Volume64.pdf (accessed on 9 April 2019).

23. Aharonian, F.A. *Very High Energy Cosmic Gamma Radiation: A Crucial Window on the Extreme Universe*; World Scientific Publishing: Singapore, 2004.

24. Stanev, T. *High Energy Cosmic Rays*; Springer: Heidelberg, Germany, 2010.

25. Kotera, K.; Olinto, A.V. The Astrophysics of Ultrahigh-Energy Cosmic Rays. *Annu. Rev. Astron. Astrophys.* **2011**, *49*, 119–153. [CrossRef]

26. Letessier-Selvon, A.; Stanev, T. Ultrahigh energy cosmic rays. *Rev. Mod. Phys.* **2011**, *83*, 907–942. [CrossRef]

27. Bykov, A.M.; Ellison, D.C.; Gladilin, P.E.; Osipov, S.M. Galactic cosmic ray origin sites: Supernova remnants and superbubbles. *AIP Conf. Proc.* **2012**, *1505*, 46–55.

28. Diehl, R. Nuclear astrophysics lessons from INTEGRAL. *Rep. Pro. Phys.* **2013**, *76*, 026301. [CrossRef]

29. Diehl, R. Gamma-ray line measurements from supernova explosions. In Proceedings of the IAU Symposium 331 "SN1987A 30 Years After", La Reunion Island, France, 20–24 February 2017; Volume 331, pp. 157–163.

30. Blasi, P. The origin of galactic cosmic rays. *Astron. Astroph. Rev.* **2013**, *21*, 70. [CrossRef]

31. Gaisser, T.K.; Engel, R.; Resconi, E. *Cosmic Rays and Particle Physics*; Cambridge University Press: Cambridge, UK, 2016.

32. Kronberg, P.P. *Cosmic Magnetic Fields*; Cambridge University Press: Cambridge, UK, 2016.

33. Amato, E.; Blasi, P. Cosmic Ray Transport in the Galaxy: A Review. *Adv. Space Res.* **2018**, *62*, 2731–2749. [CrossRef]

34. Biermann, P.L.; Becker Tjus, J.; de Boer, W.; Caramete, L.I.; Chieffi, A.; Diehl, R.; Gebauer, I.; Gergely, L.Á.; Haug, E.; Kronberg, P.P.; et al. Supernova explosions of massive stars and cosmic rays. *Adv. Space Res.* **2018**, *62*, 2773–2816. [CrossRef]

35. Aartsen, M.G.; Ackermann, M.; Adams, J.; Aguilar, J.A.; Ahlers, M.; Ahrens, M.; Altmann, D.; Andeen, K.; Anderson, T.; Ansseau, I.; et al. Astrophysical neutrinos and cosmic rays observed by IceCube. *Adv. Space Res.* **2018**, *62*, 2902–2930. [CrossRef]

36. Moskalenko, I.V.; Seo, E.-S. Preface: Origins of cosmic rays. *Adv. Space Res.* **2018**, *62*, 2729–2730. [CrossRef]

37. Seo, E.-S.; Anderson, T.; Angelaszek, D.; Baek, S.J.; Baylon, J.; Buénerd, M.; Copley, M.; Coutu, S.; Derome, L.; Fields, B.; et al. Cosmic Ray Energetics And Mass for the International Space Station (ISS-CREAM). *Adv. Space Res.* **2014**, *53*, 1451–1455. [CrossRef]

38. Seo, E.-S. Investigating Mysteries of Cosmic Rays with Space-based Experiments. 2018. Available online: http://meetings.aps.org/Meeting/MAS18/Session/G02.1 (accessed on 9 April 2019).

39. Stanev, T.; Biermann, P.L.; Gaisser, T.K. Cosmic rays IV. The spectrum and chemical composition above 10^4 GeV. *Astron. Astrophys.* **1993**, *274*, 902–908.

40. Bell, A.R.; Lucek, S.G. Cosmic ray acceleration to very high energy through the non-linear amplification by cosmic rays of the seed magnetic field. *Mon. Not. R. Astron. Soc.* **2001**, *321*, 433–438. [CrossRef]

41. Helder, E.A.; Vink, J.; Bykov, A.M.; Ohira, Y.; Raymond, J.C.; Terrier, R. Observational Signatures of Particle Acceleration in Supernova Remnants. *Space Sci. Rev.* **2012**, *173*, 369–431. [CrossRef]

42. Zirakashvili, V.N.; Ptuskin, V.S. Cosmic ray acceleration in magnetic circumstellar bubbles. *Astropart. Phys.* **2018**, *98*, 21–27. [CrossRef]

43. Jokipii, J.R. Particle drift, diffusion, and acceleration at shocks. *Astrophys. J.* **1982**, *255*, 716–720. [CrossRef]

44. Jokipii, J.R. Particle acceleration at a termination shock. I–Application to the solar wind and the anomalous component. *J. Geophys. Res.* **1986**, *91*, 2929–2932. [CrossRef]

45. Jokipii, J.R.; Morfill, G. Ultra-high-energy cosmic rays in a galactic wind and its termination shock. *Astrophys. J.* **1987**, *312*, 170–177. [CrossRef]

46. Kronberg, P.P.; Biermann, P.L.; Schwab, F.R. The continuum radio structure of the nucleus of M82. *Astrophys. J.* **1981**, *246*, 751–760. [CrossRef]

47. Kronberg, P.P.; Biermann, P.L.; Schwab, F.R. The nucleus of M82 at radio and X-ray bands–Discovery of a new radio population of supernova candidates. *Astrophys. J.* **1985**, *291*, 693–707. [CrossRef]

48. Fenech, D.M.; Muxlow, T.W.B.; Beswick, R.J.; Pedlar, A.; Argo, M.K. Deep MERLIN 5GHz radio imaging of supernova remnants in the M82 starburst. *Mon. Not. R. Astron. Soc.* **2008**, *391*, 1384–1402. [CrossRef]

49. Fenech, D.M.; Beswick, R.J.; Muxlow, T.W.B.; Pedlar, A.; Argo, M.K. Wide-field Global VLBI and MERLIN combined monitoring of supernova remnants in M82. *Mon. Not. R. Astron. Soc.* **2010**, *408*, 607–621. [CrossRef]
50. Cox, D.P. Cooling and Evolution of a Supernova Remnant. *Astrophys. J.* **1972**, *178*, 159–168. [CrossRef]
51. Biermann, P.L.; Cassinelli, J.P. Cosmic rays II. Evidence for a magnetic rotator Wolf-Rayet star origin. *Astron. Astrophys.* **1993**, *277*, 691–706.
52. Biermann, P.L.; Strom, R.G. Cosmic Rays III. The cosmic ray spectrum between 1 GeV and 10^4 GeV and the radio emission from supernova remnants. *Astron. Astrophys.* **1993**, *275*, 659–669.
53. Landau, L.D.; Lifshitz, E.M. *Fluid Mechanics*; Course of Theoretical Physics (book Series); Pergamon Press: Oxford, UK, 1959. (In English translation from the original Russian)
54. Meli, A.; Biermann, P.L. Cosmic rays X. The cosmic ray knee and beyond: Diffusive acceleration at oblique shocks. *Astron. Astrophys.* **2006**, *454*, 687–694. [CrossRef]
55. Biermann, P.L. Cosmic rays I. The cosmic ray spectrum between 10^4 GeV and $3\,10^9$ GeV. *Astron. Astrophys.* **1993**, *271*, 649–661.
56. Parker, E.N. Dynamics of the Interplanetary Gas and Magnetic Fields. *Astrophys. J.* **1958**, *128*, 664–676. [CrossRef]
57. Weber, E.J.; Davis, L., Jr. The Angular Momentum of the Solar Wind. *Astrophys. J.* **1967**, *148*, 217–227. [CrossRef]
58. Jokipii, J.R. Rate of energy gain and maximum energy in diffusive shock acceleration. *Astrophys. J.* **1987**, *313*, 842–846. [CrossRef]
59. Hillas, A.M. The Origin of Ultra-High-Energy Cosmic Rays. *Annu. Rev. Astron. Astrophys.* **1984**, *22*, 425–444. [CrossRef]
60. Völk, H.J.; Biermann, P.L. Maximum energy of cosmic-ray particles accelerated by supernova remnant shocks in stellar wind cavities. *Astrophys. J. L* **1988**, *333*, L65–L68. [CrossRef]
61. Biermann, P.L.; Becker, J.K.; Meli, A.; Rhode, W.; Seo, E.-S.; Stanev, T. Cosmic Ray Electrons and Positrons from Supernova Explosions of Massive Stars. *Phys. Rev. Lett.* **2009**, *103*, 061101. [CrossRef] [PubMed]
62. Bartel, N.; Ratner, M.I.; Rogers, A.E.E.; Shapiro, I.I.; Bonometti, R.J.; Cohen, N.L.; Gorenstein, M.V.; Marcaide, J.M.; Preston, R.A. VLBI observations of 23 hot spots in the starburst galaxy M82. *Astrophys. J.* **1987**, *323*, 505–515. [CrossRef]
63. Muxlow, T.W.B.; Pedlar, A.; Wilkinson, P.N.; Axon, D.J.; Sanders, E.M.; de Bruyn, A.G. The structure of young supernova remnants in M82. *Mon. Not. R. Astron. Soc.* **1994**, *266*, 455–467. [CrossRef]
64. Golla, G.; Allen, M.L.; Kronberg, P.P. The Starburst Nuclear Region in M82 Compared in Several Wave Bands. *Astrophys. J.* **1996**, *473*, 244–253. [CrossRef]
65. Allen, M.L.; Kronberg, P.P. Radio Spectra of Selected Compact Sources in the Nucleus of M82. *Astrophys. J.* **1998**, *502*, 218–228. [CrossRef]
66. Allen, M.L. Radio Continuum Studies of the Evolved Starburst in M82. Ph.D. Thesis, University of Toronto, Toronto, ON, Canada, 1999.
67. Kronberg, P.P.; Sramek, R.A.; Birk, G.T.; Dufton, Q.W.; Clarke, T.E.; Allen, M.L. A Search for Flux Density Variations in 24 Compact Radio Sources in M82. *Astrophys. J.* **2000**, *535*, 706–711. [CrossRef]
68. McDonald, A.R.; Muxlow, T.W.B.; Wills, K.A.; Pedlar, A.; Beswick, R.J. A parsec-scale study of the 5/15-GHz spectral indices of the compact radio sources in M82. *Mon. Not. R. Astron. Soc.* **2002**, *334*, 912–924. [CrossRef]
69. Muxlow, T.W.B.; Pedlar, A.; Beswick, R.J.; Argo, M.K.; O'Brien, T.J.; Fenech, D.; Trotman, W. Is 41.95+575 in M82 actually an SNR? *Mem. Soc. Astron. Ital.* **2005**, *76*, 586–588.
70. Gendre, M.A.; Fenech, D.M.; Beswick, R.J.; Muxlow, T.W.B.; Argo, M.K. Flux density variations of radio sources in M82 over the last three decades. *Mon. Not. R. Astron. Soc.* **2013**, *431*, 1107–1120. [CrossRef]
71. Miley, G.K. The structure of extended extragalactic radio sources. *Annu. Rev. Astron. Astrophys.* **1980**, *18*, 165. [CrossRef]
72. Parker, E.N. The dynamical state of the interstellar gas and field. *Astrophys. J.* **1966**, *145*, 811–833. [CrossRef]
73. Ames, S. Magneto-Gravitational and Thermal Instability in the Galactic Disk. *Astrophys. J.* **1973**, *182*, 387–404. [CrossRef]
74. Spitzer, L., Jr. *Physics of Fully Ionized Gases*, 2nd ed.; Wiley Interscience: New York, NY, USA, 1962.
75. Caprioli, D. Fermi acceleration at supernova remnant shocks. *AIP Conf. Proc.* **2012**, *1505*, 237–240.
76. McClements, K.G.; Dendy, R.O.; Bingham, R.; Kirk, J.G.; Drury, L.O'C. Acceleration of cosmic ray electrons by ion-excited waves at quasi-perpendicular shocks. *Mon. Not. R. Astron. Soc.* **1997**, *291*, 241–249.

77. Dieckmann, M.E.; Chapman, S.C.; McClements, K.G.; Dendy, R.O.; Drury, L.O'C. Electron acceleration due to high frequency instabilities at supernova remnant shocks. *Astron. Astrophys.* **2000**, *356*, 377–388.

78. van der Laan, H. A Model for Variable Extragalactic Radio Sources. *Nature* **1966**, *211*, 1131–1133. [CrossRef]

79. Shklovskii, I.S. Secular Variation of the Flux and Intensity of Radio Emission from Discrete Sources. *Astron. Zh.* **1960**, *37*, 256–264; translation in *Sov. Astron. A. J.* **1960**, *4*, 243–249.

80. Abramowski, A.; Aharonian, F.; Benkhali, F.A.; Akhperjanian, A.G.; Angüner, E.O.; Backes, M.; Balzer, A.; Becherini, Y.; Tjus, J.B.; Berge, D.; et al. Acceleration of petaelectronvolt protons in the Galactic Centre. *Nature* **2016**, *531*, 476–479.

81. Aab, A.; Abreu, P.; Aglietta, M.; Samarai, I.A.; Albuquerque, I.F.M.; Allekotte, I.; Almela, A.; Alvarez Castillo, J.; Alvarez-MuAiz, J.; Anastasi, G.A.; et al. Combined fit of spectrum and composition data as measured by the Pierre Auger Observatory. *J. Cosm. Astrop. Phys.* **2017**, *4*, 038. [CrossRef]

82. Aab, A.; Abreu, P.; Aglietta, M.; Al Samarai, I.; Albuquerque, I.F.M.; Allekotte, I.; Almela, A.; Castillo, J.A.; Alvarez-Muñiz, J.; Anastasi, G.A.; et al. Inferences on mass composition and tests of hadronic interactions from 0.3 to 100 EeV using the water-Cherenkov detectors of the Pierre Auger Observatory. *Phys. Rev. D* **2017**, *96*, 122003. [CrossRef]

83. Aab, A.; Abreu, P.; Aglietta, M.; Albuquerque, I.F.M.; Albury, J.M.; Allekotte, I.; Almela, A.; Castillo, J.A.; Alvarez-Muñiz, J.; Anastasi, G.A.; et al. Large-scale cosmic-ray anisotropies above 4 EeV measured by the Pierre Auger Observatory. *Astrophys. J.* **2018**, *868*, 4.

84. Abbasi, R.U.; Abe, M.; Abu-Zayyad, T.; Allen, M.; Azuma, R.; Barcikowski, E.; Belz, J.W.; Bergman, D.R.; Blake, S.A.; Cady, R.; et al. Evidence of Intermediate-scale Energy Spectrum Anisotropy of Cosmic Rays $E \geq 10^{19.2}$ eV with the Telescope Array Surface Detector. *Astrophys. J.* **2018**, *862*, 91. [CrossRef]

85. Abbasi, R.U.; Abe, M.; Abu-Zayyad, T.; Allen, M.; Azuma, R.; Barcikowski, E.; Belz, J.W.; Bergman, D.R.; Blake, S.A.; Cady, R.; et al. The Cosmic Ray Energy Spectrum between 2 PeV and 2 EeV Observed with the TALE Detector in Monocular Mode. *Astrophys. J.* **2018**, *865*, 74. [CrossRef]

86. Abbasi, R.U.; Abe, M.; Abu-Zayyad, T.; Allen, M.; Azuma, R.; Barcikowski, E.; Belz, J.W.; Bergman, D.R.; Blake, S.A.; Cady, R.; et al. Mass composition of ultra-high-energy cosmic rays with the Telescope Array Surface Detector Data. *Phys. Rev. D* **2018**, arXiv:1808.03680.

87. Abeysekara, A.U.; Alfaro, R.; Alvarez, C.; Alvarez, J.D.; Arceo, R.; Arteaga-Velázquez, J.C.; Rojas, D.A.; Solares, H.A.; Becerril, A.; Belmont-Moreno, E.; et al. Observation of Anisotropy of TeV Cosmic Rays with Two Years of HAWC. *Astrophys. J.* **2018**, *865*, 57. [CrossRef]

88. Abeysekara, A.U.; Alfaro, R.; Alvarez, C.; Arceo, R.; Arteaga-Velázquez, J.C.; Avila Rojas, D.; Belmont-Moreno, E.; BenZvi, S.Y.; Brisbois, C.; Capistrán, T.; et al. All-Sky Measurement of the Anisotropy of Cosmic Rays at 10 TeV and Mapping of the Local Interstellar Magnetic Field. *Astrophys. J.* **2019**, *871*, 96. [CrossRef]

89. Sedov, L.I. Examples of Gas Motion and Certain Hypotheses on the Mechanism of Stellar Outbursts. *Rev. Mod. Phys.* **1958**, *30*, 1077–1079. [CrossRef]

90. Hamuy, M. Observed and Physical Properties of Core-Collapse Supernovae. *Astrophys. J.* **2003**, *582*, 905–914. [CrossRef]

91. Lusk, J.A.; Baron, E. Bolometric Light Curves of Peculiar Type II-P Supernovae. *Publ. Astron. Soc. Pac.* **2017**, *129*, 044202. [CrossRef]

92. Maeda, K.; Tominaga, N.; Umeda, H.; Nomoto, K.; Suzuki, T. Supernova nucleosynthesis and stellar population in the early Universe. *Mem. Soc. Astron. Ital.* **2010**, *81*, 151–156.

93. Nakar, E.; Poznanski, D.; Katz, B. The Importance of ^{56}Ni in Shaping the Light Curves of Type II Supernovae. *Astrophys. J.* **2016**, *823*, 127. [CrossRef]

94. Utrobin, V.P.; Wongwathanarat, A.; Janka, H.-T.; Müller, E. Light-curve Analysis of Ordinary Type IIP Supernovae Based on Neutrino-driven Explosion Simulations in Three Dimensions. *Astrophys. J.* **2017**, *846*, 37. [CrossRef]

95. Kronberg, P.P.; Wilkinson, P.N. High-resolution, multifrequency radio observations of M82. *Astrophys. J.* **1975**, *200*, 430–435. [CrossRef]

96. Weliachew, L.; Fomalont, E.B.; Greisen, E.W. Radio observations of H I and OH in the center of the galaxy M82. *Astron. Astrophys.* **1984**, *137*, 335–342.

97. Todero Peixoto, C.J.; de Souza, V.; Biermann, P.L. Cosmic rays: The spectrum and chemical composition from 10^{10} to 10^{20} eV. *J. Cosm. Astrop. Phys.* **2015**, *2015*, 042; arXiv:1502.00305.

98. Thoudam, S.; Rachen, J.P.; van Vliet, A.; Achterberg, A.; Buitink, S.; Falcke, H.; Hörandel, J.R. Cosmic-ray energy spectrum and composition up to the ankle: The case for a second Galactic component. *Astron. Astrophys.* **2016**, *595*, A33. [CrossRef]

99. Biermann, P.L.; Langer, N.; Seo, E.-S.; Stanev, T. Cosmic Rays IX. Interactions and transport of cosmic rays in the Galaxy. *Astron. Astrophys.* **2001**, *369*, 269–277. [CrossRef]

100. de Boer, W.; Bosse, L.; Gebauer, I.; Neumann, A.; Biermann, P.L. Molecular Clouds as the Origin of the Fermi Gamma-Ray GeV-Excess. *Phys. Rev. D* **2017**, *96*, 043012. [CrossRef]

101. Nayerhoda, A.; Salesa Greus, F.; Casanova, S. TeV Diffuse emission from the inner Galaxy. *Front. Astron. Space Sci.* **2018**, *5*, 8. [CrossRef]

102. Aguilar, M.; Cavasonza, L.A.; Alpat, B.; Ambrosi, G.; Arruda, L.; Attig, N.; Aupetit, S.; Azzarello, P.; Bachlechner, A.; Barao, F.; et al. Observation of the Identical Rigidity Dependence of He, C, and O Cosmic Rays at High Rigidities by the Alpha Magnetic Spectrometer on the International Space Station. *Phys. Rev. Lett.* **2017**, *119*, 251101. [CrossRef] [PubMed]

103. Aguilar, M.; Cavasonza, L.A.; Ambrosi, G.; Arruda, L.; Attig, N.; Aupetit, S.; Azzarello, P.; Bachlechner, A.; Barao, F.; Barrau, A.; et al. Observation of New Properties of Secondary Cosmic Rays Lithium, Beryllium, and Boron by the Alpha Magnetic Spectrometer on the International Space Station. *Phys. Rev. Lett.* **2018**, *120*, 021101. [CrossRef]

104. Aguilar, M.; Cavasonza, L.A.; Alpat, B.; Ambrosi, G.; Arruda, L.; Attig, N.; Aupetit, S.; Azzarello, P.; Bachlechner, A.; Barao, F.; et al. Precision Measurement of Cosmic-Ray Nitrogen and its Primary and Secondary Components with the Alpha Magnetic Spectrometer on the International Space Station. *Phys. Rev. Lett.* **2018**, *121*, 051103. [CrossRef]

105. Alfaro, R.; Alvarez, C.; Álvarez, J.D.; Arceo, R.; Arteaga-Velázquez, J.C.; Rojas, D.A.; Solares, H.A.; Barber, A.S.; Becerril, A.; Belmont-Moreno, E.; et al. All-particle cosmic ray energy spectrum measured by the HAWC experiment from 10 to 500 TeV. *Phys. Rev. D* **2017**, *96*, 122001. [CrossRef]

106. Milgrom, M.; Usov, V. Possible Association of Ultra-High-Energy Cosmic-Ray Events with Strong Gamma-Ray Bursts. *Astrophys. J. L* **1995**, *449*, L37–L40. [CrossRef]

107. Milgrom, M.; Usov, V. Gamma-ray bursters as sources of cosmic rays. *Astropart. Phys.* **1996**, *4*, 365–369. [CrossRef]

108. Vietri, M. The Acceleration of Ultra-High-Energy Cosmic Rays in Gamma-Ray Bursts. *Astrophys. J.* **1995**, *453*, 883–889. [CrossRef]

109. Vietri, M. Coronal gamma-ray bursts as the sources of ultra-high-energy cosmic rays? *Mon. Not. R. Astron. Soc. Lett.* **1996**, *278*, L1–L4. [CrossRef]

110. Vietri, M. Ultrahigh Energy Neutrinos from Gamma Ray Bursts. *Phys. Rev. Lett.* **1998**, *80*, 3690–3693. [CrossRef]

111. Miralda-Escudé, J.; Waxman, E. Signatures of the Origin of High-Energy Cosmic Rays in Cosmological Gamma-Ray Bursts. *Astrophys. J. L* **1996**, *462*, L59–L62. [CrossRef]

112. Waxman, E. Cosmological Gamma-Ray Bursts and the Highest Energy Cosmic Rays. *Phys. Rev. Lett.* **1995**, *75*, 386–389. [CrossRef]

113. Waxman, E.; Bahcall, J. High Energy Neutrinos from Cosmological Gamma-Ray Burst Fireballs. *Phys. Rev. Lett.* **1997**, *78*, 2292–2295. [CrossRef]

114. Zhang, B.; Mészáros, P. Gamma-Ray Bursts: Progress, problems & prospects. *Int. J. Mod. Phys. A* **2004**, *19*, 2385–2472.

115. Piran, T. The physics of gamma-ray bursts. *Rev. Mod. Phys.* **2004**, *76*, 1143–1210. [CrossRef]

116. Abbasi, R.; Abe, M.; Abu-Zayyad, T.; Allen, M.; Anderson, R.; Azuma, R.; Barcikowski, E.; Belz, J.W.; Bergman, D.R.; Blake, S.A.; et al. Indications of intermediate-scale anisotropy of cosmic rays with energies greater than 57 EeV in the Northern sky measured with the surface detector of the Telescope Array experiment. *Astrophys. J. L* **2014**, *790*, L21. [CrossRef]

117. Abbasi, R.; Abdou, Y.; Abu-Zayyad, T.; Ackermann, M.; Adams, J.; Aguilar, J.A.; Ahlers, M.; Altmann, D.; Andeen, K.; Auffenberg, J.; et al. An absence of neutrinos associated with cosmic-ray acceleration in γ-ray bursts. *Nature* **2012**, *484*, 351–354. [CrossRef] [PubMed]

118. Albert, A.; André, M.; Anghinolfi, M.; Anton, G.; Ardid, M.; Aubert, J.J.; Avgitas, T.; Baret, B.; Barrios-Martí, J.; Basa, S.; et al. Search for high-energy neutrinos from bright GRBs with ANTARES. *Mon. Not. R. Astron. Soc.* **2017**, *469*, 906–915. [CrossRef]

119. Meli, A.; Becker, J.K.; Quenby, J.J. On the origin of ultra high energy cosmic rays: Subluminal and superluminal relativistic shocks. *Astron. Astrophys.* **2008**, *492*, 323–326. [CrossRef]

120. Rachen, J.P.; Biermann, P.L. Extragalactic ultra-high energy cosmic rays I. Contribution from hot spots in FR-II radio galaxies. *Astron. Astrophys.* **1993**, *272*, 161–175.

121. Rachen, J.P.; Stanev, T.; Biermann, P.L. Extragalactic ultra high energy cosmic rays II. Comparison with experimental data. *Astron. Astrophys.* **1993**, *273*, 377–382.

122. Aartsen, M.; Ackermann, M.; Adams, J.; Aguilar, J.; Ahlers, M.; Ahrens, M.; Al Samarai, I.; Altmann, D.; Andeen, K.; Anderson, T.; et al. Neutrino emission from the direction of the blazar TXS 0506+056 prior to the IceCube-170922A alert. *Science* **2018**, *361*, 147–151.

123. Aartsen, M.G.; Ackermann, M.; Adams, J.; Aguilar, J.A.; Ahlers, M.; Ahrens, M.; Al Samarai, I.; Altmann, D.; Andeen, K.; Anderson, T.; et al. Multimessenger observations of a flaring blazar coincident with high-energy neutrino IceCube-170922A. *Science* **2018**, *361*, eaat1378.

124. Kadler, M.; Krauß, F.; Mannheim, K.; Ojha, R.; Müller, C.; Schulz, R.; Anton, G.; Baumgartner, W.; Beuchert, T.; Buson, S.; et al. Coincidence of a high-fluence blazar outburst with a PeV-energy neutrino event. *Nat. Phys.* **2016**, *12*, 807–814. [CrossRef]

125. Kun, E.; Biermann, P.L.; Gergely, L.Á. A flat spectrum candidate for a track-type high energy neutrino emission event, the case of blazar PKS 0723-008. *Mon. Not. R. Astron. Soc. Lett.* **2017**, *466*, L34 –L38. [CrossRef]

126. Kun, E.; Biermann, P.L.; Gergely, L.Á. VLBI radio structure and core-brightening of the high-energy neutrino emitter TXS 0506+056. *Mon. Not. R. Astron. Soc. Lett.* **2019**, *483*, L42–L46. [CrossRef]

127. Halzen, F. Lectures at Erice. 2018. Available online: https://icecube.wisc.edu/~halzen/presentations.htm (accessed on 9 April 2019).

128. Caramete, L.I.; Biermann, P.L. The mass function of nearby black hole candidates. *Astron. Astrophys.* **2010**, *521*, A55. [CrossRef]

129. Gergely L.Á.; Biermann, P.L. Supermassive black hole mergers. *Astrophys. J.* **2009**, *697*, 1621–1633. [CrossRef]

130. Probing Local Sources with High Energy Cosmic Ray Electrons. Available online: https://www.mpi-hd.mpg.de/hfm/HESS/pages/home/som/2017/09/ (accessed on 9 April 2019).

131. Kardashev, N.S. Nonstationarity of Spectra of Young Sources of Nonthermal Radio Emission. *Astron. Zh.* **1962**, *39*, 393–409; reprinted in *Sov. Astron. A. J.* **1962**, *6*, 317–327.

132. Beck, R.; Brandenburg, A.; Moss, D.; Shukurov, A.; Sokoloff, D. Galactic Magnetism: Recent Developments and perspectives. *Annu. Rev. Astron. Astrophys.* **1996**, *34*, 155–206. [CrossRef]

133. Yüksel, H.; Kistler, M.D.; Stanev, T. TeV Gamma Rays from Geminga and the Origin of the GeV Positron Excess. *Phys. Rev. Lett.* 2009, *103*, 051101. [CrossRef]

134. Yüksel, H.; Stanev, T.; Kistler, M.D.; Kronberg, P.P. The Centaurus A Ultrahigh-energy Cosmic-Ray Excess and the Local Extragalactic Magnetic Field. *Astrophys. J.* **2012**, *758*, 16. [CrossRef]

135. Murphy, R.P.; Sasaki, M.; Binns, W.R.; Brandt, T.J.; Hams, T.; Israel, M.H.; Labrador, A.W.; Link, J.T.; Mewaldt, R.A.; Mitchell, J.W.; et al. Galactic Cosmic Ray Origins and OB Associations: Evidence from SuperTIGER Observations of Elements 26Fe through 40Zr. *Astrophys. J.* **2016**, *831*. [CrossRef]

136. Biermann, P.L.; Chini, R.; Haslam, C.G.T.; Kreysa, E.; Lemke, R.; Sievers, A. Evidence for heavy element dust clumps from the increasing 1300 micron emission from supernova 1987A. *Astron. Astrophys.* **1992**, *255*, L5–L8.

137. Wesson, R.; Barlow, M.J.; Matsuura, M.; Ercolano, B. The timing and location of dust formation in the remnant of SN 1987A. *Mon. Not. R. Astron. Soc.* **2015**, *446*, 2089–2101. [CrossRef]

138. Bethe, H.A. Supernova mechanisms. *Rev. Mod. Phys.* **1990**, *62*, 801–866. [CrossRef]

139. Bisnovatyi-Kogan, G.S. The Explosion of a Rotating Star As a Supernova Mechanism. *Astron. Zh.* **1970**, *47*, 813–816; translation in *Sov. A. J.* **1971**, *14*, 652–655.

140. Bisnovatyi-Kogan, G.S.; Moiseenko, S.G.; Ardelyan, N.V. Different magneto-rotational supernovae. *Astron. Rep.* **2008**, *52*, 997–1008. [CrossRef]

141. Seemann, H.; Biermann, P.L. Unstable waves in winds of magnetic massive stars. *Astron. Astrophys.* **1997**, *327*, 273–280.

142. Fischer, T.; Bastian, N.-U.F.; Wu, M.-R.; Baklanov, P.; Sorokina, E.; Blinnikov, S.; Typel, S.; Klähn, T.; Blaschke, D.B. Quark deconfinement as a supernova explosion engine for massive blue supergiant stars. *Nat. Astron.* **1954**. [CrossRef]

143. Shen, K.J.; Kasen, D.; Miles, B.J.; Broxton, J.; Townsley, D.M. Sub-Chandrasekhar-mass White Dwarf Detonations Revisited. *Astrophys. J.* **2018**, *854*, 52. [CrossRef]

144. Raddi, R.; Hollands, M.A.; Gänsicke, B.T.; Townsley, D.M.; Hermes, J.J.; Gentile Fusillo, N.P.; Koester, D. Anatomy of the hyper-runaway star LP 40-365 with Gaia. *Mon. Not. R. Astron. Soc. Lett.* **2018**, *479*, L96–L101. [CrossRef]

145. Shen, K.J.; Boubert, D.; Gänsicke, B.T.; Jha, S.W.; Andrews, J.E.; Chomiuk, L.; Foley, R.J.; Fraser, M.; Gromadzki, M.; Guillochon, J.; et al. Three Hypervelocity White Dwarfs in Gaia DR2: Evidence for Dynamically Driven Double-degenerate Double-detonation Type Ia Supernovae. *Astrophys. J.* **2018**, *865*, 15. [CrossRef]

146. Dickel, J.R.; van Breugel, W.J.M.; Strom, R.G. Radio structure of the remnant of Tycho's supernova (SN 1572). *Astron. J.* **1991**, *101*, 2151–2159. [CrossRef]

147. Chevalier, R.A. Are young supernova remnants interacting with circumstellar gas? *Astrophys. J.* **1982**, *259*, L85–L89. [CrossRef]

148. Archambault, S.; Archer, A.; Benbow, W.; Bird, R.; Bourbeau, E.; Buchovecky, M.; Buckley, J.H.; Bugaev, V.; Cerruti, M.; Connolly, M.P.; et al. Gamma-Ray Observations of Tycho's Supernova Remnant with VERITAS and Fermi. *Astrophys. J.* **2017**, *836*, 23. [CrossRef]

149. Bykov, A.M.; Ellison, D.C.; Marcowith, A.; Osipov, S.M. Cosmic Ray Production in Supernovae. *Space Sci. Rev.* **2018**, *214*, 41. [CrossRef]

150. Dwarkadas, V.V.; Chevalier, R.A. Interaction of Type IA Supernovae with Their Surroundings. *Astrophys. J.* **1998**, *497*, 807–823. [CrossRef]

151. Mazzali, P.A.; Röpke, F.K.; Benetti, S.; Hillebrandt, W. A Common Explosion Mechanism for Type Ia Supernovae. *Science* **2007**, *315*, 825–828. [CrossRef]

152. Townsley, D.M.; Jackson, A.P.; Calder, A.C.; Chamulak, D.A.; Brown, E.F.; Timmes, F.X. Evaluating Systematic Dependencies of Type Ia Supernovae: The Influence of Progenitor 22Ne Content on Dynamics. *Astrophys. J.* **2009**, *701*, 1582–1604.

153. Bell, A.R. Turbulent amplification of magnetic field and diffusive shock acceleration of cosmic rays. *Mon. Not. R. Astron. Soc.* **2004**, *353*, 550–558. [CrossRef]

154. Bell, A.R. Cosmic ray acceleration by a supernova shock in a dense circumstellar plasma. *Mon. Not. R. Astron. Soc.* **2008**, *385*, 1884–1892. [CrossRef]

155. Lucek, S.G.; Bell, A.R. Non-linear amplification of a magnetic field driven by cosmic ray streaming. *Mon. Not. R. Astron. Soc.* **2000**, *314*, 65–74. [CrossRef]

156. Raskin, C.; Scannapieco, E.; Rhoads, J.; Della Valle, M. Prompt Ia Supernovae are Significantly Delayed. *Astrophys. J.* **2009**, *707*, 74–78. [CrossRef]

157. Scannapieco, E.; Bildsten, L. The Type Ia Supernova Rate. *Astrophys. J.* **2005**, *629*, L85–L88. [CrossRef]

158. Alonso-Herrero, A.; Engelbracht, C.W.; Rieke, M.J.; Rieke, G.H.; Quillen, A.C. NGC 1614: A Laboratory for Starburst Evolution. *Astrophys. J.* **2001**, *546*, 952–965. [CrossRef]

159. Biermann, P.; Fricke, K. On the origin of the radio and optical radiation from Markarian galaxies. *Astron. Astrophys.* **1977**, *54*, 461–464.

160. Huchra, J.P. Star formation in blue galaxies. *Astrophys. J.* **1977**, *217*, 928–939. [CrossRef]

161. Huchra, J.P.; Geller, M.J.; Gallagher, J.; Hunter, D.; Hartmann, L.; Fabbiano, G.; Aaronson, M. Star formation in blue galaxies. I–Ultraviolet, optical, and infrared observations of NGC 4214 and NGC 4670. *Astrophys. J.* **1983**, *274*, 125–135. [CrossRef]

162. Aliu, E.; Archambault, S.; Aune, T.; Behera, B.; Beilicke, M.; Benbow, W.; Berger, K.; Bird, R.; Buckley, J.H.; Bugaev, V.; et al. Investigating the TeV Morphology of MGRO J1908+06 with VERITAS. *Astrophys. J.* **2014**, *787*, 166. [CrossRef]

163. Abeysekara, A.U.; Albert, A.; Alfaro, R.; Alvarez, C.; Álvarez, J.D.; Arceo, R.; Arteaga-Velázquez, J.C.; Rojas, D.A.; Solares, H.A.; Barber, A.S.; et al. Extended gamma-ray sources around pulsars constrain the origin of the positron flux at Earth. *Science* **2017**, *358*, 911–914. [CrossRef]

164. Soderberg, A.M.; Chakraborti, S.; Pignata, G.; Chevalier, R.A.; Chandra, P.; Ray, A.; Wieringa, M.H.; Copete, A.; Chaplin, V.; Connaughton, V.; et al. A relativistic type Ibc supernova without a detected γ-ray burst. *Nature* **2010**, *463*, 513–515. [CrossRef] [PubMed]

165. Seo, J.; Kang, H.; Ryu, D. The Contribution of Stellar Winds to Cosmic Ray Production. *J. Korean Astron. Soc.* **2018**, *51*, 37–48.

166. Chini, R.; Hoffmeister, V.H.; Nasseri, A.; Stahl, O.; Zinnecker, H. A spectroscopic survey on the multiplicity of high-mass stars. *Mon. Not. R. Astron. Soc.* **2012**, *424*, 1925–1929. [CrossRef]

167. Chini, R.; Nasseri, A.; Dembsky, T.; Buda, L.-S.; Fuhrmann, K.; Lehmann, H. Stellar multiplicity across the mass spectrum. In *Setting a New Standard in the Analysis of Binary Stars*; Pavlovski, K., Tkachenko, A., Torres, G., Eds.; EAS Publ. Ser., European Astron. Soc.: Geneva, Switzerland, 2013; Volume 64, pp. 155–162.

168. Heinz, S.; Sunyaev, R. Cosmic rays from microquasars: A narrow component to the CR spectrum? *Astron. Astrophys.* **2002**, *390*, 751–766. [CrossRef]

169. Mirabel, I.F. Microquasars and ULXs: Fossils of GRB Sources. In *Compact Binaries in the Galaxy and Beyond*; Tovmassian, G., Sion, E., Eds.; Rev. Mex. de Astron. Astrof. (Serie de Conf.); IAU Colloquium 2004, Volume 20, pp. 14–17. Available online: http://www.astroscu.unam.mx/~rmaa/ (accessed on 9 April 2019).

170. Mirabel, I.F. Microquasars. *Mem. Soc. Astron. Ital.* **2011**, *82*, 14–23.

171. Abeysekara, A.U.; Albert, A.; Alfaro, R.; Alvarez, C.; Álvarez, J.D.; Arceo, R.; Arteaga-Velázquez, J.C.; Rojas, D.A.; Solares, H.A.; Belmont-Moreno, E.; et al. Very high energy particle acceleration powered by the jets of the microquasar SS 433. *Nature* **2018**, *562*, 82–85. [CrossRef] [PubMed]

172. Vulic, N.; Hornschemeier, A.E.; Wik, D.R.; Yukita, M.; Zezas, A.; Ptak, A.F.; Lehmer, B.D.; Antoniou, V.; Maccarone, T.F.; Williams, B.F.; et al. Black Holes and Neutron Stars in Nearby Galaxies: Insights from NuSTAR. *Astrophys. J.* **2018**, *864*, 150.

173. Abbott, B.P.; Abbott, R.; Abbott, T.D.; Acernese, F.; Ackley, K.; Adams, C.; Adams, T.; Addesso, P.; Adhikari, R.X.; Adya, V.B.; et al. Multi-messenger Observations of a Binary Neutron Star Merger. *Astrophys. J. L* **2017**, *848*, L12. [CrossRef]

174. Abbott, B.P.; Abbott, R.; Abbott, T.D.; Acernese, F.; Ackley, K.; Adams, C.; Adams, T.; Addesso, P.; Adhikari, R.X.; Adya, V.B.; et al. GW170817: Observation of gravitational waves from a binary neutron star inspiral. *Phys. Rev. Lett.* **2017**, *119*, 161101. [CrossRef]

175. Albert, A.; André, M.; Anghinolfi, M.; Ardid, M.; Aubert, J.J.; Aublin, J.; Avgitas, T.; Baret, B.; Barrios-Martí, J.; Basa, S.; et al. Search for High-energy Neutrinos from Binary Neutron Star Merger GW170817 with ANTARES, IceCube, and the Pierre Auger Observatory. *Astrophys. J. Lett.* **2017**, *850*, 35. [CrossRef]

176. Pshirkov, M.S.; Tinyakov, P.G.; Kronberg, P.P.; Newton-McGee, K.J. Deriving the Global Structure of the Galactic Magnetic Field from Faraday Rotation Measures of Extragalactic Sources. *Astrophys. J.* **2011**, *738*, 192. [CrossRef]

177. Mao, S.A.; McClure-Griffiths, N.M.; Gaensler, B.M.;Brown, J.C.; van Eck, C.L.; Haverkorn, M.; Kronberg, P.P.; Stil, J.M.; Shukurov, A.; Taylor, A.R.; et al. New Constraints on the Galactic Halo Magnetic Field Using Rotation Measures of Extragalactic Sources toward the Outer Galaxy. *Astrophys. J.* **2012**, *755*, 21. [CrossRef]

178. Kronberg, P.P. *ISVHECRI 2012—XVII Int. Symp. on Very HE CR Interactions*; Gensch, U., Walter, M., Eds.; EPJ Web of Conferences: Berlin, Germany, 2013; Volume 52, p. 06004.

179. Heesen, V.; Krause, M.; Beck, R.; Adebahr, B.; Bomans, D.J.; Carretti, E.; Dumke, M.; Heald, G.; Irwin, J.; Koribalski, B.S.; et al. Radio haloes in nearby galaxies modelled with 1D cosmic ray transport using SPINNAKER. *Mon. Not. R. Astron. Soc.* **2018**, *476*, 158–183. [CrossRef]

180. Krause, M.; Irwin, J.; Wiegert, Th.; Miskolczi, A.; Damas-Segovia, A.; Beck, R.; Li, J.-T.; Heald, G.; Müller, P.; Stein, Y.; et al. CHANG-ES. IX. Radio scale heights and scale lengths of a consistent sample of 13 spiral galaxies seen edge-on and their correlations. *Astron. Astrophys.* **2018**, *611*, A72. [CrossRef]

181. Miskolczi, A.; Heesen, V.; Horellou, C.; Bomans, D.-J.; Beck, R.; Heald, G.; Dettmar, R.-J.; Blex, S.; Nikiel-Wroczyński, B.; Chyży Bomans, D.-J.; et al. CHANG-ES XII: A LOFAR and VLA view of the edge-on star-forming galay NGC 3556. *Astron. Astrophys.* **2018**, *622*, A9. [CrossRef]

182. Rossa, J.; Dettmar, R.-J. An Hα survey aiming at the detection of extraplanar diffuse ionized gas in halos of edge-on spiral galaxies. I. How common are gaseous halos among non-starburst galaxies? *Astron. Astrophys.* **2003**, *406*, 493–503. [CrossRef]

183. Merten, L.; Bustard, C.; Zweibel, E.G.; Becker Tjus, J. The Propagation of Cosmic Rays from the Galactic Wind Termination Shock: Back to the Galaxy? *Astrophys. J.* **2018**, *859*, 63. [CrossRef]

184. Albert, A.; André, M.; Anghinolfi, M.; Ardid, M.; Aubert, J.J.; Aublin, J.; Avgitas, T.; Baret, B.; Barrios-Martí, J.; Basa, S.; et al. Joint constraints on Galactic diffuse neutrino emission from ANTARES and IceCube Neutrino Telescopes. *Astrophys. J. Lett.* **2018**, *868*, L20. [CrossRef]

185. Nath, B.B.; Gupta, N.; Biermann, P.L. Spectrum and ionization rate of low-energy Galactic cosmic rays. *Mon. Not. R. Astron. Soc. Lett.* **2012**, *425*, L86–L90. [CrossRef]

186. Mirabel, I.F. The Formation of Stellar Black Holes. *New Astron. Rev.* **2017**. [CrossRef]
187. Mirabel, I.F. Stellar progenitors of black holes: Insights from optical and infrared observations. In *New Frontiers in Black Hole Astrophysics*; Cambridge University Press: Cambridge, UK, 2017; Volume 324, pp. 27–30.
188. Mirabel, I.F. Black holes formed by direct collapse: Observational evidences. In *New Frontiers in Black Hole Astrophysics*; Cambridge University Press: Cambridge, UK, 2017; Volume 324, pp. 303–306.

    MDPI

Article

Theoretical Discourse on Producing High Temporal Yields of Nuclear Excitations in Cosmogenic ^{26}Al with a PW Laser System: The Pathway to an Astrophysical Earthbound Laboratory

Klaus Michael Spohr [1,*], Domenico Doria [1] and Bradley Stewart Meyer [2]

[1] Extreme Light Infrastructure (ELI-NP) & Horia Hulubei National Institute for R & D in Physics and Nuclear Engineering (IFIN-HH), Str. Reactorului No. 30, P.O. Box MG-6, 077125 Bucharest– Măgurele, Romania; domenico.doria@eli-np.ro

[2] Department of Physics and Astronomy, Clemson University, Clemson, SC 29634-0978, USA; mbradle@clemson.edu

* Correspondence: klaus.spohr@eli-np.ro; Tel.: +40-21-404-2301

Received: 7 November 2018; Accepted: 14 December 2018; Published: 26 December 2018

Abstract: The development of the 10 PW laser system at the Extreme Light Infrastructure is a crucial step towards the realization of an astrophysical Earthbound laboratory. The interaction of high-power laser pulses with matter results in ultrashort (fs-ps) pulses of 10s of MeV ions and radiation that can create plasma and induce nuclear reactions therein. Due to the high fluxes of reaction-driving beam pulses, high yields of radioactive target nuclei in their ground and excited states can be provided in situ on short time scales. Cosmogenic ^{26}Al, which is of pronounced astrophysical interest, is a prime candidate for evaluating these new experimental possibilities. We describe how, for a short duration of $\Delta t \sim 200$ ps, laser-driven protons with energies above $E_p \sim 5$ MeV can induce the compound nucleus reaction ^{26}Mg$(p, n)^{26}$Al leading to high and comparable yields of the three lowest-lying states in ^{26}Al including the short-lived, $t_{1/2} = 1.20$ ns state at 417 keV. In the aftermath of the reaction, for a short duration of $t \sim$ ns, the yield ratios between the ground and the two lowest-lying excited states will resemble those present at thermodynamic equilibrium at high temperatures, thus mimicking high ^{26}Al entropies in cold environments. This can be seen as a possible first step towards an investigation of the interplay between those states in plasma environments. Theory suggests an intricate coupling of the ground state ^{26}Al$_{g.s.}$ and the first excited isomer ^{26m}Al via higher-lying excitations such as the $J = 3^+$ state at 417 keV resulting in a dramatic reduction of the effective lifetime of ^{26}Al which will influence the isotope's abundance in our Galaxy.

Keywords: laser-induced nuclear reactions; high-power laser systems; laser plasma; nuclear astrophysics; effective lifetime; ^{26}Al

1. Introduction

Laboratory measurements of nuclear reaction rates are made on target nuclei in their ground states. However, in stellar plasma conditions which govern the evolution of abundances in the Universe, the reacting nuclei are distributed among their excited states. Without providing these conditions, a proper experimental accounting of stellar models remains elusive. Nonetheless, direct experiments on the interplay of ground and exited states relevant for nuclear astrophysics have so far escaped scientific investigation since such studies are extremely difficult to realize in Earthbound laboratories driven by existing RF-based accelerator systems. This is mainly due to the short-lived, fs–ns nature of the relevant excited nuclear states in conjunction with the overall rather low instantaneous beam intensities deliverable by RF-based accelerator systems during that short timescales which result in low spontaneous yields for the promptly decaying excited states. As of today, reaction rate evaluations on excited states have not been measured and theory relies therefore on Hauser-Feshbach calculations [1] in which an assumption of a thermal population of excited states is undertaken. This approach is deemed to be very crude in general and is certainly incorrect for a handful of isotopes that have longer-lived isomers and for which internal thermalization does not therefore easily occur. For such nuclei, the equilibration of the longer-lived isomer and the ground state (g.s.) occurs only indirectly via upper-lying levels at high temperatures in the MK to GK regime. The cosmogenic ^{26}Al, the subject of this investigation, is perhaps the most prominent case in nuclear astrophysics for this kind of nuclei.

In the following discourse, we depict how the new generation of all-optical, high-power laser systems (HPLSs) based on chirped pulse amplification (CPA) invented by Strickland and Mourou [2] can address the problem of providing short-lived states with considerably high yields and hence can promote astrophysical research in the future.

2. Production and Decay of Cosmogenic ^{26}Al

2.1. Spontaneous Yields

Core to the depicted concept is the high spontaneous flux of ion-driven reactions as provided by a single short pulse from a HPLS which leads to high temporal yields $Y_i^d(t)$ of excited states by direct population (d) in a nucleus. Assuming that all nuclear levels are numbered subsequently according to their energy E_i, we find for the direct population of the i-th state;

$$Y_i^d(t) = N \cdot \sigma_i \cdot \phi \cdot (1/\lambda_i) \cdot \begin{cases} (1 - \exp(-\lambda_i \cdot t)), & \text{for } t < t_{\text{irrad}}, \\ [1 - \exp(-\lambda_i \cdot t_{\text{irrad}})] \cdot \exp(-\lambda_i \cdot (t - t_{\text{irrad}})), & \text{for } t \geq t_{\text{irrad}}; \end{cases} \quad (1)$$

wherein N is the number of atoms exposed to the beam in the irradiated sample, σ_i the energy dependent cross-section for the direct population of the i-th state in barns, ϕ the beam flux and λ_i the total decay rate of the i-th level in s^{-1}. The parameter t_{irrad} represents the irradiation time during which the beam of incoming particles triggers the reaction. For simplification, all the aforementioned parameters are seen to be time-independent and thus constant during t_{irrad}. The factor $(t - t_{\text{irrad}})$ is the time elapsed after the production of a specific nucleus and its states via an irradiating beam has stopped. Equation (1) is valid as long as $\sigma_i \cdot \phi \ll \lambda_i$ which is the case even for the high, instantaneous beam fluxes, ϕ that are achievable with a HPLS. Equation (1) only describes the direct population of the i-th state by the reaction-driving pulse and needs to be extended if an indirect population (feeding) of this state via higher-lying energetic states occurs. For this more general case, the total yield $Y_i(t)$ for the i-th state, assuming that $1 \leq i < k \leq N_{\text{tot}}$, can be calculated via,

$$Y_i(t) = Y_i^d(t) + \sum_{k>i}^{N_{\text{tot}}} f_{ki} \cdot \int_0^t |A_k(t')| \, dt' \quad (2)$$

in which

$$A_k(t') = -\frac{dY_k(t')}{dt'} \quad (3)$$

is the instantaneous activity of the feeding state k with half-life λ_k and f_{ki} the partial branching ratio of the k-th state to the i-th state. The variable N_{tot} describes the total number of states in the nucleus which feed into the level at E_i. For complex systems in which a long decay chain with constant initial activities $A_k(0)$ exists, Equation (2) will result in the Bateman equation for the i-th state [3]. To include the possibility that the feeding state at E_k is also directly populated by the flux of particles with a specific cross-section of σ_k, the instantaneous activity $A_k(t')$ defined in Equation (3) was used in Equation (2) for the calculations in this work. We also exclude the possibility of any induced upward transitions from E_i to the higher-lying states at E_k via photo-absorption as no extremely hot equilibrated plasma condition can be provided by a HPLS for which those photo-absorption processes would appear with high enough yields to change Y_i.

Equation (1) shows Y_i scales linear with ϕ. Experiments show that $\sim 10^{8-9}$ times higher values for the spontaneous beam fluxes ϕ are achievable with HPLS if compared to typical values associated RF-based technology. Though these extremely high fluxes can only be delivered for ultrashort irradiation time spans t_{irrad} which are in the order of the lifetime of short-lived states, the resulting enhancement in the presented case, $t_{1/2} = 1.20\,\text{ns}$ level at 417 keV in ^{26}Al, will still be in the order of 10^{7-8} larger when compared to a DC or RC beam from a conventional accelerator system, since the crucial factor $[1 - \exp(-\lambda \cdot t_{irrad})]$ is still around ~ 0.1. The other important factor in Equation (1) is the exponential decline after t_{irrad} which is given by the term $\exp(-\lambda \cdot (t - t_{irrad}))$. This term expresses the fact that any time-resolved probing of the interplay of states including short-lived isomers must occur within a short timeframe itself, which can only be realized by a generically short probing pulse as provided by a HPLS. Therefore, to allow such kinds of studies at e.g., the Extreme Light Infrastructure for Nuclear Physics (ELI-NP), two synchronized laser beamlines are implemented. The possibility of ultrafast probing is another unique feature of HPLS-technology.

In addition to a direct reaction-induced production of a nuclear state, a population enhancement of states of interest can often be achieved via the feeding from states at higher energies as shown in Equation (2). This circumstance, which will become especially interesting once mono-energetic beams, can be delivered by a HPLS.

It is also important to note that Equation (1) only applies for a single pulse of short duration and the overall total reaction yields driven by a HPLS are some orders of magnitude smaller than those achievable with most of the RF-based systems over a prolonged experimental campaign. However, in summary, it is clear that if t_{irrad} is in the order of the lifetimes of short-lived excited states ($\tau_{short} = 1/\lambda_{short}$), their yields, relative to longer living, e.g., few seconds, or even stable states, can be optimized for time spans in the order of τ_{short}.

2.2. Yield Distributions of Excited States and Temperature Equivalents

The yield ratios between the ground and exited states can be mapped to a corresponding Maxwell-Boltzmann distribution describing a thermal equilibrium represented by a single temperature T via,

$$\frac{Y_2(t)}{Y_1(t)} = \frac{(2 \cdot J_2 + 1)}{(2 \cdot J_1 + 1)} \cdot \exp{-\left(\frac{E_2 - E_1}{k_B T}\right)}, \tag{4}$$

in which Y_1 and Y_2 describe the state populations at a given time, J_1 and J_2 are the spin values of the nuclear states at the energies E_1 and E_2. Equation (4) has been used for the matching of state populations to temperature values of evaporation fragments emitted from a compound nucleus after a fusion reaction. Thus, the appliance of this concept to laser-induced nuclear compound reactions seems more than adequate, especially as the time spans involved are similar to the typical interaction times which can be provided by a HPLS. The formation stage for the compound nucleus takes a period of time approximately equal to the time interval for the bombarding particle to travel across the diameter of the target nucleus which is $\sim 1 \times 10^{-21}$ s. After a relatively long period during which the compound system has thermally equilibrated which can extend up to 1×10^{-15} s, it disintegrates, usually into an ejected small particle and a larger evaporation fragment nucleus which inherits the compound's temperature represented by the corresponding value of $k_B T$. Thus, after a correction for the energy given to the rotational motion of the residuals is applied, one can deduce on the internal temperature of a reaction residual from its measured population ratio. The validity of this approach was e.g., demonstrated by Morrissey et al. [4] who proved that the population ratio does represent the internal temperature of a residual nucleus for a wide range of internal excitation energies. Hence, $k_B T$ will represent the temperature of a nucleus if equilibrated in hot plasma. Obviously, in an experiment no hot plasma conditions can be provided in the matter surrounding the compound nucleus sustaining an equilibrium in a secondary reaction target. Therefore, the high value for $k_B T$ within the nucleus will only prevail for the fleeting time spans associated with the shortest lifetimes of the decaying excited states that were populated. High-power laser accelerators can create this special state of synthetic entropy condition where we will find a gamut of hot nuclei in their short-lived excited states in a surrounding cold environment after a reaction-triggering laser pulse. During the associated short time spans the temperature, which was present in the nuclear matter at time of the evaporation, is normally equivalent to $k_B T$ values in the MeV regime which equates to hot plasma temperatures in the GK region. It is worth noting that of course one can produce excited states in nuclei, for example via nuclear reactions or beta decay, or in-flight whereby γ-rays are detected with arrays of germanium detectors around a target position with RF-based technology. However, it is the provision of high spontaneous yields which makes a HPLS an interesting tool for selected cases of nuclei in which the equilibrium studies on thermalized conditions are particularly hard to achieve due to the interplay of long- and short-lived isomers. Moreover, the high temporary yields of hot evaporation residues can, in principle, be probed by a secondary ion- or γ-beam originating from a second synchronized laser pulse. We again must stress the fact that no real plasma environment at hot temperatures will exist in such a hypothetical experiment; it is just the population distributions of the excited states in plasma which is mimicked by the inheritance of the compound nucleus' temperature.

The cosmogenic ^{26}Al, was chosen for this evaluation due to its importance for astrophysics and some already existing or soon to be published experimental work [5,6]. Our work here will give an answer to the long-standing questions: "What is a unique feature of all-optical PW accelerators which cannot be achieved with traditional RF-based accelerator technology and how can the the the CPA technology invented by Strickland and Mourou be best applied to enrich fundamental nuclear physics research?". We believe that we can answer this question satisfactorily as we depict the unique ability of high-intensity laser acceleration to allow the production of high temporal yields of ultrashort-lived nuclear species in their ground and excited states, which is a crucial step towards the realization of an astrophysical laboratory. The supply of high yields of short-lived states positively distinguishes HPLS-driven accelerators from standard RF-based technology. In the future, these states can be may subjected to further probing by e.g., creating a low-temperature plasma environment such as WDM during the reaction processes or using an ultraintense probing X-ray beam in coincidence. Moreover, our work will show that already existing TW and PW laser-driven systems can be used to measure hitherto elusive in situ nuclear reaction rates in plasma to underpin astrophysical theory. As such, the presented discourse promotes future nuclear research at facilities such as ELI-NP and elucidates how they will help understanding of the complex chemical evolution within our Galaxy. In this respect one needs to point out that it is e.g., already possible to create warm dense-matter conditions in the laboratory [7,8]. Even the study of exotic fusion-fission reactions in the quest to produce very neutron-rich isotopes to understand the production of heavy nuclei in the Universe [9] is envisaged at ELI-NP.

2.3. The Importance of ^{26}Al for Astrophysics

To understand the astrophysical abundances in our Galaxy, the cosmogenic isotope ^{26}Al is of outstanding astrophysical significance as it is a core isotope for γ-ray astronomy [10]. Its presence in the Galaxy was discovered by Mahoney et al. [11]. Theoretically, the isotope derives from explosive helium, carbon, oxygen, and silicon burning cycles [12], as well as novae or supernovae explosions, Wolf-Rayet Stars, red giants, and supermassive stars [13,14]. Moreover, the correlated excesses of its daughter product ^{26}Mg in comparison with stable ^{27}Al provide important constraints on the formation time of primitive meteorites and components therein [15–17]. The total galactic abundance of ^{26}Al is estimated to be around $2\,M_\odot$ to $3\,M_\odot$. State-of-the-art theories only account for a fraction of this value. A study of the effective lifetime τ_{eff} of ^{26}Al is henceforth necessary. Any related experimental investigation must account for the rather unique and complex decay pattern of the isotope which is sketched in Figure 1.

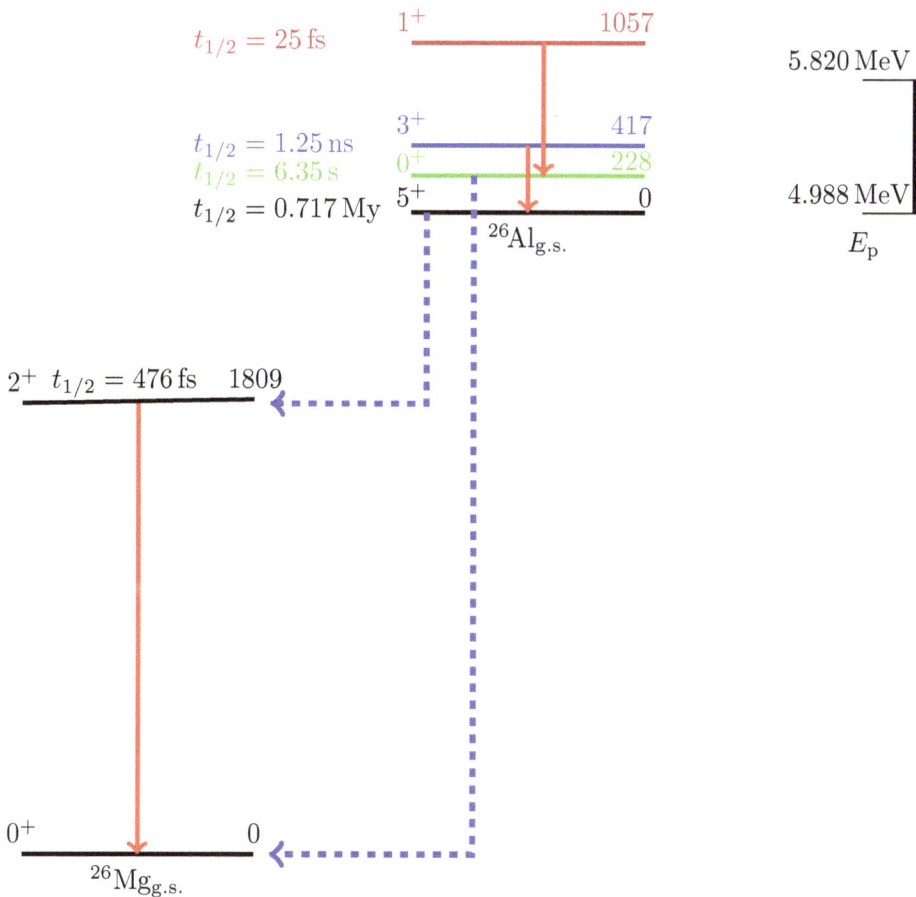

Figure 1. Simplified decay pattern of ^{26}Al including the lowest four states. The two β^+ decay branches are depicted with the blue dotted lines. The γ-transitions between the states are indicated with red arrows, such as the 417 keV state which solely decays into the ground state ^{26}Al$_{g.s.}$ To the right of the level scheme, the proton energy range 4.988 MeV $\leq E_p \leq$ 5.820 MeV surveyed for the reaction ^{26}Mg(p,n)^{26}Al by Skelton et al. is depicted [18].

In cold environments, ^{26}Al decays via two prominent β^+ decay routes from its ground and first excited state which are characterized by very different values of their degree of forbiddenness. This special scenario complicates the evaluation of the effective ^{26}Al decay rate λ_{eff} in a stellar plasma as one cannot assume the first two states existed in an equilibrated thermal distribution. Specifically, the decay from its $J = 5^+$ ground state is a second-order forbidden β^+ decay with a corresponding long half-life of $t_{1/2} = 0.717$ My leading to a branching of 97.24% to the first excited state in ^{26}Mg at 1809 keV. The β^+ decay is followed by a quasi-prompt emission ($t_{1/2} = 476$ fs) of the corresponding γ-ray in ^{26}Mg. It was this transition which was discovered by Mahoney et al. with the HEAO-3 satellite surveying the radiation background of the Galaxy in 1984 [11]. On the other hand, the first excited state of ^{26}Al at 228 keV (^{26m}Al) decays via a rather fast ($t_{1/2}(^{26m}$Al$) = 6.35$ s) super-allowed $0^+ \longrightarrow 0^+$ β^+ decay directly into the stable ground state ^{26}Mg$_{g.s.}$. Due to the high spin difference, a direct electromagnetic transition from the 228 keV level to the ground state $0^+ \longrightarrow 5^+$ in ^{26}Al has an extremely low transition probability and, as such, one is forced to treat the decay of ^{26}Al$_{g.s.}$ and ^{26m}Al as different species in theoretical evaluations. To shed light on the overall destruction of ^{26}Al one needs to estimate the interplay between these two states in hot environments to be able to calculate the effective decay rate λ_{eff} in stellar plasma.

Theory predicts that a thermal equilibrium between the long-lived 5^+ ground state and the first isomeric 0^+ state at 228 keV will occur at high temperatures (1×10^6 K to 1×10^9 K) via a manifold of interlinking high-energy transitions which induce a population of short-lived (fs–ns) high-lying energy levels. This will result in a much shorter τ_{eff} for the ^{26}Al in hot environments. The groundwork of this interpretation was done by Ward and Fowler [19], followed by a more complex analysis by Coc et al. in 1999 [20]. Both efforts are cited in the most elaborated work to date, which focuses on the decay of ^{26}Al in thermodynamic equilibrium by Gupta and Meyer [21]. In that work a new, holistic approach that introduces a dissection of the structure of ^{26}Al into two different ensembles e_1 and e_2 is based on whether the most prominent decay path feeds into the ground state (e_1) or the first excited state (e_2) at cold temperatures. They then calculate in detail how these two ensembles start to merge into a single Maxwell-Boltzmann distribution. In detail, the authors evaluated the Einstein coefficients of induced absorption between the lowest-lying 64 states in ^{26}Al which allows the determination of the effective transition rate between the two assumed species represented by e_1 and e_2, labelled λ_{12} for ^{26}Al as a function of temperature T. Their results show that below $T \sim 0.4$ GK the ^{26m}Al metastable state has no chance to equilibrate with the ground state before decaying via ^{26m}Al. The scenario dramatically changes above $T \sim 0.4$ GK where internal equilibration of ^{26}Al competes with the β^+ decay from the metastable ^{26m}Al thus influencing the transition between the two species. This equilibration leads to a much-enhanced λ_{eff}. The increase of λ_{eff} between 0.1 GK to 10 GK is around 25 orders of magnitude (1×10^{-10} s^{-1} to 1×10^{15} s^{-1}) thus underpinning the need to find experimental pathways for investigations. It is, however, clear that a technical realization of such a hot plasma in the above-cited temperature regime will experimentally remain elusive for the foreseeable future. However, the production of the short-lived states via the decay of a hot nuclear compound results in the production of sufficiently high yields of the excited 417 keV 3^+ state, comparable to those of the first two states in ^{26}Al for a duration of several ns (see Equation (1)). According to Gupta and Meyer, the aforementioned 2nd excited state plays the most crucial role in steering the equilibration between the two ensembles. As such, the provision of the short-lived 417 keV isomer with high yield can be seen as a first step towards measurements at short-lived excited states and their interplay with longer-lived states. This scenario will become particularly interesting if, in future, high fluxes of fs-long probing X-rays can be delivered in coincidence with the production of the excited states or warm dense-matter conditions (WDM) with the driving proton beams. Again, in both cases, no extremely hot plasma conditions will exist but still such investigations must be seen as experimental stepping stones on the pathway to an astrophysical laboratory.

2.4. Cross-Section of ^{26}Al and its Production in Laser Plasma Experiments

The most effective way to produce ^{26}Al in situ is by harvesting laser-induced proton acceleration. In the associated compound nucleus reaction ^{26}Mg(p,n)^{26}Al in which the formed compound nucleus ^{27}Al is excited by the energy above the reaction threshold of the impacting proton. The emittance of a neutron in the exit channel leads to excited states in ^{26}Al as well as directly to the ground state ^{26}Al$_{g.s.}$. Due to the prominence of this reaction, precise measurements of the individual cross sections for ^{26}Al$_{g.s.}$, ^{26m}Al and the 3^+ level at 417 keV, ^{26}Al$_{417}$, have been carried out over three decades ago by Skelton et al. [18]. Their experimental campaign facilitated an efficient neutron detector, the graphite-cube neutron-detection system (GCNDS) at the California Institute of Technology in Pasadena, CA, US. The cross-section measurements spanned over a proton energy range of almost 1 MeV from the reaction threshold at $E_p = 4.988$ MeV to $E_p = 5.820$ MeV (see Figure 1). A very precise average energy resolution for the cross sections of $\delta E = 2.5$ keV was achieved by measuring a total of

328 different energy values [22]. In detail they examined three different physical quantities to conclude on the reaction cross sections for the first three states in ^{26}Al. Details are displayed in Table 1. Therein n_{tot} refers to all neutrons emitted in the reaction process while n_{228} and n_{417} represent neutron channels that lead to the population of the first or second excited state at 228 keV and 417 keV respectively.

Based on the published data, the integral cross sections σ^{int} and their uncertainties for the three lowest-lying states in ^{26}Al were calculated in this work. The relative uncertainty, $\delta\sigma^{int}_{g.s.}/\sigma^{int}_{g.s.}$, of the integral cross-section for ^{26}Al$_{g.s.}$, was found to be higher than those of the other two integral cross sections, as the integral cross section for the population of the ground state $\sigma^{int}_{g.s.}$ was not measured, but calculated from the difference of the measured total neutron yield and the sum of the yields for n_{417} and n_{228}. In the experiment it was not possible to directly measure the produced amount of ^{26}Al$_{g.s.}$, an unfortunate restriction which will also apply for future high-power laser-based investigations. This is due to the overall low total production yield which will not allow a direct quantification of $\sigma^{int}_{g.s.}$. Firstly, the very long lifetime of ^{26}Al$_{g.s.}$ does not permit the measurement of its decay with any radiation detector. Moreover, the expected total yields for the ground state will be in the order of $\sim 1 \times 10^5$ to $\sim 1 \times 10^7$ per pulse for a thick target which is still far below the sensitivity of conventional mass separators. The installation of an online isotope separation system including a Penning-trap; as e.g., depicted in [23] would allow to measure smallest amounts of ^{26}Al$_{g.s.}$ that would emerge from a thin target after a laser pulse; could potentially address this issue. However, to our knowledge, such an online isotope separation system is not foreseen to be implemented at any upcoming HPLS worldwide in the near future.

Table 1. Calculated integral cross sections σ^{int} and methods used to measure $\sigma(E)$ for the lowest three states in ^{26}Al.

State	σ^{int}/MeVmb	Method Used in [18]
$\sigma^{int}_{g.s.}$	4.9(9)	Total neutron yield from the ^{26}Mg$(p, n_{tot})^{26}$Al reaction measured by the GCNDS. Note, due to the very long half-life of ^{26}Al$_{g.s.}$, no delayed 511 keV annihilation radiation yield will be measurable emerging from the associated ground state decay.
σ^{int}_{228}	30.2(3)	Delayed 511 keV annihilation radiation yield from the ^{26}Mg$(p, n_{228})^{26m}$Al reaction measured by two NaI(Tl) detectors in coincidence. An irradiation-count cycle with $t_{irrad} = 6$ s and $t_{count} = 30$ s was applied.
σ^{int}_{417}	19.3(9)	Prompt 417 keV γ-ray yield from the ^{26}Mg$(p, n_{417})^{26}$Al using an ultrathin target with a thickness of 8.5 µgcm^{-2}. The prompt radiation was measured by a 60 cm^3 Ge(Li) detector.

We assume that integral cross sections will be valid for the high instantaneous fluxes ϕ of protons at a HPLS as plasma conditions will not prevail in the secondary target. The precise measurements as depicted by Skelton et al. [18] cannot be repeated at a HPLS for the foreseeable future. First and foremost, the systems are not capable of delivering high-flux pulses of reaction-driving mono-energetic protons with high frequency. Hence, overall yields and associated measurement statistics derived from a laser plasma experiment will lack behind the results obtainable with RF-based technology. Indeed, laser accelerators are best described as high-intensity low-repetition-rate systems if compared to RF-based technology. In addition, the most widely applied ion acceleration mechanism is the target normal sheath acceleration (TNSA) scheme as described by e.g., Wilks et al. [24]. Although proton energies of up to 100 MeV are achievable with TNSA [25], their proton energy distribution dN/dE_p inherits the Maxwell-Boltzmann energy distribution of the initially accelerated electrons and is therefore not mono-energetic, hence HPLS will practical allow only the measurement of integral cross sections at present.

Moreover, a direct efficient way of a measurement of neutrons in the way presented by Skelton et al. [18] is also not feasible for the time being at any existing PW laser site as they used a highly efficient neutron detector surrounding the relatively small target chamber. For comparison, the target chambers used in high-power laser research will have a volume in the order of several m^3 and hence cannot be surrounded by a 4π detector system.

Additionally, during any proton pulse, an in situ measurement of the 417 keV transition is also not possible, as the initially produced proton beam coincides with intense soft and hard X-ray background radiation producing a strong electromagnetic pulse (EMP). Therefore, the only observable in a laser-driven ^{26}Al experiment will be the 511 keV delayed annihilation line, from which $\sigma_{228}^{\mathrm{int}}$ can be derived. Nonetheless, we will be able to scale the presumed yields for the ground state and second excited state at 417 keV based on the calculated integral cross sections $\sigma_{\mathrm{g.s.}}^{\mathrm{int}}$ and $\sigma_{417}^{\mathrm{int}}$ using the measured value of $\sigma_{228}^{\mathrm{int}}$ as reference.

An inaugural *prima facie* experiment was undertaken with the Vulcan laser at the Rutherford Appleton Laboratory (RAL) in which the ^{26}Mg(p, n$_{228}$)$^{26\mathrm{m}}$Al reaction was studied with laser accelerated protons created by 300 TW pulses at a rate of $\sim f = 2 \times 10^{-4}$ Hz. Protons well above the reaction threshold at $E_{\mathrm{p}} = 4.988$ MeV could be produced. A delayed γ-radiation line measured in situ at 511 keV with $t_{1/2}^{\mathrm{exp}} = 6.6(3)$ s was unambiguously identified as the annihilation radiation associated with the β^+ decay of $^{26\mathrm{m}}$Al. The total yield per pulse measured was as high as 5×10^5 of $^{26\mathrm{m}}$Al. They produced nuclei were concentrated within a volume of ~ 0.08 cm^3 in the secondary MgO reaction target. For more details see [5,6].

3. Theoretical Evaluation of Achievable Population Distributions for the First Three Excited States in ^{26}Al in Laser Induced Plasma Experiments

3.1. General Considerations

Based on Equations (1) and (2) and the values of the integrated cross sections σ^{int} from Table 1, as well as the results of the inaugural experimental study, the achievable population yields of the lowest three states in ^{26}Al were theoretically estimated. A short proton irradiation period, $t_{\mathrm{irrad}} = 200$ ps was assumed. To adjust such a short t_{irrad} in an experiment, one needs to place the ultrathin plastic sheets used as primary targets for proton production in very close proximity to the secondary reaction target, consisting of ^{26}Mg or ^{26}MgO in isotopically enriched form. This was applied in the aforementioned inaugural test experiment ah RAL. In such close geometry one is able to minimize the spread in the time-of-flight, δt_{TOF}, between the most energetic protons produced and those that will be just above the reaction threshold at $E_{\mathrm{p}} = 4.988$ MeV [6]. It is worth pointing out that the minimization of δt_{TOF} is crucial to provide a short t_{irrad}. Distances between the primary production target and the secondary reaction target of less than 1 cm are technical feasible to date. Choosing a much closer distance could mean that the secondary production target maybe damaged or even destroyed by a single laser pulse at high intensity. Obviously, the original 25 fs duration of the initial laser pulse which drives the proton production cannot be sustained, but with distances in the cm regime, δt_{TOF} values below 100 ps can be achieved. At any experiment one ideally would maximize the intensity of protons in the relevant energy rather than producing higher energetic protons with e.g., $E_{\mathrm{p}} \geq 9$ MeV. Also, the high magnitude of protons with energies below the reaction threshold maybe facilitated in future experiments to create conditions of WDM in the volume within the secondary target where the reaction takes place. That alone is worth a consideration, as such scenarios cannot be achieved with conventional RF-based technology.

Isotopically enriched ^{26}Mg target comes at a high cost. Natural magnesium cannot be taken due to a series of intruder reaction channels which would make the identification of the $t_{1/2} = 6.35$ s decay of $^{26\mathrm{m}}$Al impossible. A crucial parameter is also the thickness of the ^{26}Mg target. To enhance the yield, one would install a bulky target which would extend to thicknesses in the sub-mm to mm regime. In the *prima facie* experiment at RAL a thick target consisting of MgO powder was used in which all protons were stopped [5]. This optimization of the target thickness was based on SRIM [26] calculations in which the condition was set that even the most energetic protons produced in the experiment with $E_p \sim 9\,\text{MeV}$ are deaccelerated to energy values below the reaction threshold within the secondary target. A thickness of 1 mm was derived. SRIM can also be used to calculate the corresponding time t_{stop} by which all protons are reduced to kinetic energies below $E_p = 4.988\,\text{MeV}$. It was found that within a duration of $t_{\text{stop}} = 100\,\text{ps}$ the electronic and nuclear stopping has reduced all impacting protons with $E_p \sim 9\,\text{MeV}$ to an energy below the reaction threshold in the chosen thick target. With respect to t_{stop} it is important to mention that the stopping of the fast protons occurs already during the time less energetic, hence slower, reaction-driving protons arrive at the target. Hence, a crude estimate of a maximal irradiation time of $t_{\text{irrad}} = 200\,\text{ps}$ ($\delta_{\text{TOF}} + t_{\text{stop}}$) can be assumed. Most importantly, for the case study on ^{26}Al, is the fact this value is still well below the lifetime of the crucial short-lived second excited state at 417 keV with $t_{1/2} = 1.20\,\text{ns}$. With optimized geometry and targetry the assumed 200 ps value for t_{irrad} can easily be halved in future experiments with PW systems.

It is important to mention that HPLS will eventually be capable of producing macroscopic sheets of protons in which collective stopping effects will lead to a much-increased stopping power and hence reduced stopping times. In this work we do not consider these potential effects, but, as well as minimizing t_{irrad}, an enhanced stopping will result in higher plasma temperatures than those e.g., characterizing WDM conditions, offering new pathways for astrophysical investigations in the future.

3.2. Simulations of Population Yields for ^{26}Al obtained by a Short Pulsed Laser Proton Beam of High Intensity

Figure 2 shows the calculated distributions of $Y_{\text{g.s.}}$, Y_{228} and Y_{417} deduced from Equations (1) and (2) assuming $t_{\text{irrad}} = 200\,\text{ps}$. The extracted values for the associated integral cross sections σ^{int} in the region between $E_p = 4.988\,\text{MeV}$ and $E_p = 5.820\,\text{MeV}$ were taken from Table 1. We assume the proton energy distribution, dN/dE_p, within this range to be constant as a first order approximation. The time zero $t_0 = 0$ which is truncated in the logarithmic display, refers to the time of the first reaction induced by the fastest protons arriving in the secondary production target. Although the yield depicted on the $y-$abscissa is in arbitrary units, a.u., the value is aligned with obtainable levels of concentration in units of $1\,\text{cm}^{-3}$ deduced from the initial measurements of ^{26m}Al at the RAL. The values can be seen as a minimum estimate for achievable yield concentrations with a PW laser system. We assume all consecutive pulses to impact on the same volume within the secondary production target, hence the yield for $^{26}\text{Al}_{\text{g.s.}}$, $Y_{\text{g.s.}}$, builds up continuously after each pulse due to the very long lifetime of the ground state, $t_{1/2} = 0.717\,\text{My}$. In addition, the ground state is populated by direct feeding from the 417 keV level (Equation (2)) which decays via an E2 transition (see Figure 1). As ^{26}Al is cosmogenic we can assume its original concentration in Mg or MgO, before the first shot, in the secondary target probe to be $Y_{\text{g.s.}}^0 \sim 0$. Moreover, we consider that two sequential proton pulses are separated by an interval of $t = 100\,\text{s}$ for simplicity which reflects roughly the time interval between two consecutive pulses from the 10 PW system at ELI-NP. As a result, the yield for ^{26m}Al, Y_{228}, does not build up between two consecutive shots which would be the case for the 1 PW and 0.1 PW systems at ELI-NP as they operate at higher repetition rates $\sim$ Hz. Most interestingly is the fact that $\sigma_{\text{g.s.}}^{\text{int}}$ describing the direct population of the ground state, is significantly lower than for the excited states but builds up continuously in the active secondary target volume after each pulse due to its long lifetime. We calculated these conditions for a series of consecutive laser pulses, assuming the target to remain undamaged during the impact of all subsequent pulses. Crucially for the understanding of temporal yield evolution is the fact that the irradiation time $t < 200\,\text{ps}$ is still lower than the lifetime of the 417 keV state which allows this state to accrue comparable yields with regard to the first excited state at 228 keV per single pulse. For the first couple of pulses this is even true for a comparison with $Y_{\text{g.s.}}$ as can be derived from Figure 2. The two yields Y_{228} and Y_{417} remain unchanged for the two consecutive pulses shown in Figure 2 and for any additional subsequent pulse. This is a very special condition only achievable with the ultrashort pulses provided by an all-optical accelerator system. Consecutive pulses will only lead to higher values for $Y_{\text{g.s.}}$ in the same target volume.

For the near future, a total of several 100 pulses with a 10 PW laser system can be seen as a reasonable estimation for the number of total shots deliverable in a single experiment. With optimized target geometry, the appliance of proton focusing and PW induced proton pulses, potentially of mono-energetic nature, one can expect yields to be several orders of magnitude higher than those sketched in this work based on the RAL experiment. As shown in Table 1, the decay of the 511 keV delayed annihilation radiation will be the only measurable entity in any future laser-driven experiment while the other intensities would have to be calculated in reference to the precise cross-section data of Skelton et al. [18]. It is also important to point out that, if the irradiated volume is changed on a pulse-by-pulse base, each consecutive pulse will impact on a region where no measurable yield for ^{26}Al$_{g.s.}$ existed before. The importance of that scenario and associated technical aspects will be depicted in the aforementioned forthcoming publication [6]. However, it is clear that due to the high costs of isotopically enriched ^{26}Mg an experiment relating on a constant supply of non-irradiated ^{26}Mg will be not feasible for budget and resourcing reasons.

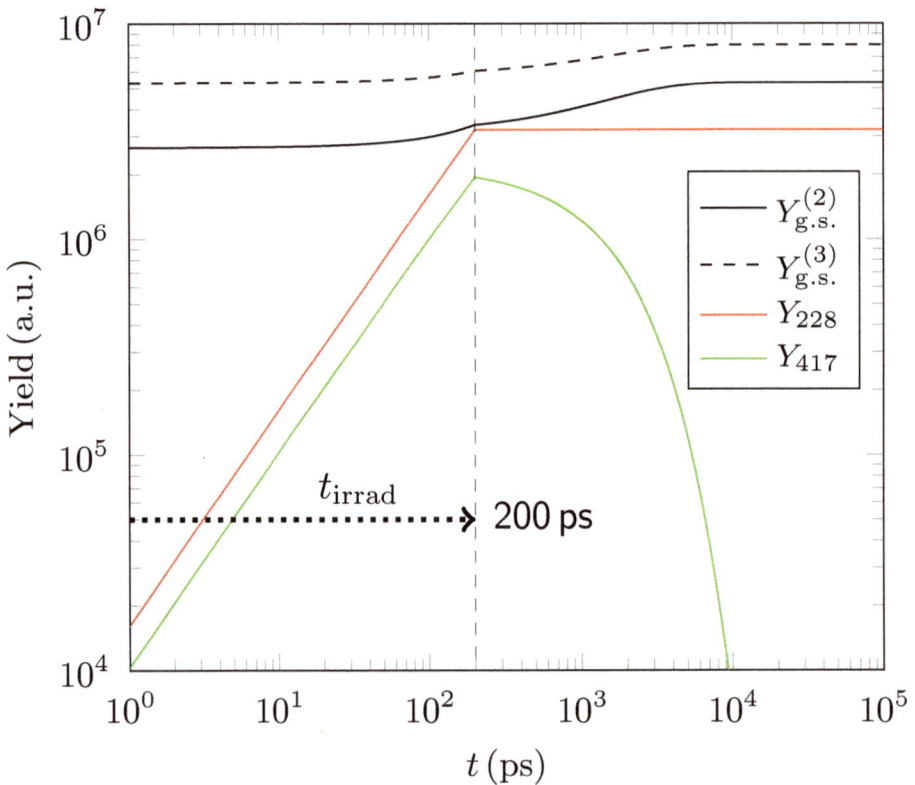

Figure 2. Calculated yields Y_{228} (red) and Y_{417} (green) as function of time t per proton pulse with $t_{\mathrm{irrad}} = 200$ ps (vertical dashed line) and $4.988\,\mathrm{MeV} \leq E_p \leq 5.820\,\mathrm{MeV}$. The accumulated yields $Y_{g.s.}$ for the ground state are superimposed for the 2nd (solid black) and 3rd (dashed black) consecutive proton pulses. We assume those pulses to impact on the same target volume in the secondary production target. Note the influence of the direct feeding from the 417 keV level to the ground state from the non-linear enhancement of $Y_{g.s.}$ during t_{irrad}. A time interval of 100 s between the 2nd and 3rd pulse was assumed.

3.3. Thermodynamic Temperature Equivalents from the Distribution of the First Three Excited States in ^{26}Al in Laser Plasma Experiments

Using the Maxwell-Boltzmann distribution as depicted in Equation (4) the ratios of the two yield distributions; $Y_{228}/Y_{g.s.}$ and $Y_{417}/Y_{g.s.}$ were mapped into temperature equivalents to emphasize the unique ability of laser-driven experiments to facilitate the short, reaction-driving, ion pulses to mimic temperature-equivalent scenarios as present in hot interstellar conditions. In absence of a real hot plasma this interpretation exploits the fact that the hot state of the ^{26}Al compound residue is conserved for fleeting lifetimes after the reaction-driving pulse has impacted. The results for the consecutive pulses numbered, 1, 2, 3, 5, 10, 25, 50 and 100 are displayed in Figure 3.

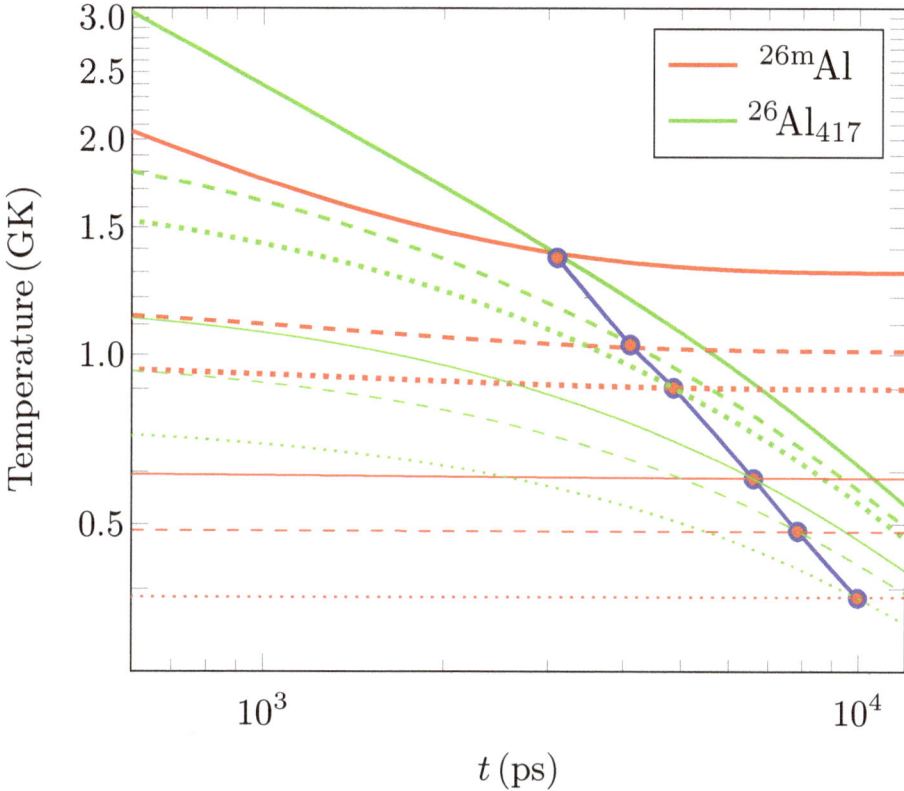

Figure 3. Calculated T equivalents for $Y_{228}/Y_{g.s.}$ (red lines) and $Y_{417}/Y_{g.s.}$ (green) from yield distributions according to Maxwell-Boltzmann distribution, for consecutive pulse numbers 1 (solid thick), 2 (dashed thick), 3 (dotted thick), 5 (solid thin), 25 (dashed thin) and 100 (dotted thin). The protons are considered to irradiate the same volume in the secondary target. The times t_{equi} for each pulse at which the two yield ratios converge to resemble one temperature value T are indicated by the blue & red circles which are connected to a blue line to guide the eye.

Theory states that to infer a temperature between the ground state and the first excited state in the case of ^{26}Al is not a straight forward concept, as those states are only connected via higher-lying excited states in low to middle temperature plasma. Nevertheless, the temperature equivalents between these two states, based on their yield population, remain in the GK temperature regime due to the short times of the populating laser proton pulse. At such high temperatures, thermalization can be assumed in stellar environments [21]. Most crucially, for any successive pulse there exist a defined time t_{equi}, where the T equivalents of the first two excited states are equal, mimicking the conditions of a thermal equilibrium distribution in the absence of real hot temperature conditions in the production target. For the calculated cases, this regime stretches from 3.10 ns in which the temperature equivalent is 1.3 GK for the first pulse to 9.95 ns with $T = 0.4$ GK for the 100th pulse. Any probing X-ray flash or secondary ion beam would be best timed around t_{equi} to exploit that artificially stellar-like astrophysical scenario mimicking an equilibrium in hot plasma. It is worth noting that for consecutive runs, this particular time increases to almost 10 ns while the temperature-equivalent value decreases. The fact represents the natural buildup of the long-lived ground state.

Using RF-based technology the realization of such conditions characterized by yield distributions resembling high temperatures at short time intervals after reaction-triggering pulses is not possible. It is the shortness of the fastest transition that demands a short triggering high beam flux ϕ to achieve such scenarios. Moreover, the assumed high yields achievable with a HPLS system are multiple orders of magnitude higher than those of current RF-technology which summarizes the essence of the contribution these systems can provide for future astrophysical studies. Again, one needs to emphasize that we are not referring here to real hot plasma temperatures. It can be envisaged that, once mono-energetic laser proton pulses are achieved, the depicted calculation can be extended to include any of the 64 states below the proton separation energy $S_p = 6.306$ MeV threshold of ^{26}Al. Such a calculation will help in the understanding of the complex interplay between the short-lived states belonging to the e_1 and e_2 ensembles and the longer-lived low lying states as function of the temperature T thus allowing to study the evolution of the effective lifetime τ_{eff} for ^{26}Al.

3.4. Considerations about Measurability at HPLSs

The above work sketches the unique experimental reaction scenarios which can be established with the short pulses of high-power laser-driven system. Core to the considerations is the fact that short-lived states can be produced with high yields, which exhibit, in the depicted case of ^{26}Al a 1.9×10^{22} times shorter lifetime compared to that of the ground state. It is however important to consider the measurability of those scenarios in this work. First and foremost, in the case of ^{26}Al, the only measurable parameter at a high-power laser system will be, for the foreseeable future, the delayed 511 keV radiation. A measurement of the prompt radiation will stay elusive mainly to the EMP pulse which will not allow a direct deduction of the yield of the $t_{1/2} = 1.20$ ns at 417 keV. Current γ-ray detector systems placed at PW are maybe capable of resolving ms isomers at best [27]. Moreover, it is unlikely for any forthcoming HPLS system to host a sophisticated, efficient neutron detector system for inaugural experiments. To deduce yield changes in ^{26}Al independently of pulse-to-pulse variations, one could refer to a second reaction channel with a precisely measured cross-section that is triggered in coincidence with the production of ^{26}Al. The *prima facie* experiment therefore used MgO to compare changes in the yield ratio between ^{26m}Al and ^{13}N obtained via ^{16}O(p, α)^{13}N. Since the β^+ half-life of the resulting ^{13}N nucleus is 9.965 min the corresponding yield could be clearly distinguished from the super-allowed β^+ decay of ^{26m}Al with $t_{1/2} = 6.35$ s [6].

The beamlines at the ELI-NP facility are ideally suited to spearhead any related measurement. In Table 2 shows the core features of the ELI-NP HPLS characterized by their power P. The ELI-NP laser system has two equal arms allowing the synchronization of the two beams. The energy E_{pul} and pulse duration t_{pul} depicted for the laser pulse represent the average values as required from the THALES Group, the laser manufacturer in France. At the highest power of 10 PW a repetition rate f of 1 pulse per minute is foreseen. The intensity I_{pul} for the 10 PW is estimated to reach values in excess of 1×10^{23} Wcm^{-2}. At this intensity simulations provide a maximum proton energy E_p^{max} of a few 100 s of MeV, to be conservative. The corresponding values for the 0.1 PW and 1 PW systems are taken from published work [25]. Most importantly it can be seen that the least energetic 100 TW beamline can deliver protons above the reaction threshold, having additionally the advantage of a comparatively high repetition rate. It is however important to note that the mean beam currents achievable with any of the HPLS are still a few orders of magnitude smaller than those achievable with the conventional

RF-accelerator technology. The values supplied for I_p^{max} describe the estimated ranges of the electrical peak currents induced by the reaction-driving protons from a single high-power laser pulse.

Table 2. Characteristic features of the high-power laser beamlines to be implemented at ELI-NP.

P/PW	E_{pul}/J	t_{pul}/fs	I_{pul}/Wcm^{-2}	f/Hz	E_p^{max}/MeV	I_p^{max}
10	250	25	10^{22-23}	0.017	>200	kA-MA
1	25	25	10^{21-22}	1	~100	A-kA
0.1	2.5	25	10^{20-21}	10	~30	mA-A

Our presented paper should be seen as a guideline for development, pointing out the possibilities, rather than an experimental proposal form. It is however hoped that with ever increasing fluxes ϕ, optimized geometry and high laser-to-proton energy conversion obtainable by the next generation of multi-PW laser plasma accelerators the total obtainable yields for the production of ^{26}Al will provide densities in the order of 1×10^{10} cm^{-3} and possibly even much higher. With improved values for laser energy-to-γ-conversion, a second laser beam in coincidence could therefore be used to probe the coupling of excited states by photo-excitation transitions by supplying a secondary, intense, X-ray beam. This will allow, potentially, the characterization of a yield change of Y_{228} in the presence of a hot photon bath. An enhancement of Y_{228} in such a case would be a first indication of the onset of an equilibrium between the ground and first excited state via the coupling of the e_1 and e_2 ensembles in ^{26}Al [21]. Moreover, such a study could enable to understand the intricate interplay between the excited states which govern the λ_{eff}. As current theories only account for a fraction of the abundance of ^{26}Al one may expect that a more elaborate evaluation of the effective lifetime will better reflect the production and decay mechanisms in actual astrophysical scenarios explaining the measured surplus of ^{26}Al in our Galaxy.

From the depicted scenarios, it is clear that an online isotope separator facility which uses a Penning-trap would be an ideal extension to improve any related experiment as it would allow a direct identification of ^{26}Al$_{g.s.}$ as well as supporting the measurement of ^{26m}Al. It is hoped to implement such a system eventually at the experimental stations dedicated to nuclear reaction research at ELI-NP.

4. Conclusions

We presented a pathway for high-power laser physics at the interface of nuclear astrophysics. The example of ^{26}Al showed that laser systems can provide short-lived states in high yields, comparable with those states that exhibit much longer lifetimes. This feature is unique to HPLS due to the ability of all-optical accelerator systems to deliver high temporal fluxes ϕ of reaction-driving protons for the fleeting time spans in the order of the lifetimes of the shortest-lived excited states. Exploiting the fact that the high temperatures within the residual ^{26}Al nuclei prevail for some ultrashort duration after its creation by a compound reaction, high entropy conditions in cold plasma environments can be achieved and potentially be tested with a second radiation or particle beam. In the case of ^{26}Al the yield ratios, $Y_{228}/Y_{g.s.}$ and $Y_{417}/Y_{g.s.}$, interpreted in the framework of the Maxwell-Boltzmann statistics, were found to mimic thermal equilibrium at defined times t_{equi} in the range of 300 ps to 10 ns resembling thermalization in a fictitious GK plasma environment. This circumstance can help in future experiments in the quest to create a nuclear astrophysical laboratory. In this respect, future investigations could be the application of coinciding X-ray pulses of high intensities, or an additional proton beam creating WDM. The study of other isotopes and the interplay of their excited states maybe envisaged with, e.g., ^{34}Cl, ^{115}In and ^{180}Ta being other very interesting cases for future investigation [20,21].

Author Contributions: K.M.S. conceived and designed the idea and theoretical planning; D.D. suggested and scrutinized the laser systems abilities regarding the paper; B.S.M. supported and revised the theoretical aspects regarding the entropy conditions as stated.

Funding: This research received no external funding.

Conflicts of Interest: The authors declare no conflict of interest.

Abbreviations

The following abbreviations are used in this manuscript:

CPA	Chirped pulse amplification
ELI-NP	Extreme Light Infrastructure-Nuclear Physics
EMP	Electromagnetic pulse
GCNDS	Graphite-cube neutron-detection system
HPLS	High-power laser system
RAL	Rutherford Appleton Laboratory
TNSA	Target normal sheath acceleration
TOF	Time-of-flight
WDM	Warm dense matter

References

1. Hauser, W.; Feshbach, H. The Inelastic Scattering of Neutrons. *Phys. Rev.* **1952**, *87*, 366–373. [CrossRef]
2. Strickland, D.; Mourou, G. Compression of amplified chirped optical pulses. *Opt. Commun.* **1985**, *56*, 219–221. [CrossRef]
3. Bateman, H. The solution of a system of differential equations occurring in the theory of radioactive transformations. *Proc. Camb. Philos. Soc.* **1910**, *15*, 423–427.
4. Morrissey, D.J.; Bloch, C.; Benenson, W.; Kashy, E.; Blue, R.A.; Ronningen, R.M.; Aryaeinejad, R. Thermal population of nuclear excited states. *Phys. Rev. C* **1986**, *34*, 761–763. [CrossRef]
5. Denis-Petit, D.; Ledingham, K.W.D.; McKenna, P.; Mascali, D.; Tudisco, S.; Spohr, K.M.; Tarisien, M. Nuclear Reactions in a Laser-Driven Plasma Environment. In *Applications of Laser-Driven Particle Acceleration*; Bolton, P., Parodi, K., Schreiber, J., Eds.; CRC-Press, Taylor & Francis Group: Boca Raton, FL, USA, 2018; Chapter 22, pp. 339–352.
6. Spohr, K.M.; Ledingham, K.W.D.; Clarke, R.; Carroll, D.C.; Doria, D.; Gray, R.J.; Hassan, M.; Labiche, M.; McCanny, T.; McKenna P.; et al. Unpublished work, 2019.
7. Patel, P.K.; Mackinnon, A.J.; Key, M.H.; Cowan, T.E.; Foord, M.E.; Allen, M.; Price, D.F.; Ruhl, H.; Springer, P.T.; Stephens, R. Isochoric Heating of Solid-Density Matter with an Ultrafast Proton Beam. *Phys. Rev. Lett.* **2003**, *91*, 125004. [CrossRef] [PubMed]
8. Ping, Y.; Correa, A.; Ogitsu, T.; Draeger, E.; Schwegler, E.; Ao, T.; Widmann, K.; Price, D.; Lee, E.; Tam, H.; et al. Warm dense matter created by isochoric laser heating. *High Energy Density Phys.* **2010**, *6*, 246–257. [CrossRef]
9. Thirolf, P.G.; Habs, D.; Gross, M.; Allinger, K.; Bin, J.; Henig, A.; Kiefer, D.; Ma, W.; Schreiber, J. Fission-Fusion: A new reaction mechanism for nuclear astrophysics based on laser-ion acceleration. *AIP Conf. Proc.* **2011**, *1377*, 88–95.
10. Oginni, B.; Iliadis, C.; Champagne, A. Theoretical evaluation of the reaction rates for ^{26}Al(n, p)^{26}Mg and ^{26}Al(n, α)^{23}Na. *Phys. Rev. C* **2011**, *83*, 025802. [CrossRef]
11. Mahoney, W.A.; Ling, J.C.; Wheaton, W.A.; Jacobson, A.S. HEAO$_3$ discovery of ^{26}Al in the interstellar medium. *Astrophys. J.* **1984**, *286*, 578–585. [CrossRef]
12. Arnould, M.; Norgaard, H.; Thielemann, F.K.; Hillebrandt, W. Synthesis of ^{26}Al in explosive hydrogen burning. *Astrophys. J.* **1980**, *237*, 931–950. [CrossRef]
13. Arnett, W.D. Explosive Nucleosynthesis in Stars. *Astrophys. J.* **1969**, *157*, 1369. [CrossRef]
14. Diehl, R.; Halloin, H.; Kretschmer, K.; Lichti, G.; Schönfelder, V.; Strong, A.; von Kienlin, A.; Wang, W.; Jean, P.; Knödlseder, J.; et al. Radioactive ^{26}Al from massive stars in the Galaxy. *Nature* **2006**, *439*, 45–47. [CrossRef] [PubMed]
15. Lee, T.; Papanastassiou, D.; Wasserburg, G. Demonstration of ^{26}Mg excess in Allende and evidence for ^{26}Al. *Geophys. Res. Lett.* **1976**, *3*, 41–44. [CrossRef]
16. Villeneuve, J.; Chaussidon, M.; Libourel, G. Homogeneous Distribution of ^{26}Al in the Solar System from the Mg Isotopic Composition of Chondrules. *Science* **2009**, *325*, 985–988. [CrossRef] [PubMed]

17. Kita, N.T.; Yin, Q.Z.; MacPherson, G.J.; Ushikubo, T.; Jacobsen, B.; Nagashima, K.; Kurahashi, E.; Krot, A.N.; Jacobsen, S.B. ^{26}Al-^{26}Mg isotope systematics of the first solids in the early solar system. *Meteorit. Planet. Sci.* **2013**, *48*, 1383–1400. [CrossRef]

18. Skelton, R.; Kavanagh, R.; Sargood, D. ^{26}Mg(p, n)^{26}Al and ^{23}Na(α, n)^{26}Al Reactions. *Phys. Rev. C* **1987**, *35*, 45–54. [CrossRef]

19. Ward, R.A.; Fowler, W.A. Thermalization of long-lived nuclear isomeric states under stellar conditions. *Astrophys. J.* **1980**, *238*, 266–286. [CrossRef]

20. Coc, A.; Porquet, M.G.; Nowacki, F. Lifetimes of ^{26}Al and ^{34}Cl in an astrophysical plasma. *Phys. Rev. C* **1999**, *61*, 015801. [CrossRef]

21. Gupta, S.; Meyer, B. Internal equilibration of a nucleus with metastable states: ^{26}Al as an example. *Phys. Rev. C* **2001**, *64*, 025805. [CrossRef]

22. Dunford, C.; Burrows, T. *Online Nuclear Data Service*; Report IAEA-NDS-150; NNDC Informal Report NNDC/ONL-95/10; BNL National Nuclear Data Center: Brookhaven, NY, USA, 1996; p. 015801.

23. Äystö, J.; Eronen, T.; Jokinen, A.; Kankainen, A.; Moore, I.; Penttilä, H. An IGISOL portrait. *Hyperfine Interact.* **2013**, *223*, 1–3. [CrossRef]

24. Wilks, S.C.; Langdon, A.B.; Cowan, T.E.; Roth, M.; Singh, M.; Hatchett, S.; Key, M.H.; Pennington, D.; MacKinnon, A.; Snavely, R.A. Energetic proton generation in ultra-intense laser–solid interactions. *Phys. Plasmas* **2001**, *8*, 542–549. [CrossRef]

25. Higginson, A.; Gray, R.J.; King, M.; Dance, R.J.; Williamson, S.D.R.; Butler, N.M.H.; Wilson, R.; Capdessus, R.; Armstrong, C.; Green, J.S.; et al. Near-100 MeV protons via a laser-driven transparency-enhanced hybrid acceleration scheme. *Nat. Commun.* **2018**, *9*, 724. [CrossRef] [PubMed]

26. Ziegler, J.F.; Ziegler, M.D.; Biersack, J.P. SRIM—The stopping and range of ions in matter. *Nucl. Instrum. Methods Phys. Res. B* **2010**, *268*, 1818–1823. [CrossRef]

27. Negoita, F.; Roth, M.; Thirolf, P.G.; Tudisco, S.; Hannachi, F.; Moustaizis, S.; Pomerantz, I.; McKenna, P.; Fuchs, J.; Spohr, K.M.; et al. Laser Driven Nuclear Physics at ELI-NP. *Rom. Rep. Phys.* **2016**, *68*, S37–S144.

MDPI

St. Alban-Anlage 66

4052 Basel

Switzerland

Tel. +41 61 683 77 34

Fax +41 61 302 89 18

www.mdpi.com

Galaxies Editorial Office

E-mail: galaxies@mdpi.com

www.mdpi.com/journal/galaxies